Relativistic Effects in Heavy-Element Chemistry and Physics

Wiley Series in Theoretical Chemistry

Series Editors

Thermodynamics of Irreversible Processes: Applications to Diffusion and Rheology
Gerard D. C. Kuiken
Published 1994, ISBN 0 471 94844 6

Modelling Molecular Structures
Alan Hinchliffe
Published 1995, ISBN 0 471 95921 9 (cloth), ISBN 0 471 95923 5 (paper)
Published 1996, ISBN 0 471 96491 3 (disk)

Molecular Interactions
Edited by Steve Scheiner
Published 1997, ISBN 0 471 97154 5

Density-Functional Methods in Chemistry and Materials Science
Edited by Michael Springborg
Published 1997, ISBN 0 471 96759 9

Theoretical Treatments of Hydrogen Bonding
Dušan Hadži
Published 1997, ISBN 0 471 97395 5

The Liquid State – Applications of Molecular Simulations
David M. Heyes
Published 1997, ISBN 0 471 97716 0

Quantum-Chemical Methods in Main-Group Chemistry
Thomas M. Klapötke and Axel Schulz
Published 1998, ISBN 0 471 97242 8

Electron Transfer in Chemistry and Biology: An Introduction to the Theory
Alexander M. Kuznetsov and Jens Ulstrup
Published 1998, ISBN 0 471 96749 1

Metal Clusters
Edited by Walter Ekardt
Published 1999, ISBN 0 471 98783 2

Methods of Electronic-Structure Calculations From Molecules to Solids
Michael Springborg
Published 2000, ISBN 0 471 97975 9 (cloth), 0 471 97976 7 (paper)

Modelling Molecular Structures, Second Edition
Alan Hinchliffe
Published 2000, ISBN 0 471 62380 6

Relativistic Effects in Heavy-Element Chemistry and Physics

Edited by

Bernd A. Hess
Friedrich-Alexander-Universität
Erlangen-Nürnberg, Germany

Phone (+44) 1243 779777

Email (for orders and customer service enquiries): cs-books@wiley.co.uk
Visit our Home Page on www.wiley.co.uk or www.wiley.com

Other Wiley Editorial Offices

John Wiley & Sons Inc., 111 River Street, Hoboken, NJ 07030, USA

Jossey-Bass, 989 Market Street, San Francisco, CA 94103-1741, USA

Wiley-VCH Verlag GmbH, Boschstr. 12, D-69469 Weinheim, Germany

John Wiley & Sons Australia Ltd, 33 Park Road, Milton, Queensland 4064, Australia

John Wiley & Sons (Asia) Pte Ltd, 2 Clementi Loop #02-01, Jin Xing Distripark, Singapore 129809

John Wiley & Sons Canada Ltd, 22 Worcester Road, Etobicoke, Ontario, Canada M9W 1L1

Wiley also publishes its books in a variety of electronic formats. Some content that appears in print may not be available in electronic books.

Library of Congress Cataloguing-in-Publication Data

Relativistic effects in heavy element chemistry and physics/edited by Bernd Hess.
p. cm. - (Wiley series in theoretical chemistry)
ISBN 0-470-84138-9 (cloth : alk. Paper)
1. Quantum chemistry. 2. Relativistic quantum theory. 3. Heavy elements. I. Hess, Bernd. II Series.

QD462.6.R2 R45 2002
541.2'8-dc21

British Library Cataloguing in Publication Data

A catalogue record for this book is available from the British Library

ISBN 0 470 84138 9

Produced from LaTeX files supplied by the authors, typeset by T&T Productions Ltd, London.
Printed and bound in in Great Britain by T J International Ltd, Padstow, Cornwall.
This book is printed on acid-free paper responsibly manufactured from sustainable forestry in which at least two trees are planted for each one used for paper production.

SERIES PREFACE

Theoretical chemistry is one of the most rapidly advancing and exciting fields in the natural sciences today. This series is designed to show how the results of theoretical chemistry permeate and enlighten the whole of chemistry together with the multifarious applications of chemistry in modern technology. This is a series designed for those who are engaged in practical research. It will provide the foundation for all subjects which have their roots in the field of theoretical chemistry.

How does the materials scientist interpret the properties of the novel doped fullerene superconductor or a solid-state semiconductor? How do we model a peptide and understand how it docks? How does an astrophysicist explain the components of the interstellar medium? Where does the industrial chemist turn when he wants to understand the catalytic properties of a zeolite or a surface layer? What is the meaning of 'far-from-equilibrium' and what is its significance in chemistry and in natural systems? How can we design the reaction pathway leading to the synthesis of a pharmaceutical compound? How does our modelling of intermolecular forces and potential energy surfaces yield a powerful understanding of natural systems at the molecular and ionic level? All these questions will be answered within our series which covers the broad range of endeavour referred to as 'theoretical chemistry'.

The aim of the series is to present the latest fundamental material for research chemists, lecturers and students across the breadth of the subject, reaching into the various applications of theoretical techniques and modelling. The series concentrates on teaching the fundamentals of chemical structure, symmetry, bonding, reactivity, reaction mechanism, solid-state chemistry and applications in molecular modelling. It will emphasize the transfer of theoretical ideas and results to practical situations so as to demonstrate the role of theory in the solution of chemical problems in the laboratory and in industry.

D. Clary, A. Hinchliffe, D. S. Urch and M. Springborg
June 1994

Contents

List of Contributors

Martin Diefenbach
Institut für Chemie, Technische Universität Berlin,
Straße des 17. Juni 135, D-10623 Berlin, Germany

Michael Dolg
Institut für Physikalische und Theoretische Chemie, Rheinische
Friedrich-Wilhelm-Universität Bonn, Wegelerstr. 12,
D-53115 Bonn, Germany

Reiner Dreizler
Institut für Theoretische Physik, Johann-Wolfgang-Goethe-Universität Frankfurt,
Robert Mayer Straße 8–10, D-60054 Frankfurt am Main, Germany

Hubert Ebert
Department Chemie, Ludwig-Maximilians-Universität München, Butenandtstraße
11, D-81377 München, Germany

Eberhard Engel
Institut für Theoretische Physik, Johann-Wolfgang-Goethe-Universität Frankfurt,
Robert Mayer Straße 8–10, D-60054 Frankfurt am Main, Germany

Burkhard Fricke
Fachbereich Physik, Universität Kassel, Heinrich-Plett-Straße 40,
D-34109 Kassel, Germany

Eberhard K. U. Gross
Institut für Theoretische Physik, Universität Würzburg, Am Hubland,
D-97074 Würzburg, Germany

Bernd A. Hess
Lehrstuhl für Theoretische Chemie, Friedrich-Alexander-Universität
Erlangen–Nürnberg, Egerlandstraße 3, D-91058 Erlangen, Germany

Jürgen Hinze
Fakultät für Chemie, Universität Bielefeld, D-33615 Bielefeld, Germany

Jens Volker Kratz
Institut für Kernchemie, Johannes-Gutenberg-Universität Mainz,
Fritz-Straßmann-Weg 2, D-55128 Mainz, Germany

Valeria Pershina
Gesellschaft für Schwerionenforschung, Planckstraße 1,
D-64291 Darmstadt, Germany

Günter Plunien
Institut für Theoretische Physik, Technische Universität Dresden,
Zellescher Weg 17, D-01062 Dresden, Germany

Markus Reiher
Lehrstuhl für Theoretische Chemie, Friedrich-Alexander-Universität
Erlangen–Nürnberg, Egerlandstraße 3, D-91058 Erlangen, Germany

Annette Schier
Institut für Anorganische Chemie, Technische Universität München,
Lichtenbergstr 4, D-85747 Garching, Germany

Hubert Schmidbaur
Institut für Anorganische Chemie, Technische Universität München,
Lichtenbergstr 4, D-85747 Garching, Germany

Detlef Schröder
Institut für Chemie, Technische Universität Berlin,
Straße des 17. Juni 135, D-10623 Berlin, Germany

Helmut Schwarz
Institut für Chemie, Technische Universität Berlin,
Straße des 17. Juni 135, D-10623 Berlin, Germany

Gerhard Soff
Institut für Theoretische Physik, Technische Universität Dresden,
Zellescher Weg 17, D-01062 Dresden, Germany

Sven Varga
Fachbereich Physik, Universität Kassel, Heinrich-Plett-Straße 40,
D-34109 Kassel, Germany

Foreword

In his book, *Mr Tompkins in Wonderland*, the physicist, George Gamow, gives a popular description of two of the most important scientific developments of the 20th century: quantum theory and the theory of relativity. Gamow's book is split into two parts, in which we follow the adventures of the main character, Mr Tompkins, in a universe where some of the physical laws have been modified. In the first part, the speed of light has been greatly reduced, while in the second Planck's constant has been greatly increased. In this way, the typical length scales of the theory at hand have been changed to those of our daily life in order to make the differences between the classical-mechanical description of our customary life and the relativistic or quantum-mechanical description clear. In the first case (in which the theory of relativity is popularized), this corresponds to a reduction in the distance light travels in one time unit by orders of magnitude, whereas in the second case (i.e. the description of quantum theory) the typical length of objects experiencing quantum effects has been increased by orders of magnitude.

This book testifies to a common belief that the theory of relativity is to be applied to one class of physical objects, while quantum theory is relevant for another class of physical objects. The exceptions are exotic and most often not relevant for a scientist working in physics or chemistry, at least as long as he or she is studying the properties of real materials that could also be of interest to a nonscientist.

Thus, the most common assumption was that a material's properties are governed by quantum theory and that relativistic effects are mostly minor and of only secondary importance. Quantum electrodynamics and string theory offer some possible ways of combining quantum theory and the theory of relativity, but these theories have only very marginally found their way into applied quantum theory, where one seeks, from first principles, to calculate directly the properties of specific systems, i.e. atoms, molecules, solids, etc. The only place where Dirac's relativistic quantum theory is used in such calculations is the description of the existence of the spin quantum number. This quantum number is often assumed to be without a classical analogue (see, however, Dahl 1977), and its only practical consequence is that it allows us to have two electrons in each orbital.

However, relativistic effects have a much more profound influence on a material's properties. Thus, the existence of a spin quantum number allows for the existence of magnetism. Moreover, as shown in most standard textbooks on physical chemistry, the phenomenon of phosphorescence can only be explained through the existence of relativistic effects. Phosphorescence involves transitions between, for example, singlet and triplet states which are only possible if some spin-operating effects exist, e.g. spin–orbit couplings. Furthermore, several experimental techniques are indirectly based on exploiting relativistic effects. These include, for example, electron-spin and NMR spectroscopies.

Relativistic effects are more important for heavier elements. Moreover, the different electronic orbitals are influenced differently by these effects, which means that they ultimately may modify the material's properties differently for different elements. A famous example of this is the colour of gold, which is atypical when compared with the colours of copper and silver but which can be explained by incorporating relativistic effects.

These very few examples should demonstrate that in many cases relativistic effects are important and cannot be either simply ignored or, better, treated at a low-accuracy level. Nevertheless, the combination of quantum theory and the theory of relativity at a level that makes qualitative and quantitative predictions possible is very far from trivial and has not yet reached a mature level, but is a currently active field of dispute, discussion and progress.

There exist many methods for electronic-structure calculations (see, for example, Springborg 2000) that are becoming increasingly important in both science and industry. Some of these methods have led to commercially available programs that are used worldwide. With these methods we can study many different systems but relativistic effects are at best treated only approximately. As indicated above, this is fine for many properties and for systems with lighter atoms, but in many cases it is precisely the relativistic effects that are responsible for important details. Therefore, there is a strong need for reliable methods for electronic-structure calculations that include relativistic effects and the treatment of those is a topic of intensive worldwide research.

In this context, Europe is playing a central role, so that a relatively large part of the frontier research in this field has taken place there. And within Europe, Germany is particularly strong, with two major research programs, one of which was supported by the European Science Foundation and the other by the German Research Council.

The purpose of this book, edited by Bernd A. Hess, is twofold. On the one hand, the different theoretical tools for including relativistic effects in quantum-theoretical calculations (at very many different levels of theory and sophistication) developed recently are presented. And on the other hand, the results obtained with these methods for both simple and complicated (i.e. many-atom) systems are presented.

The book concentrates largely on the results of the German research activities (with, however, contributions from other groups outside Germany) and is the result of the above-mentioned German research effort. It is interdisciplinary with first of all contributions from chemistry and physics and has been written by scientists who are among the leaders in this field. Since the authors have had close contact over several years, the present book is not solely a collection of largely independent parts but the different parts are strongly interrelated.

Michael Springborg
Physical Chemistry, University of Saarland,
66123 Saarbrücken, Germany
m.springborg@mx.uni-saarland.de

Preface

In August 1991, seven scientists from seven European Countries, working in the field of relativistic electronic structure theory, met in Strasbourg as guests of Dr Manfred Mahnig, who was at this time a representative of the European Science Foundation. They gathered in order to discuss the state of the art in relativistic electronic structure theory of atoms and molecules and to explore the possibilities of promoting the field at a European scale. The initiative for this meeting can be traced back to correspondence between Dr Mahnig and Professor Pekka Pyykkö, which can therefore be taken as the germ of the various REHE programmes promoting the field of relativistic effects in heavy-element chemistry and physics. The first was the REHE programme of the European Science Foundation, which was current in the years 1993–1998 and proved extremely effective by providing the opportunity for many scientists to take advantage of short visits and also longer visits lasting up to several weeks to the most active laboratories in the field all over Europe. In addition, a series of Euroconferences and workshops ensured a rapid exchange of ideas. The newsletter of the program[1] provided rapid exchange of information between the participating groups. The programme was directed by a steering committee comprising E. J. Baerends (Amsterdam), J. P. Daudey (Toulouse), K. Faegri (Oslo), I. P. Grant (Oxford), B. Hess (Bonn, Vice-Chairman), J. Karwowski (Torun), P. Pyykkö (Helsinki, Chairman), K. Schwarz (Vienna) and A. Sgamellotti (Perugia). The success of this programme is documented by an impressive list of papers which received funding by REHE.[2]

The European REHE programme fostered rapid development of the area, and soon a Collaborative Research Program ('Schwerpunkt') programme of the German Science Foundation (DFG) was approved, which in the years 1994–2000 provided funding for about 30 research groups, mostly in the form of positions for graduate students and, in some cases, postdoctoral researchers. This funding was granted based on applications of individual research groups, which were reviewed in a collaborative context every second year by an international committee of referees. The collaborative aspect was strengthened by reports by the groups at yearly meetings, which soon led to new collaborations and exchanges of ideas across the participating groups.

---

[1] http://www.chemie.uni-erlangen.de/hess/html/esf/nl.html

[2] http://www.chemie.uni-erlangen.de/hess/html/esf/papers.html

The topic of the REHE programmes was relativistic effects in heavy-element chemistry, but what are relativistic effects? In principle, the answer is easy. A relativistic effect is any phenomenon which can be traced back to the fact that the velocity of light is a universal, finite constant in all frames of reference, even those moving with some unchanging velocity with respect to each other. Thus, a relativistic effect relies on a comparison with a fictitious world where the velocity of light is infinite and a 'nonrelativistic' description applies.

There is still discussion, in particular, in experimentally oriented papers, about whether relativistic effects 'really exist' and are measurable, or if they are an artefact of a 'wrong theory', namely, the nonrelativistic one, and cannot be measured because in reality there are no nonrelativistic atoms. However, since the notion of relativistic effects is well defined, I claim that they can even be measured in favourable cases directly from the behaviour of a simple function of the atomic number Z. Consider the binding energies of the $1s$ electron of hydrogen-like atoms, which are well accessible to measurement. Obviously, this quantity depends on Z, and we attribute the dependence on Z beyond second order to relativity, since nonrelativistic theory predicts that there are no nonvanishing Taylor coefficients beyond second order. The relativistic effect therefore can in this particular case be measured as the deviation of E as a function of Z from parabolic behaviour, a very simple prescription indeed.

It is remarkable that the Dirac theory of the relativistic electron perfectly describes this deviation, and the difference to the reference (the nonrelativistic value) is unusually well defined by the limit of a single parameter (the velocity of light) at infinity. The special difficulty encountered in 'measuring' relativistic effects is that relativistic quantum mechanics is by no means a standard part of a chemist's education, and therefore the theory for interpreting a measurement is often not readily at hand. Still, a great many of the properties of chemical substances and materials, in particular, 'trends' across the periodic system of elements, can be understood in terms of relativistic effects without having to consider the details of the theory.

Needless to say, many-electron atoms and molecules are much more complicated than one-electron atoms, and the realization of the nonrelativistic limit is not easily accomplished in these cases because of the approximations needed for the description of a complicated many-particle system. However, the signature of relativistic effects (see, for example, Chapter 3 in this book) enables us to identify these effects even without calculation from experimental observation. Two mainly experimentally oriented chapters will report astounding examples of relativistic phenomenology, interpreted by means of the methods of relativistic electronic structure theory. These methods for the theoretical treatment of relativistic effects in many-electron atoms and molecules are the subject of most of the chapters in the present volume, and with the help of this theory relativistic effects can be characterized with high precision.

The present book serves a twofold purpose. On one hand, the book was designed to serve as a final report on the work done in the Collaborative Research Programme ('Schwerpunkt') of the German Science Foundation on Relativistic Effect in Heavy-Element Chemistry and Physics. I apologize that for that reason it is certainly biased towards the work of the groups who had participated in the last period of this programme, and as a consequence our account will certainly be found to have missed important contributions to the field. On the other hand, to some extent it should also give an account of the worldwide progress made in the last decade as far as the methodology of calculating relativistic effects in heavy-element chemistry and physics and their interpretation are concerned. Thus, we have made some effort to review the work done in Europe, in particular, in the framework of the REHE programme, and all over the world in the field of relativistic electronic structure calculations.

On behalf of the participants of the 'German REHE Schwerpunkt', I thank the German Science Foundation for their generous and highly effective funding. The professional guidance through the technical aspects of the programme by Dr Carnell, Dr Kuchta and Dr Mahnig, who were the officers of the DFG in charge of the 'Schwerpunkt', was instrumental to its success. Since Dr Mahnig also guided the early stages of the 'European REHE' at the European Science Foundation, it is fair to say that without his help and initiative in the early 1990s the REHE programmes would not exist, nor would the present book. An important instrument in the design of a 'Schwerpunkt' is the refereeing process, carried out by an international committee of experts, all of them with illustrious reputations in the field. Their presence at the meetings was highly appreciated and without doubt provided many stimuli which found their way into the work of the groups refereed by them. I express my sincerest gratitude for the time and the work they have devoted to the programme.

Finally, I also thank the participants of the last term of this programme for supplying a large amount of the material which has been used by the authors of the seven chapters of this book in order to accomplish the task of providing a report on six years of research in a fascinating area of modern science.

Bernd Hess
Chair of Theoretical Chemistry, University of Erlangen–Nuremberg,
91058 Erlangen, Germany
bernd.hess@chemie.uni-erlangen.de

1 Basic Theory and Quantum Electrodynamics in Strong Fields

Günter Plunien and Gerhard Soff
Institut für Theoretische Physik, Technische Universität Dresden

1.1 Introduction

Theoretical and experimental investigations of relativistic and QED effects in atomic physics and chemistry have increased continuously during the last decade. As a consequence of this interest in various relativistic phenomena and in their empirical manifestations a diverse field of research has developed linking together widespread activities ranging from high-energy heavy-ion collision physics, atomic or molecular physics and chemistry of heavy elements to solid-state physics.

Giving a rigorous account of relativistic effects is now an important goal in theoretical and experimental studies because of recent progress made in experimental techniques and because of the accuracy currently achievable in measurements, e.g. in atomic and molecular spectroscopy, or in view of newly available laser techniques. Present accessible energies in heavy-ion accelerators allow a new generation of experiments with ultrarelativistic ions, which, for example, enable us to probe the structure of the vacuum via the electromagnetic particle–antiparticle pair creation.

In particular, the consideration of relativistic and QED effects of electronic systems (i.e. free electrons, electronic ions, atoms or molecules) in strong external electromagnetic fields provides various appropriate scenarios for sensitive tests of our understanding of the underlying interactions. Theories of fundamental interactions, such as quantum electrodynamics (QED) or the standard model of electroweak interactions can be tested conclusively by studying QED radiative corrections and parity-violating effects (PNC) in the presence of strong fields.

This chapter partly reviews current developments in studies of relativistic phenomena that occur under the influence of the strongest accessible electromagnetic fields realized in nature and in the laboratory:

(1) electrons in external electromagnetic fields of superintense lasers;

(2) lepton pair creation in time-dependent external fields generated in peripheral ultrarelativistic collisions of heavy ions;

Relativistic Effects in Heavy-Element Chemistry and Physics
Edited by B. A. Hess

(3) relativistic and QED effects in strong external electric and magnetic fields of highly charged ions.

A proper framework for a relativistic description of the various scenarios indicated above is based on Dirac's theory of the electron and QED as the quantum field theory of leptons and photons.

A rigorous approach requires us to perform exact, i.e. all-order, calculations with respect to the coupling to the strong external fields. Laser field intensities of the order of 10^{21} W cm^{-2} corresponding to so-called *ponderomotive energies* comparable with the electron rest mass or effective coupling constants $Z\alpha \sim \mathcal{O}(1)$, where Z denotes the nuclear charge number of heavy ions, reveal the inherent nonperturbative nature of the problems under consideration. Perturbation theory is thus inadequate and can at best only serve as a rough estimate.

On the level of quantum mechanics we are faced with the problem of solving numerically the Dirac equation governing the time-evolution of an electron state $|\Psi(t)\rangle$ under the influence of a space-time-dependent (classical) electromagnetic field $A^{\mu}_{\text{ext}}(\boldsymbol{r}, t)$ including the binding nuclear potential $A^{\mu}_{\text{nuc}}(\boldsymbol{r})$:

$$H(t)|\Psi(t)\rangle = \mathrm{i}\partial_t|\Psi(t)\rangle,$$
$$H(t) = \boldsymbol{\alpha} \cdot (\boldsymbol{p} - e\boldsymbol{A}_{\text{ext}}(\boldsymbol{x}, t)) + eA^0_{\text{ext}}(\boldsymbol{x}, t) + \beta m.$$

Employing simplifying conditions, this task can in principle be achieved for particular scenarios. To develop efficient computer codes for solving the time-dependent Dirac equation in four dimensions with arbitrary external electromagnetic potentials is the task of numerous theoretical investigations based on different numerical approaches. Once this goal is achieved it will facilitate applications ranging from the description of dynamical processes in atoms or molecules, from studies of the behaviour of electrons or atoms in time-dependent intense laser fields to nonperturbative calculations of lepton pair production in the time-varying electromagnetic field generated in ultrarelativistic heavy-ion collisions.

A general quantum field-theoretical framework for treating all the different problems outlined above may be specified in the interaction representation. Let the total system under consideration be described by the time-dependent Hamiltonian

$$H(t) = H_0 + H_{\text{I}}(t).$$

This decomposition suggests that at least in principle the H_0-problem can be solved. The additional interaction Hamiltonian $H_{\text{I}}(t)$ is assumed to be localized in time, i.e. $H_{\text{I}}(|t| \to \infty) = 0$. The explicitly time-independent Hamiltonian H_0 can be diagonalized through a proper definition of creation/annihilation operators $a_i^{(s)\dagger}/a_i^{(s)}$ corresponding to noninteracting particles of the species (s) and characterized by a set of quantum numbers denoted by a label i:

$$H_0 = \sum_{s,i} \varepsilon_i^{(s)} a_i^{(s)\dagger} a_i^{(s)}.$$

The vacuum $|\mathbf{0}\rangle \equiv \bigotimes_s |0_s\rangle$ defined as the ground state of the system is annihilated according to $a_i^{(s)}|\mathbf{0}\rangle = 0$. Time-dependent operators obey the equation of motion

$$\mathrm{i}\frac{\mathrm{d}O}{\mathrm{d}t} = -[H_0, O] + \mathrm{i}\frac{\partial O}{\partial t},$$

while the dynamics of any given quantum state $|\Psi(t)\rangle$ is governed by

$$\mathrm{i}\frac{\mathrm{d}}{\mathrm{d}t}|\Psi(t)\rangle = H_\mathrm{I}(t)|\Psi(t)\rangle.$$

This equation of motion for the quantum state $|\Psi(t)\rangle$ is formally integrated by means of the (unitary) time-evolution operator U

$$\begin{aligned}|\Psi(t)\rangle &= U(t,t')|\Psi(t')\rangle = T\exp\left(-\mathrm{i}\int_{t'}^{t}\mathrm{d}t_1\,H_\mathrm{I}(t_1)\right)|\Psi(t')\rangle \\ &= \sum_{n=0}^{\infty}\frac{(-\mathrm{i})^n}{n!}\int_{t'}^{t}\mathrm{d}t_1\cdots\int_{t'}^{t}\mathrm{d}t_n\,T[H_\mathrm{I}(t_1)\cdots H_\mathrm{I}(t_n)]|\Psi(t')\rangle.\end{aligned}$$

Whether or not perturbation theory with respect to the interaction described by the interaction Hamiltonian is meaningful depends on the problem under consideration. Otherwise, we have to search for appropriate, nonperturbative approximations for the time-evolution operator U.

Strongly bound electrons in highly charged ions experience permanently the external electromagnetic field generated by the nucleus and interact permanently as well with the quantized radiation field. Typical binding energies of K-shell electrons in heavy, hydrogen-like ions, which, for example, in the uranium system amount to about 132 keV, become comparable in magnitude with the electron rest energy. In the same system the dominant QED-radiative corrections contribute to the 1s-Lamb shift already at the level of 10^{-4}. A value for the effective coupling constant $Z\alpha \sim 0.6$ for uranium indicates the breakdown of perturbation theory with respect to the nuclear Coulomb field. Moreover, K-shell electrons have considerable overlap with the nucleus and with its nearby region, where the local energy density of the electric field reaches the critical Schwinger value. Spontaneous electron–positron pair creation could occur if this region extended over a spatial volume of the order of the electron Compton wavelength. These qualitative remarks strongly support the expectation that relativistic and QED effects will become relevant, for example, for a quantitative explanation of recent spectroscopic data.

When dealing with few-electron ions, corrections due to the interelectronic interaction have to be properly taken into account. While this can be achieved with sufficient accuracy perturbatively for helium- or lithium-like ions, a rigorous accounting of electron correlation effects requires the ambitious task of solving the interacting many-electron system. Various all-order approaches—such as relativistic Dirac–Hartree–Fock (RDHF), relativistic many-body perturbation theory (RMBPT), relativistic configuration interaction methods or coupled cluster expansions—are available employing specific approximations to the exact many-body problem, such as, for example, projecting out the negative-energy Dirac states (no-pair approximation), etc. Accordingly, theoretical predictions of QED and PNC effects in atoms are still limited by uncertainties set by the present capability to account for all relevant electron correlation. Because of such inherent problems within the methods used in relativistic atomic-structure calculations it is desirable to search for a consistent framework that treats simultaneously electron-correlation and QED effects on an equal footing from the very beginning. Attempts to merge QED and many-body approaches on a fundamental level also represent a challenging task in future research.

1.2 Electrons in Superintense Laser Fields

Investigation of the various aspects related to the problem of the interactions of fast electrons with intense laser fields has become a very active field of current research (Bula *et al.* 1996; Connerade and Keitel 1996; Grochmalicki *et al.* 1990; Hartemann and Kerman 1996; Hartemann and Luhmann 1995; Hartemann *et al.* 1995; Keitel and Knight 1995; Meyerhofer *et al.* 1996; Moore *et al.* 1995), strongly motivated by the potential applications discussed in the recent literature (Hussein *et al.* 1992; Körmendi and Farkas 1996; Maine *et al.* 1988; Patterson and Perry 1991; Patterson *et al.* 1991)—such as the generation of X-rays via scattering intense radiation with fast free electrons (Körmendi and Farkas 1996) and the acceleration of charged particles by powerful lasers (Hussein *et al.* 1992)—that have now become possible in present highly intense pulsed and focused laser fields (Maine *et al.* 1988; Patterson *et al.* 1991). To appreciate intensities of the order of $I = 10^{21}$ W cm^{-2} of currently generated laser light we should compare this number with the intensity $I_a = 1$ a.u. $= 3.5 \times 10^{16}$ W cm^{-2} (a.u. = atomic units) corresponding to the binding Coulomb field experienced by an electron in the 1s ground state of hydrogen.

Pulsed laser fields may be represented by a classical external vector potential $\boldsymbol{A}(\eta)$ considered as a function of the invariant phase $\eta = \omega t - \boldsymbol{k} \cdot \boldsymbol{r}$. A natural scale for characterizing the intensities is provided by the dimensionless parameter

$$q = \sqrt{\frac{2e^2\langle \boldsymbol{A}\rangle^2}{(mc^2)^2}} = \frac{ea}{mc^2} = \sqrt{\frac{2r_0 I \lambda^2}{\pi mc^3}} = \sqrt{\frac{4U_\mathrm{p}}{mc^2}}, \tag{1.1}$$

which relates characteristic properties of the electron, i.e. charge e, mass m and its classical radius r_0, with those of the radiation fields, such as the speed of light c, wavelength λ, intensity I and the peak amplitude a, respectively. The averaging $\langle \boldsymbol{A}\rangle^2$ is taken over a time interval that is short compared with the pulse duration but long enough to contain many field cycles. The so-called ponderomotive energy $U_\mathrm{p} = e^2 I/(4m\omega_0^2)$ becomes comparable with the rest energy of the electron. As a consequence, the motion of the electron can become relativistic even in the case of optical photons possessing nonrelativistic energies, for example, $\hbar\omega_0 \simeq 2$ eV. Fields characterized by values $q \geqslant 1$ are called superintense. Accordingly, with the advent of superintense lasers we have entered the regime where relativistic effects become relevant and thus have to be taken into account. Already at intensities of about $I = 10^{18}$ W cm^{-2} the electron can gain energies of the order of its rest energy from the electric-field mode of the laser. Under those conditions the dynamics of the electron is dominated by relativistic effects. We should mention, for instance, the effect of self-focusing and acceleration of electrons up to energies of about 100 MeV (Wagner *et al.* 1997). Achieving the goal of a consistent relativistic description of the atom–laser interaction has also been the subject of recent research (Blasco *et al.* 2001; Faisal and Radozycki 1993; Salamin and Faisal 1996, 1997, 1999, 2000).

One important issue among other interesting topics to which current theoretical studies are devoted is the examination of the influence of superintense laser fields on the process of coherent emission of high-energy photons with frequencies ω equal to high multiples of the fundamental frequency ω_0 of the incident laser field. The generation of such photons, commonly called high-harmonic generation (HHG), has been first observed experimentally in intense but nonrelativistic laser fields (Ferray *et al.* 1988; McPerson *et al.* 1897). Intensive nonperturbative calculations employing the formalism of Floquet states (see, for example, Gavrila 1992; Muller and Gavrila 1993, and references therein) or solving numerically the time-dependent Schrödinger equation on the lattice (Kulander *et al.* 1991; Muller 1999) have contributed important results to the elucidation of atom–laser interactions. In these computationally intensive studies, the laser field is treated in the dipole approximation and thus magnetic and retardation effects are usually neglected.

Similarly, for superintense fields we could attempt to solve the corresponding time-dependent Dirac problem. It turns out, however, that unlike the nonrelativistic Schrödinger case a full numerical integration of the Dirac equation in four dimensions for realistic scenarios is faced with additional technical difficulties connected with the discretization of spinor equations of motion. Accordingly, many of the approaches which work nicely in the Schrödinger case cannot be applied straightforwardly to the solving of the relativistic Dirac problem. An inherent difficulty obviously follows from the specific nature of the atom–laser interaction. Depending on the particular system and process under consideration, it could be appropriate to employ the Furry picture, where the bound-electron problem is solved exactly (Coulomb states) and the laser field is treated as a perturbation, or, conversely, the problem is solved exactly with respect to the interaction with the external laser field (Volkov states) and the Coulomb potential plays the role of a minor perturbation. Both interaction pictures, however, may become inadequate in the scenario of highly charged ions in superintense laser fields.

Exact and nonperturbative numerical evaluations are thus severely constrained. As a first step towards a fully numerical integration, we can investigate an analogous dimensionally reduced model scenario, for example, in $1 + 1$ dimensions (one spatial and time direction). Such simulations still allow us to elucidate some generic features of relativistic effects on the electron dynamics. Calculations along this line have been performed recently providing new insight into the process of HHG and atomic ionization under the influence of superintense fields (Blase 2001). It turns out that at such high intensities the electron, although initially bound, will become ionized even before the maximum (peak) field strength of the incident laser fields sweeps over the atom. Hence, at the time when the superintense field takes effect, the atom–laser interaction becomes essentially a problem of a free electron in the presence of a superintense field. With this picture in mind and placing particular emphasis on the relativistic aspects, we could try to simulate the problem of HHG from free electrons in superintense laser fields employing semiclassical or even classical relativistic electrodynamics in four dimensions, treating the electron as classical, relativistic point particle (Salamin and Faisal 1996). In the following we discuss the result of such model simulations on HHG and above-threshold-ionization (ATI) spectra in superintense fields.

1.2.1 Model simulations

For the $(1 + 1)$-dimensional model system, the numerical integration of the corresponding Dirac equation has been performed (Blase 2001) employing an explicit finite-difference method. Numerical studies revealed that, for example, the time-discretization interval δt has to be chosen about three orders of magnitude smaller than the corresponding interval δx of the spatial grid. This is basically connected with the finite velocity of light, which amounts to $c = 137.036$ a.u. To be explicit, for the calculations performed for a hydrogen atom $(Z = 1)$ in an intense laser field (Blase 2001) numerical stability required a choice of $\delta x = 0.05$ a.u. for the spatial and $\delta t = 0.000\,02$ a.u. for the temporal grid.

We can employ the results of such simulations for both the Dirac and Schrödinger equations in order to calculate the HHG as well as the ATI spectra for the same laser parameters. This allows us to estimate the relativistic effects. An important observable is the multiharmonic emission spectrum $S(\omega)$. It can be represented as the temporal Fourier transform of the expectation value of the Dirac (Schrödinger) current density operator $j(t)$ according to

$$S(\omega) = \frac{\alpha^3}{3\pi}\left|\int_{-\infty}^{+\infty} \mathrm{d}t\, \langle\Psi(t)|j(t)|\Psi(t)\rangle\right|^2.$$

It describes the radiated power, while the atomic electron undergoes the dynamics. Another important physical quantity of interest is the ATI spectrum, which corresponds to the energy density spectrum of the electron in the continuum. It can be calculated by projecting the simulated, time-dependent wave function $|\Psi(t)\rangle$ that evolves from the initially unperturbed ground state of the atom $|\Psi^0\rangle = |\Psi(t \to -\infty)\rangle$ under the influence of the external laser pulse onto free-electron continuum states $|\Phi_n\rangle$. The transition amplitudes involved are obtained via

$$\langle\Phi_n|U(t, -\infty)|\Psi^0\rangle = \left\langle\Phi_n\middle|T\exp\left\{\int_{-\infty}^{t} \mathrm{d}t'\, H_\mathrm{I}(t')\right\}\middle|\Psi^0\right\rangle,$$

where the interaction Hamiltonian contains the time-dependent laser pulse.

Generic features of the results for both HHG and ATI spectra obtained in such model simulations have been recently provided (Blase 2001). In particular, the simulations

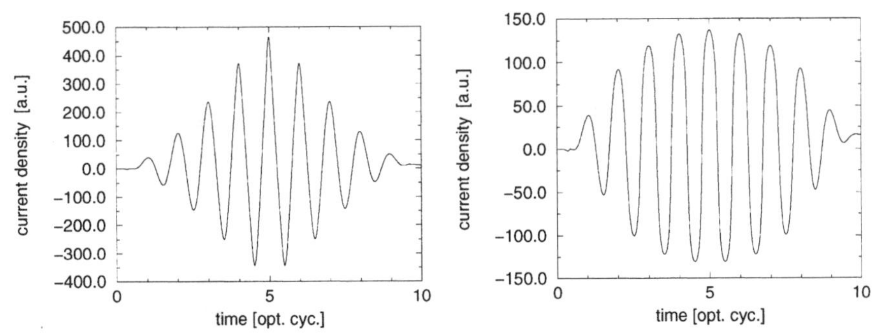

Figure 1.1 Electron current density as a function of time. The laser pulse is characterized by a frequency $\omega = 27.21$ eV and an intensity $I = 3.51 \times 10^{21}$ W cm^{-2} $\equiv 10^5$ a.u. Nonrelativistic (left part) and relativistic (right part) results are shown.

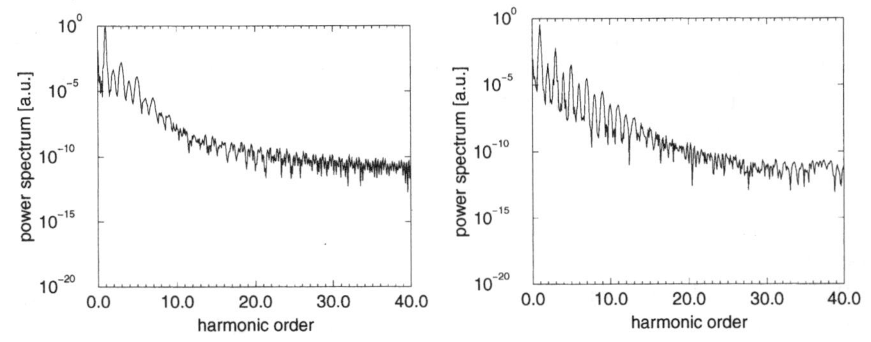

Figure 1.2 Generated HHG spectrum after 10 optical cycles. Parameters of the laser pulse are the same as in the previous figure. Nonrelativistic (left) and relativistic case (right).

have been carried out for hydrogen. The atomic system is considered as being initially in the 1s ground state. The laser pulse was modelled by means of a $\sin^2(\frac{1}{2}\omega t)$-shaped external field assuming a laser frequency $\omega = 27.2$ eV and a pulse duration between 6 and up to 20 cycles. In order to investigate the role of relativistic effects results were obtained for laser intensities I varying between $I = 1$ a.u. $\simeq 3.51 \times 10^{16}$ W cm^{-2} and $I = 10^5$ a.u. $\simeq 3.51 \times 10^{21}$ W cm^{-2}. For laser intensities $I = 1$ a.u., apart from slight details, the electron-current density and the HHG as well as the ATI spectra turn out to be more or less the same in both simulations, i.e. in the nonrelativistic Schrödinger and the relativistic Dirac case. For the considered laser pulse with an intensity $I = 3.51 \times 10^{16}$ W cm^{-2}, no influence of relativistic effects can be observed. However, the situation changes significantly when the intensity of the laser pulse is increased. In Figures 1.1–1.3 the corresponding results for the same quantities are displayed for a much higher intensity $I = 3.51 \times 10^{21}$ W cm^{-2} $= 10^5$ a.u., which is five orders of magnitude larger than typical atomic intensities.

While in the case of low-intensity laser pulses, $I = 1$ a.u., we can hardly recognize any differences between the Schrödinger and the Dirac simulations, for large inten-

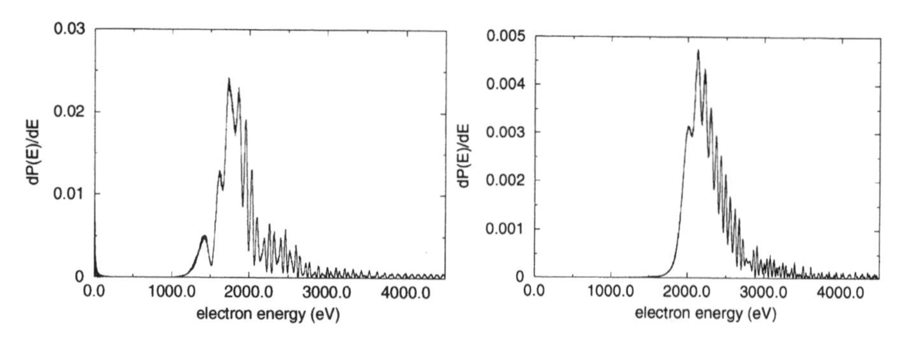

Figure 1.3 ATI spectrum after 10 optical cycles. Otherwise the same as in the previous figure.

sities, $I = 10^5$ a.u., nonrelativistic and relativistic simulations can now be clearly distinguished. Looking, for example, at the densities displayed in Figure 1.1, we observe that the maximum of the nonrelativistic electron-current density (left part) amounts to about 500 a.u. According to the classical interpretation, we can relate the current density j and the probability density ρ to a classical electron velocity v via $j = \rho v$, which exceeds the velocity of light c at the maximum nonrelativistic value. This consideration may already indicate the breakdown of the nonrelativistic Schrödinger theory and that the relativistic Dirac equation has to be used in order to provide a consistent description of the behaviour of atomic electrons in very intense laser fields. Accordingly, the relativistic simulation (right part) in which the maximum of the envelope of the current density flattens out and thus does not give rise to an electron velocity greater than c at any time. Figure 1.2 reveals that in very intense laser fields harmonic generation of very large order becomes possible. This conclusion may already be drawn from the occurrence of a rich peak structure in the emitted power spectrum up to high photon energies. Note that although the nonrelativistic (left) and relativistic simulations (right) show significant qualitative differences in the details, the spectra appear to be quite similar. As shown in Figure 1.2 the corresponding ATI spectra exhibit a remarkable difference. In comparison with the nonrelativistic simulation (left) the relativistic ATI spectrum (right) the peak is shifted towards lower energies. Note that there is a nonvanishing probability that the electron can be excited into continuum states with energies of several keV.

It is also interesting to examine the situation of the time-evolution of bound electrons under the influence of intense laser pulses in hydrogen-like ions with considerably higher nuclear charge numbers Z. In order to provide some knowledge about the generic effect of strong electron binding, we discuss some results for hydrogen-like neon ($Z = 10$). In this case, the initially bound 1s electron has a binding energy nearly 100 times larger than that in the hydrogen atom. The electron wave function of such tightly bound electrons strongly localized in the vicinity of the nucleus. Accordingly, as mentioned above, the number of spatial and temporal grid points has to enlarged. To give an example, the corresponding results of relativistic simulations of the wave function for $Z = 10$ are shown in Figure 1.4 (Blase 2001).

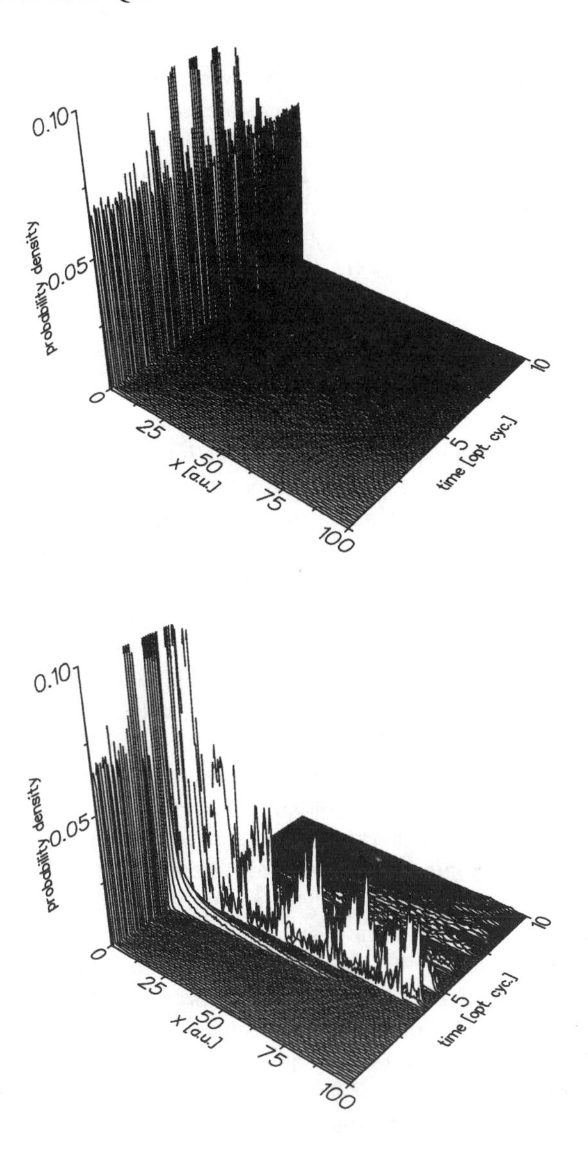

Figure 1.4 Relativistic simulation of the time-evolution of the ground-state electron density of hydrogen-like neon during 10 optical cycles of the laser field ($\omega = 54.4$ eV). Upper part: assuming an intensity $I = 3.51 \times 10^{19}$ W cm^{-2}, the electron remains tightly bound and the probability for ionization is negligible. Lower part: assuming an intensity $I = 3.51 \times 10^{20}$ W cm^{-2}, the electron becomes ionized.

Figure 1.4 displays the simulated probability-density distribution for finding the ground-state electron at a finite distance x (a.u.) away from the nucleus after a time (measured in units of optical cycles) corresponding to 10 laser-field cycles. The important feature to be pointed out here is that for a laser intensity of $I = 10^3$ a.u. a tightly bound electron remains near to the vicinity of the nucleus during the duration of the laser pulse (upper part of Figure 1.4). This indicates that the ionization probability is negligible. However, the situation changes drastically with increasing intensity; for example, as indicated in the lower part of Figure 1.4, a further increase in the intensity by one order of magnitude now has the consequence that after four and a half cycles the electron density has already propagated far out from the nucleus and reached the edge of the simulation grid ($x > 100$ a.u.).

1.2.2 Laser–electron interaction from classical electrodynamics

From the considerations of the previous subsection, it is evident that a fully relativistic description of the laser–electron interaction within the framework of QED still represents a challenge for present and future research efforts. In view of this situation it is also interesting to consider the problem of laser–electron interaction within the framework of classical electrodynamics. It is known that semiclassical approaches in nonrelativistic quantum mechanics allow for a description of atomic physics processes whenever large quantum numbers are involved. In laser–electron interactions the laser field can be described as a classical external field. Electronic excitation processes taking place in strong fields imply large energy scales and quantum numbers as well. To a good approximation, the atomic electron interacting with very intense laser fields can be considered as being 'born' freely in a continuum state. Furthermore, classical electrodynamics is capable of accounting for relativistic effects because of its Lorentz invariance. In order to check whether or not the behaviour of the ionized electron interacting with intense laser fields can be described in the framework of classical electrodynamics, the so-called *ponderomotive scattering* of electrons emitted via ionization from neutral atom near the focus of a very intense laser pulse has been investigated theoretically by several authors (Corkum *et al.* 1992; Reiss 1990, 1996; Salamin and Faisal 1997). Thereby, the ionized electron is scattered by the force derived as the gradient of the ponderomotive potential of the laser field. Provided the electron is ionized with sufficiently high energy, it may penetrate the laser beam and undergo a scattering process (Bucksbaum *et al.* 1987; Bula *et al.* 1996; Hartemann *et al.* 1995; Kibble 1966; Meyerhofer *et al.* 1996; Moore *et al.* 1995). In recent experiments (Meyerhofer *et al.* 1996; Moore *et al.* 1995) that used electrons generated by ionization near the focus of a circularly polarized laser pulse, the ponderomotive scattering has been measured. The experimental results have been found to be consistent with the assumption that the ionized electron has initially almost zero kinetic energy before it gets highly accelerated and undergoes ponderomotive scattering in the intense laser field. However, it is important to analyse the dependence of the measured results on the initial conditions.

Salamin and Faisal (1996, 1997) analysed harmonic generation and the ponderomotive scattering of electrons in intense laser fields based on a classical relativistic

Hamilton–Jacobi theory of charged point particles in external electromagnetic fields. For laser pulses with arbitrary intensity, polarization and pulse shape under rather general initial conditions, exact expressions for the ponderomotive scattering angles of electrons have been derived. This also allowed earlier investigations to be generalized (Eberly 1969; Sarachik and Schappert 1970). The results obtained within this classical approach have been employed in the analysis of the measurements (Meyerhofer *et al.* 1996).

Let us briefly recall the essentials of the relativistic Hamilton–Jacobi approach (we refer to Salamin and Faisal (1996, 1997) for a detailed discussion). Assume that initially the electron is incident on the focal region of the laser pulse with an arbitrary velocity $\boldsymbol{v}_0 = c\boldsymbol{\beta}_0$, where c denotes the velocity of light. Accordingly, the electron possesses an initial energy $E_0 = \gamma_0 mc^2 = (1 - \beta_0^2)^{-1/2}$ and a canonical momentum $\boldsymbol{\pi}_0 = \gamma_0 mc\boldsymbol{\beta}_0$. Accordingly, the kinetic energy of the electron K is obtained from the energy E by subtracting its rest energy $K = (\gamma - 1) mc^2$. Assuming elliptical polarization, the laser pulse may be modelled by a transversal vector potential $(\nabla \cdot \boldsymbol{A} = 0)$

$$\boldsymbol{A}(\eta) = a(\boldsymbol{\epsilon}_1 \, \delta \cos \eta + \boldsymbol{\epsilon}_2 \sqrt{1 - \delta^2} \, \sin \eta) \, g(\eta).$$

Here a denotes the maximum field amplitude, η is the ellipticity together with the pulse-shape function $g(\eta)$, which depends on the phase $\eta = \omega t - \boldsymbol{k} \cdot \boldsymbol{r}$. The laser beam is characterized by the frequency ω and the wave vector $\boldsymbol{k}$ with $c\boldsymbol{k} = \omega$. The transversality condition implies $\boldsymbol{k} \cdot \boldsymbol{A} = 0$. For a charged point particle interacting with this external electromagnetic field, the Hamilton–Jacobi equation reads

$$\left(\frac{\partial S}{\partial t}\right)^2 = c^2 \left(\nabla S + \frac{e}{c} \boldsymbol{A}\right)^2 + (mc^2)^2, \tag{1.2}$$

where S denotes Hamilton's principal function. The latter may be assumed to be of the form

$$S(\boldsymbol{r}, t) = \boldsymbol{s} \cdot \boldsymbol{r} + \xi ct + F(\eta).$$

The constant vector $\boldsymbol{s}$ and the constant ξ are determined by the initial conditions. Insertion of this ansatz into (1.2) together with the transversality condition for the vector potential allows for the solution (Sarachik and Schappert 1970)

$$F(\eta) = \tfrac{1}{2}(\boldsymbol{s} \cdot \boldsymbol{k} + \xi k)^{-1} \int_{\eta_0}^{\eta} \mathrm{d}\eta' \left[s^2 - \xi^2 + (mc)^2 + 2\frac{e}{c} \, \boldsymbol{s} \cdot \boldsymbol{A}(\eta') + \frac{e^2}{c^2} \, \boldsymbol{A}^2(\eta') \right].$$

The initial conditions for the electron motion mentioned above are consistent with the choice $\boldsymbol{s} = \gamma_0 mc \, \boldsymbol{\beta}_0$ and $\xi = -\gamma_0 mc$, respectively (Salamin and Faisal 1996). Given this result for S the energy of the electron can be derived via $E(\eta) = -\partial S / \partial t$. Subtracting the rest energy the kinetic energy of the electron can be expressed as

$$K(\eta) = K_0 + \frac{\gamma_0 mc^2}{1 - \hat{\boldsymbol{k}} \cdot \boldsymbol{\beta}_0} \left[\frac{1}{2} \left(\frac{e A(\eta)}{\gamma_0 mc^2} \right)^2 + \frac{e \boldsymbol{A}(\eta) \cdot \boldsymbol{\beta}_0}{\gamma_0 mc^2} \right].$$

Here $\boldsymbol{k} = (\omega/c)\hat{\boldsymbol{k}}$ defines the propagation direction of the laser beam. Employing the relation between the canonical momentum, the principal function S and the kinetic momentum of the electron, i.e. $\boldsymbol{\pi} = \nabla S = \boldsymbol{p} - (e/c)\boldsymbol{A}$, yields

$$\boldsymbol{p}(\eta) = \gamma_0 mc \left\{ \boldsymbol{\beta}_0 + \frac{e\boldsymbol{A}(\eta)}{\gamma_0 mc^2} + \frac{\hat{\boldsymbol{k}}}{1 - \hat{\boldsymbol{k}} \cdot \boldsymbol{\beta}_0} \left[\frac{1}{2} \left(\frac{e\boldsymbol{A}(\eta)}{\gamma_0 mc^2} \right)^2 + \frac{e\boldsymbol{A}(\eta) \cdot \boldsymbol{\beta}_0}{\gamma_0 mc^2} \right] \right\}.$$

The energy and the kinetic momentum of the electron derived above are both consistent with the initial conditions.

The initial velocity $\boldsymbol{\beta}_0$ of the electron may be represented with respect to the local polar coordinate system with the origin at the electron–laser crossing beam

$$\boldsymbol{\beta}_0 = \beta_0 \, (\sin\theta_0 \cos\phi_0 \, \boldsymbol{\epsilon}_1 + \sin\theta_0 \sin\phi_0 \, \boldsymbol{\epsilon}_2 + \cos\theta_0 \, \hat{\boldsymbol{k}}),$$

where the angles θ_0, ϕ_0 determine the initial orientation of the electron velocity. With respect to the same coordinate system we can describe the ponderomotive scattering of the electron in terms of the angles θ and ϕ, respectively. It is now straightforward to derive explicit expressions for the ponderomotive scattering angles for arbitrary initial conditions as function of the electron kinetic energy and the parameters of the laser pulse:

$$\begin{aligned} \tan\theta &= \frac{\sqrt{(\boldsymbol{p} \cdot \boldsymbol{\varepsilon}_1)^2 + (\boldsymbol{p} \cdot \boldsymbol{\varepsilon}_2)^2}}{\boldsymbol{p} \cdot \hat{\boldsymbol{k}}} \\ &= \frac{\sqrt{\beta_0^2 \sin\theta_0 + 2(1 - \beta_0 \cos\theta_0)(K - K_0)/\gamma_0 mc^2}}{\beta_0 \cos\theta_0 + (K - K_0)/\gamma_0 mc^2} \end{aligned}$$

and

$$\tan\phi = \frac{\beta_0 \sin\theta_0 \sin\phi_0 + ae\, g(\eta)/(mc^2\gamma_0)\, \sqrt{1 - \delta^2} \sin\eta}{\beta_0 \sin\theta_0 \cos\phi_0 + ae\delta\, g(\eta)/(mc^2\gamma_0)\, \cos\eta}.$$

Note that in the equation for θ all reference to the radiation field has been eliminated in favour of the kinetic-energy difference $K - K_0$ of the electron. By this means the scattering angle θ becomes a functional of the laser-field amplitude $\boldsymbol{A}(\eta)$ and its phase η. The predicted curve for the angle θ as function of the escape kinetic energy K holds for any laser field. It is the scattering angle ϕ that reflects specific features of the laser pulse.

The results of this classical relativistic calculations (Salamin and Faisal 1997) are presented in Figure 1.5 and compared with experimentally deduced ponderomotive scattering angles θ (Meyerhofer *et al.* 1996; Moore *et al.* 1995) as a function of the escape kinetic energy K of the electron.

Various initial conditions for the emitted electron are depicted and it can be concluded that the measured results are well reproduced by the classical theory based on the assumption that the electrons are 'born' in the laser field with small kinetic energy K_0 and move essentially in the direction of the polarization of the field, i.e.

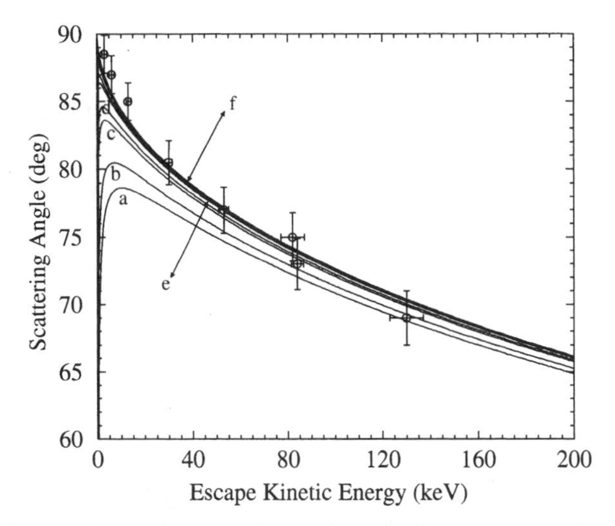

Figure 1.5 The ponderomotive scattering angle θ (in degrees) versus the escape kinetic energy of the electron (in keV) for initial kinetic energies $K_0 = 0.1, 1, 10$ and 100 eV and for initial scattering angles $\theta_0 = 0$, $\pi/4$ and $\pi/2$. Experimental data have been taken from (Meyerhofer *et al.* 1996). The various curves for a given pair of initial conditions (θ_0, K_0) are as follows: a, (0, 100); b, $(\pi/4, 100)$; c, (0, 10); d, $(\pi/4, 10)$; e, (0, 1) and $(\pi/4, 1)$; and f, (0, 0.1), $(\pi/4, 0.1)$, $(\pi/2, 0.1)$, $(\pi/2, 1)$, $(\pi/2, 10)$, $(\pi/2, 100)$ and (0, 0).

$\theta_0 \sim \pi/2$. It is then accelerated to higher energies by the field. In very intense laser fields the electron can already appear in a region of lower intensity before the pulse maximum arrives and it can then be accelerated by the high-intensity peak.

As another consequence the interaction between ionized (free) electrons with very intense laser fields results in a highly nonlinear Thomson scattering of the incident laser photons. This gives rise to a variation in the angular distribution of the emitted (scattered) photon of the same energy as that of the incident laser photon. Without intending to review the theoretical description (see Salamin and Faisal 1996, for detailed derivations), we mention only the following. For the derivation of a cross-section it is convenient to start from the (time-) averaged scattered radiant power $\mathrm{d}P_{\mathrm{R}}$ per unit solid angle $\mathrm{d}\Omega_{\mathrm{R}}$ with respect to the averaged rest frame (R) of the electron and to Lorentz-transform the quantities into the laboratory (L) frame. The total scaled cross-section for the observed radiated power (summed over all harmonic contributions n) is obtained by dividing the averaged power $\mathrm{d}P_{\mathrm{L}}$ observed per solid angle $\mathrm{d}\Omega_{\mathrm{L}}$ by the incident laser-beam intensity $I = (eq\omega_{\mathrm{L}}^{(0)})^2/(8\pi c r_0^2)$:

$$\frac{1}{r_0^2}\frac{\mathrm{d}\sigma_{\mathrm{L}}}{\mathrm{d}\Omega_{\mathrm{L}}} \equiv \frac{1}{r_0^2 I}\frac{\mathrm{d}P_{\mathrm{L}}}{\mathrm{d}\Omega_{\mathrm{L}}}.$$

Here $\omega_{\mathrm{L}}^{(0)}$ denotes the observed (L-frame) frequency of the incident laser frequency. Figure 1.6 indicates the generic features based on the classical approach (Salamin and Faisal 1996). The calculated angular distribution of the scattering cross-section

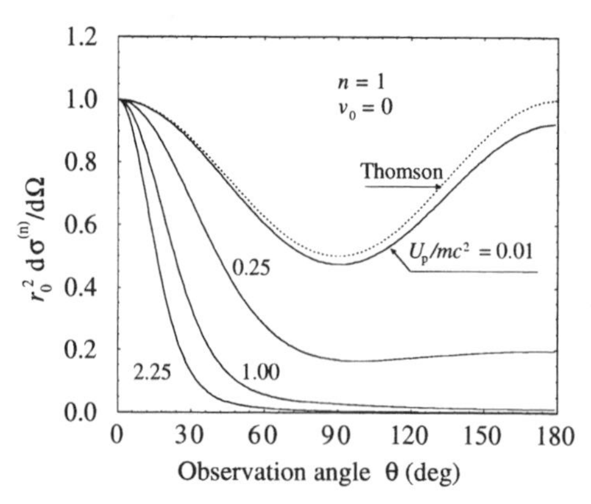

Figure 1.6 Angular distribution of the scaled differential cross-section $(1/r_0^2)(\mathrm{d}\sigma^{(1)}/\mathrm{d}\Omega)$ of the first harmonic $n = 1$. Vanishing initial velocity $\beta_0 = 0$ is assumed and $r_0 = 2.82 \times 10^{-13}$ cm denotes the classical electron radius. The dotted line corresponds to the ordinary Thomson scattering cross-section.

of the first harmonic ($n = 1$) is plotted versus the observation angle θ with respect to the laboratory frame.

The different curves correspond to various values of the ponderomotive energy measured in terms of $U_\mathrm{p}/(mc^2)$ (see Equation (1.1)). For a comparison, the ordinary Thomson cross-section is also plotted in order to demonstrate the effect of high-intensity laser fields. Increasing the laser intensity (measured in $U_\mathrm{p}/(mc^2)$), but keeping the laser frequency fixed, strong deviations in the cross-section occur, with the general tendency that it becomes highly asymmetrical with respect to the direction perpendicular to the incident laser light. Thus, the well-known $(1+\cos^2\theta)/2$ behaviour characteristic of the linear regime of Thomson scattering changes drastically. For increasing intensity the cross-section becomes peaked towards $\theta = 0$ and the backward scattering diminishes. In other words, in very intense laser fields the forward–backward symmetry, which is characteristic of the Thomson regime, is broken. Radiation due to Thomson scattering occurs only in the forward direction. The spectrum of emitted radiation can be analysed in terms of multiple Compton harmonics of the incident photons scattered off from the free electron due to the presence of intense laser fields (see also Salamin and Faisal 1999, 2000). A typical sequence of spectra measured in terms of scaled emission cross-sections $r_0^{-2}\,\mathrm{d}\sigma_\mathrm{L}^{(n)}/\mathrm{d}\Omega_\mathrm{L}$ for increasing intensities $U_\mathrm{p}/(mc^2)$ is plotted in Figure 1.7 versus the order of the harmonic n.

For low harmonics of radiation, the spectra exhibit a nonlinear dependence even at a logarithmic scale. Moreover, in this region the emission cross-section is found to be larger for higher harmonics, sometimes at the expense of contributions of lower order (e.g. as for $n = 1$). It is also seen that with increasing intensities the harmonics have comparable emission strengths that can extend to very large harmonic orders. Furthermore, the semi-logarithmic plot reveals a remarkably simple asymptotic dependence in the sense that it turns over into straight lines for large orders of n. This pro-

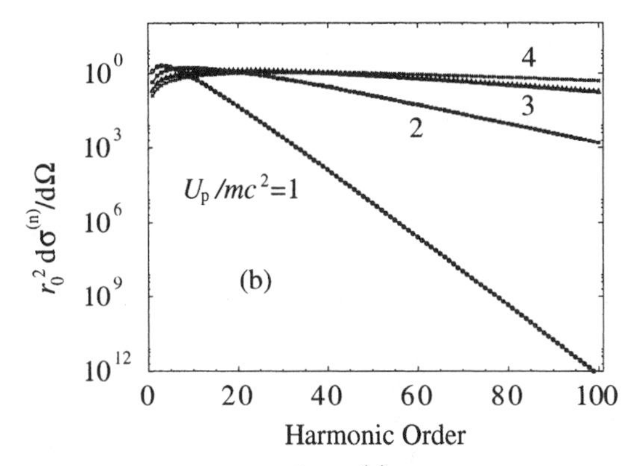

Figure 1.7 Differential cross-section $(1/r_0^2)(d\sigma_L^{(n)}/d\Omega_L)$ against the harmonic order n for ponderomotive potential $U_p/(mc^2) = 1, 2, 3, 4$ and a fixed observation angle $\theta = 40°$.

vides the possibility of obtaining the contributions from even higher harmonics by linear extrapolation. The classical electrodynamical approach turns out to be capable of analysing important features of the electron–laser interaction in very intense fields. However, we should emphasize that investigations of relativistic effects in the electron–laser interaction for realistic scenarios require a fully relativistic quantum mechanical description.

1.3 Electron–Positron Pair Creation in Relativistic Heavy-Ion Collisions

The investigation of electron–positron pair production due to the strong time-varying electromagnetic fields that occur during relativistic collisions of highly charged heavy ions allows us to probe quantum electrodynamics at small distances as well as to determine the interaction between fast projectiles and matter. Electron–positron (lepton–antilepton) pair production in relativistic heavy-ion collisions has received great attention in recent years. Due to the development of heavy-ion accelerators, numerous calculations have been devoted to this subject employing both perturbative and nonperturbative methods (see Bertulani and Baur 1988; Eichler and Meyerhof 1995; Soff 1980; Wu *et al.* 1999, and references therein). A review of recent developments can also be found in Becker *et al.* (1987), Thiel *et al.* (1994) and Tenzer *et al.* (2000a). Lepton-index pair production in relativistic collisions of highly charged ions at small impact parameters is a nonperturbative process. This is evident in view of the very strong electromagnetic fields generated during the collision. For systems with large nuclear charge numbers Z the effective coupling constant $Z\alpha$ is not small compared with one. Accordingly, a proper description requires a nonperturbative numerical solution of the Dirac equation in the strong, time-dependent external field generated by the colliding ions. Being able to provide efficient numerical codes to generate exact solutions of the time-dependent Dirac equation allows for a description of various dynamical processes, such as electronic excitations, ionization and charge transfer as well. Here we report on recent developments and achievements.

Basically, one distinguishes between pair creation, where the electron and positron are produced in free (continuum) states and pair creation with an electron in a bound state of one of the ions, for example, in a K-shell state. The latter process is also called bound-free pair production. Pair creation with electron capture into a bound state of the ion changes the charge state of the ion represents one of the major loss processes affecting the stability of ion beams in relativistic heavy-ion colliders. The corresponding cross-section for electron capture is of the order of 100 barn for RHIC energies in $U^{92+}(100\ \mathrm{GeV/u}) + U^{92+}(100\ \mathrm{GeV/u})$ collisions and determines the lifetime of ionic charge states in the beam and then the luminosity.

Electron–positron pair creation is usually calculated within the semiclassical approximation. Within this approach the motion of the ions is treated classically and described by constant velocities and straight-line trajectories. The electromagnetic fields generated by the ions are considered as classical fields, while the electron–positron field is quantized. It obeys the Dirac equation in the presence of a time-dependent external electromagnetic field. The field strengths that occur during the collision depend on the nuclear charge numbers Z of the ions and on the impact parameter. Depending on the particular scenario, simplifying approximations can be employed. In the case of lower Z and larger impact parameters, time-dependent perturbation theory of first or second order can be applied yielding reasonable probabilities for the pair creation. In the ultrarelativistic regime of extremely high incident energies, taking the limit $\gamma \to \infty$ for the Lorentz factor is legitimate. For this limiting case an exact solution of the Dirac equation has been found (Baltz 1997) that allows the calculation of cross-sections in the light-cone approximation. We shall return to this situation below.

Most of the difficulties occur in the regime of moderate relativistic energies together with impact parameters smaller than the reduced Compton wavelength, i.e. $b \leqslant \lambda\!\!\!^{-} = \hbar/(mc) = 386\ \mathrm{fm}$, and for highly charged heavy ions. This characterizes the regime, where the time-dependent perturbation theory fails and the electron–positron pair creation becomes an inherently nonperturbative process. Results on electron–positron pair production have been obtained within the framework of two approaches for solving the Dirac equation nonperturbatively. The first method is based on the solution of coupled-channel equations employing momentum eigenfunctions (Tenzer *et al.* 2000b) and the second one carries out a direct integration with the help of the finite-element method (Busic *et al.* 1999a).

1.3.1 Theoretical framework

Let us first summarize the results of the field-theoretical description of pair creation. For details we may refer to Eichler and Meyerhof (1995) and Strayer *et al.* (1990). Electron–positron pair creation can be viewed as an excitation of an electron from the (occupied) negative energy continuum into a positive energy bound or continuum state leaving a hole in the Dirac sea. The latter is reinterpreted as a free positron. Assuming that the (vacuum) ground state is a many-body state of noninteracting electrons, the total Hamiltonian may be decomposed into an unperturbed (time-independent) part H_0 and a time-dependent interaction part $H_{\mathrm{I}}(t)$. This decomposition of the Hamiltonian is not unique and may be performed suitably depending on the particular situation under consideration and with respect to physical initial conditions. However, we may assume that the interaction part vanishes for asymptotic times, i.e. $H_{\mathrm{I}}(|t| \to \infty) = 0$. The proper definition of (quasi-)particles is provided by diagonalization of the unperturbed Hamiltonian H_0.

With respect to the rest frame of the target ion we can write

$$H_0 = c\boldsymbol{\alpha} \cdot \boldsymbol{p} + \beta mc^2 + V_{\mathrm{T}}(r) \tag{1.3}$$

with the static Coulomb potential of the target nucleus. We can specify the one-particle states of the target Hamiltonian as Coulomb–Dirac wave functions

$$\psi_{n,\pm}(\boldsymbol{r}, t) = \phi_{n,\pm}(\boldsymbol{r})\, \mathrm{e}^{-\mathrm{i}E_{n,\pm}t}. \tag{1.4}$$

The subscripts '$(n, +)$' label positive-energy levels (bound and upper continuum states) and '$(n, -)$' refer to negative-energy continuum states, respectively. The time-dependent interaction H_{I} is due to the projectile ion giving rise to a time-dependent external electromagnetic four-potential $(A^0_{\mathrm{P}}(\boldsymbol{r}, t), \boldsymbol{A}_{\mathrm{P}}(\boldsymbol{r}, t))$. Assuming the projectile to move with a constant velocity v along a straight line (e.g. parallel to the z-axis) with impact parameter b, the time-dependent interaction Hamiltonian H_{I} is described by the Lorentz-boosted Coulomb potential of the projectile ion. The total Hamiltonian of the scattering system reads

$$H(t) = c\boldsymbol{\alpha} \cdot \boldsymbol{p} + \beta mc^2 + V_{\mathrm{T}}(r) + H_{\mathrm{I}},$$

where

$$H_{\mathrm{I}} = -e^2 \gamma\, Z_{\mathrm{P}} \frac{1 - v/c\, \alpha_z}{(x^2 + (y - b)^2 + \gamma^2 (z - vt)^2)^{1/2}}.$$

This total Dirac Hamiltonian possesses time-dependent solutions

$$H(t)\, \chi_{n,\pm}(\boldsymbol{r}, t) = \mathrm{i}\hbar \frac{\partial}{\partial t} \chi_{n,\pm}(\boldsymbol{r}, t) \tag{1.5}$$

with the initial condition that in the infinite past the solutions $\chi_{n,\pm}$ refer to a solution $\psi_{n,\pm}$ of the unperturbed Hamiltonian with a defined sign of energy and same quantum numbers n, i.e.

$$\lim_{t \to -\infty} \chi_{n,\pm}(\boldsymbol{r}, t) = \phi_{n,\pm}(\boldsymbol{r})\, \mathrm{e}^{-\mathrm{i}E_{n,\pm}t}.$$

The field operator can be represented at first in terms of eigenfunctions (1.4)

$$\psi(\boldsymbol{r}, t) = \sum_{n,+} a_{n,+}(t)\, \phi_{n,+}(\boldsymbol{r})\, \mathrm{e}^{-\mathrm{i}E_{n,+}t} + \sum_{n,-} b^{\dagger}_{n,-}(t)\, \phi_{n,-}(\boldsymbol{r})\, \mathrm{e}^{-\mathrm{i}E_{n,-}t}$$

with time-evolved (electron) annihilation operators $a_{n,+}$ and (positron) creation operators $b^{\dagger}_{n,-} = a_{n,-}$, respectively. Equivalently, we may decompose the (Heisenberg) field operator with respect to the (one-particle) basis set (1.5)

$$\psi(\boldsymbol{r}, t) = \sum_{n,+} \bar{a}_{n,+}\, \chi_{n,+}(\boldsymbol{r}, t) + \sum_{n,-} \bar{b}^{\dagger}_{n,-}\, \chi_{n,-}(\boldsymbol{r}, t)$$

with time-independent (one-particle) operators $\bar{a}_{n,+}$ and $\bar{b}^{\dagger}_{n,-} = \bar{a}_{n,-}$. The above relations have the following physical interpretation. Under the influence of the time-dependent interaction with the projectile ion, any positive-energy state $\phi_{n,+}$ considered at asymptotic time $t \to -\infty$ will become an admixture of positive- and negative-energy states $\chi_{n,\pm}$ after the time-evolution to $t \to +\infty$. Conversely, any positive-energy state $\chi_{n,\pm}$ at $t \to +\infty$ appears as admixture of positive- and negative-energy initial states $\phi_{n,\pm}$. Accordingly, the number of electrons $N_{n,+}$ created in the state $(n, +)$ and equivalently the number of positrons $N_{n,-}$ created referring to the initially occupied negative-energy state $(n, -)$ are given by

$$N_{n,+} = \sum_{m,-} |\langle \phi_{n,+} | \chi_{m,-}(+\infty) \rangle|^2, \qquad N_{n,-} = \sum_{m,+} |\langle \phi_{n,-} | \chi_{m,+}(+\infty) \rangle|^2.$$

Employing time-reversal symmetry, the number of generated electrons can be rewritten as

$$N_{n,+} = \sum_{m,-} |\langle \phi_{m,-} | \chi_{n,+}(+\infty) \rangle|^2. \tag{1.6}$$

This expression is much more convenient for practical evaluations since only one state has to be evolved in time, namely the state $\chi_{n,+}$ referring to the final electron state under consideration. By these means the equation above may describe pair production with electron capture. It also yields ionization and electron-transfer probabilities of a hydrogen-like target. These probabilities are obtained by projecting $\chi_{n,+}$ onto the target and projectile states. Non-perturbative calculations of such processes require a fully numerical time-evolution of the Dirac equation.

1.3.2 Coupled-channel calculations

Let us briefly discuss the idea of and the results obtained from coupled-channel calculations with momentum eigenfunctions as it has been employed recently (Tenzer *et al.* 2000b). The collision is considered in a coordinate system, where the scattering ions have equal but opposite velocities $\boldsymbol{v} = \pm v_0 \boldsymbol{e}_z$. With respect to this system the total time-dependent Hamiltonian H is decomposed into the unperturbed (free) Dirac Hamiltonian

$$H_0 = c\boldsymbol{\alpha} \cdot \boldsymbol{p} + \beta mc^2,$$

Table 1.1 Cross-sections for free electron–positron pair production. Collision energies are given in units of GeV per nucleon (GeV/u).

		cross-section (barns)				
System	E_{lab} (GeV/u)	Exp.	coupled channels	Born approx.	PT (a)	PT (b)
$La^{57+} + Cu^{29+}$	1.3	0.30	0.54			
$La^{57+} + Ag^{47+}$	1.3	0.80	1.16	0.23		
$La^{57+} + Au^{79+}$	1.3	2.64	3.56	0.43		
$Au^{79+} + Cu^{29+}$	10.8	42	43.4	13.9	12	15
$Au^{79+} + Ag^{47+}$	10.8	85	92.2	41.3	31	40
$Au^{79+} + Au^{79+}$	10.8	180	212.0	91.2	87	113

and a time-dependent interaction part, which now appears as the sum of the Lorentz-boosted Coulomb potentials of both target and projectile ions,

$$H_I(t) = -e^2\gamma\left[\frac{Z_T}{r_+}\left(1+\frac{v_0}{c\,\alpha_z}\right)+\frac{Z_P}{r_-}\left(1-\frac{v_0}{c\,\alpha_z}\right)\right], \tag{1.7}$$

with

$$r_\pm = \sqrt{(x\pm\tfrac{1}{2}b)^2+y^2+\gamma^2(z\pm v_0t)^2}.$$

Target and projectile ions have the charge numbers Z_T and Z_P, respectively, b denotes the impact parameter and $\gamma = (1-(v_0/c)^2)^{-1/2}$ the Lorentz factor. The corresponding Dirac equation is solved by expanding the electron–positron field ψ into eigenstates of the free Hamiltonian H_0. Since these eigenstates are free continuum states with a momentum $\boldsymbol{p}$, it is convenient to discretize the positive and negative continuum leading to a system of coupled equations for the expansion coefficients of the electron–positron field ψ.

In Table 1.1 experimental and calculated cross-sections for free electron–positron pair production in collisions of La^{57+}(1.3 GeV/u) and Au^{79+}(10.8 GeV/u) on Cu^{29+}, Ag^{47+} and Au^{79+} are listed. The experimental data for the cross-sections are taken from Belkacem *et al.* (1997) and A. Belkacem (1999, personal communication) and compared with various theoretical predictions. The results obtained with the described coupled-channel method are also compared with predictions based on first-order Born approximation (Steih 1999) and with corresponding data obtained by means of perturbation theory as reported by Becker *et al.* (1986) (PT (a)) and by Ionescu and Eichler (1993) (PT (b)), respectively. In the latter case the numbers given are deduced from the scattering system $U^{92+}+U^{92+}$ by scaling with a factor $(Z_T Z_P)^2$. The perturbative calculations yield cross-sections that are too small. This fact indicates the necessity of using nonperturbative methods, e.g. coupled-channel procedures, for the calculation of cross-sections for electron–positron pair production.

1.3.3 Finite-element method

In a recent work (Busic *et al.* 1999a), the time-dependent Dirac equation has been solved by a finite-element method. For convenience, the rest frame of target nucleus has been chosen. For calculating the electron–positron pair production probabilities with electron capture into the K-shell of the target ion the time-reversal symmetry has been employed. According to (1.6) the $1s_{1/2}$ wave function has to be developed in time under the influence of the projectile ion. In order to obtain the proper asymptotic behaviour for the wave function at large distances we can make use of the gauge freedom and transform the wave function by a phase transformation according to Eichler and Meyerhof (1995). This leads to potentials falling off faster than the Coulomb potential for large internuclear distances. The projectile is assumed to move on a straight line parallel to the z-axis in the (y, z)-plane and the electron spin is quantized with respect to the x-axis. A three-dimensional, Cartesian lattice with equidistant grid points is introduced. Then the Dirac equation is transformed into a matrix equation by using basis functions written as a product $\phi_i(x)\phi_j(y)\phi_k(z)$ of three one-dimensional linear finite-element functions depending on a Cartesian coordinate. Each function fulfils the following condition $\phi_i(x_j) = \delta_{ij}$ at the grid points x_j. The time-evolution of the resulting matrix equation is carried out in discrete time-steps Δt, where higher powers of Δt are taken into account. This method has been applied to the study of the scattering systems (Busic *et al.* 1999a)

$$U^{92+}(\gamma = 1.5 \mathrel{\widehat{=}} 466\ \mathrm{MeV/u}) + U^{91+}$$

and

$$Au^{79+}(\gamma = 2 \mathrel{\widehat{=}} 930\ \mathrm{MeV/u}) + U^{91+}.$$

For a Lorentz factor $\gamma = 1.5$, the probability for pair creation turns out to be rather small. However, in this energy region we can study excitation, ionization and charge transfer into the ground state of the projectile. For collisions of

$$U^{92+}(\gamma = 1.5) + U^{91+},$$

typical results for the corresponding probabilities are displayed in Figure 1.8 in comparison with results obtained by perturbation theory. The spin is quantized with respect to a direction perpendicular to the scattering plane. Note that the perturbative ionization probability exceeds unity for small impact parameters. In Figure 1.9 the density distribution in the scattering plane at a time $t = 4780$ fm/c after the collision is given. The target is at rest in the right corner at $(y = 0, z = 0)$, where the residual probability of the ground state is located. The projectile appears underneath the second maximum in the front left corner. The charge transfer is clearly visible in the density distribution indicating that the electron has been torn away partly by the projectile ion.

In the case of the collision system $Au^{79+}(\gamma = 2) + U^{91+}$ calculations have been carried out with high precision. The time-step Δt is set to 25.4 fm/c. In order to achieve an error in the norm of the wave function less than 10^{-12}, the time-evolution

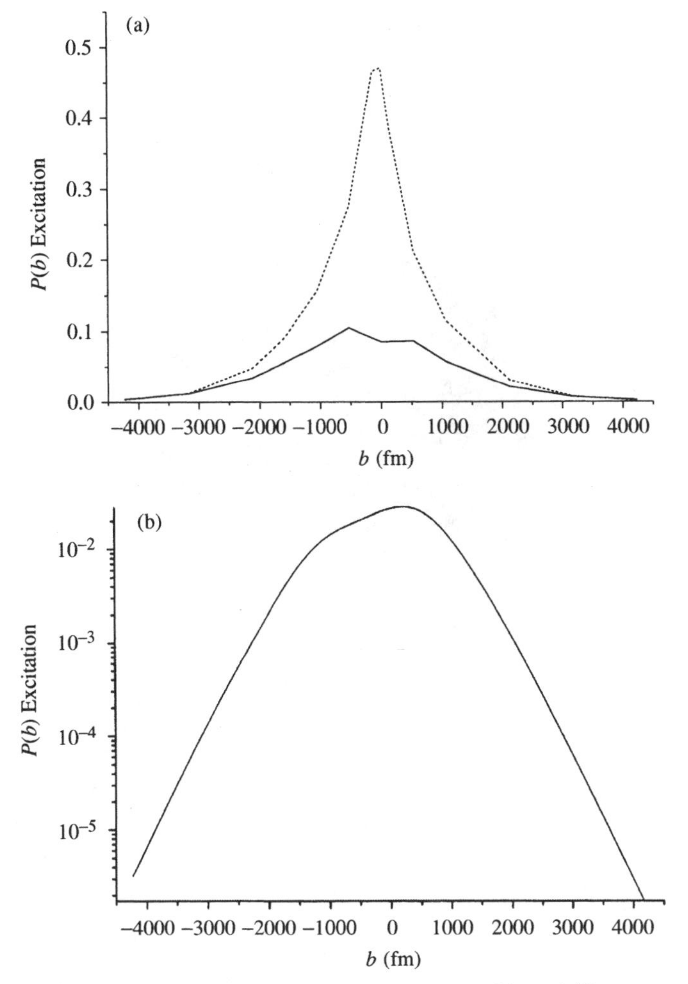

Figure 1.8 (a) Probability for excitation of the $1s_{1/2}$ state of U^{91+} in $U^{92+}(\gamma = 1.5)+U^{91+}$ collisions as a function of the impact parameter. The solid curve is calculated with the finite-element method, the dashed curve with perturbation theory. (b) The same for charge transfer into the ground state of the projectile ion.

operator has been expanded up to the 25th order in the Hamiltonian. The probabilities for pair production with electron capture are calculated by projecting the final wave function onto states of the negative energy continuum with angular momentum quantum numbers $|\kappa| \leqslant 8$. A probability for free-pair production of 3.1×10^{-4} at an impact parameter $b = 0$ is predicted in comparison with a value of 3.9×10^{-4} reported by Momberger *et al.* (1996). The lattice calculations yield 1.3 barn for the cross-section. However, this value disagrees considerably with the experimental value of 2.19 barn (Belkacem *et al.* 1993).

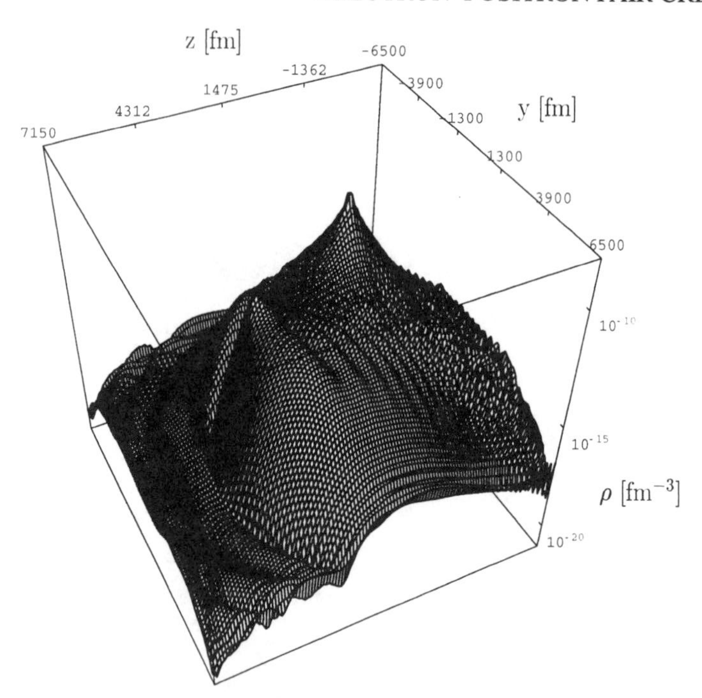

Figure 1.9 Probability density of the time-evolved ground state of U^{91+} in $U^{92+}(\gamma = 1.5) + U^{91+}$ collisions in the scattering plane after the collision for $b = 1060$ fm.

At high Lorentz factors γ, the effective width of the Lorentz-contracted projectile potentials becomes smaller than the lattice spacing and, therefore, the potentials are poorly described on a grid which is fixed in the target system. Hence, it is more convenient to introduce a coordinate frame, where both target and projectile move with equal but opposite velocities. Furthermore, the z-coordinate in the direction of motion is transformed to $z' = \gamma z$ to ensure that the lattice spacing is sufficiently small for a proper representation of the electromagnetic potentials. To give an example, we present calculations for $U^{92+}(\gamma = 10\,000) + U^{91+}$ collision at an impact parameter of $b = 530$ fm with the finite-element method (Busic *et al.* 1999a) in Table 1.2. The results for the probabilities for the excitation of five bound states are compared with those obtained in the ultrarelativistic limit $\gamma \to \infty$ (Baltz 1997) and with predictions based on perturbation theory. The values provided by the finite-element method are already in fair agreement with those calculated in the limit of infinite γ. As shown in Figure 1.10 the results obtained for ionization and pair-creation probabilities display a similar behaviour.

Some aspects of the so-called fermion doubling and how to avoid them in standard finite-element or finite-difference methods have been addressed recently by Busic *et al.* (1999b).

The examples presented above demonstrate that the finite-element method can be applied successfully to the investigation of processes such as excitation, ioniza-

Table 1.2 Probabilities for the excitation of the lowest bound states of U^{91+} in collisions $U^{92+}(\gamma = 10\,000) + U^{91+}$ at $b = 530$ fm.

state	limit $\gamma \to \infty$	finite elements	PT
$1s_{1/2}(+\frac{1}{2})$	0.524	0.528	—
$2s_{1/2}(+\frac{1}{2})$	4.565×10^{-2}	4.497×10^{-2}	8.628×10^{-2}
$2p_{1/2}(-\frac{1}{2})$	9.329×10^{-4}	9.765×10^{-4}	1.848×10^{-3}
$2p_{3/2}(-\frac{1}{2})$	5.049×10^{-3}	5.15×10^{-3}	1.169×10^{-2}
$2p_{3/2}(+\frac{3}{2})$	7.049×10^{-3}	6.949×10^{-3}	1.609×10^{-2}

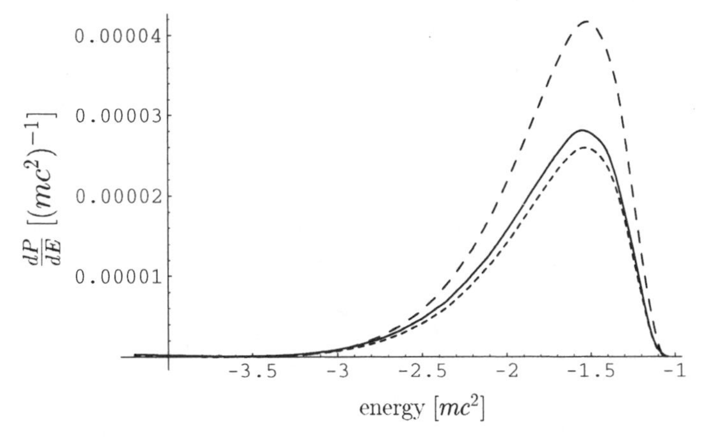

Figure 1.10 Spectrum of positrons in the $p_{1/2}$ ($m = -1/2$) state at $b = 530$ fm calculated at the ultrarelativistic limit (solid curve), with the finite-element method (short dashes) and with perturbation theory (longer dashes).

tion, charge transfer and electron–positron pair production in relativistic collisions of highly charged heavy ions. The results reveal that the considered reactions can only be correctly described by nonperturbative methods. However, further investigations with respect to the analysis of the states in the negative electron continuum have to be carried out in order to provide definite statements on the probabilities for electron–positron production.

1.3.4 Electromagnetic pair production: the ultrarelativistic limit

As we have mentioned above it is possible to evaluate the electromagnetic lepton pair production in the limiting case of infinite Lorentz factors γ. One interesting aspect among others is that peripheral heavy-ion collisions at ultrarelativistic energies offer

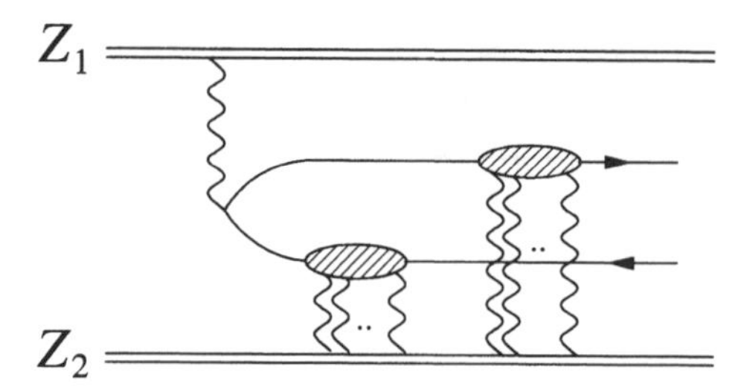

Figure 1.11 Type of diagrams contributing to pair production.

great opportunities for testing QED in the strongest, time-dependent, electromagnetic fields accessible in experiments.

Focusing on electron–positron pair creation in collisions of highly charged heavy ions, it is well known that in such strong external fields—for point-like nuclei the result dates back to the early days of QED (Landau and Lifshitz 1934; Racah 1937)—according to which the total pair creation cross-section, which behaves as $\ln^3 \gamma$, has to be extended in two main directions. First we have to consider multiple pair production and vacuum diagrams to restore the unitarity-violation of this result, and, secondly, any process considered has to be evaluated to all orders in the expansion parameter $Z\alpha$ since for heavy collision systems this effective coupling is of order 1. Assuming, however, the independence of multiple pair production, the single pair production probability may be interpreted as the mean number of created pairs (Best *et al.* 1992; Hencken *et al.* 1995; Rhoades-Brown and Weneser 1991). In this respect it is sufficient to consider only the impact of higher orders of this process and not to care about the unitarization procedure leading to the mentioned inclusion of multiple-pair production and vacuum diagrams.

The attempt to calculate pair production via the crossing invariance of the exact scattering amplitude has led to the conclusion that the full consideration of higher-order corrections to the cross-section vanishes (Baltz and McLerran 1998; Eichmann *et al.* 1999; Segev and Wells 1998). On the other hand, the straight calculation starting directly from pair-production diagrams results in rather large higher-order corrections (Ivanov *et al.* 1999).

In the following we briefly discuss both approaches in more detail. It has been shown that the calculation of the amplitude for lepton scattering in the field of ultrarelativistic colliding nuclei can be performed exactly (Eichmann 2000; Eichmann *et al.* 2000a). The interaction part of the Dirac Hamiltonian (1.7) for two colliding nuclei, referred to by the subscripts '(1)' and '(2)', at infinite energy $\gamma \to \infty$ reads

$$H_{\mathrm{I}}(t) = 2\pi[-\alpha_{-}\delta(t-z)V_{\perp}^{(1)}(\boldsymbol{r}_{\perp}) - \alpha_{+}\delta(t+z)V_{\perp}^{(2)}(\boldsymbol{r}_{\perp})], \tag{1.8}$$

where $\alpha_{\pm} = 1 \pm \alpha_z$ defines the light-cone components of the Dirac matrices. $V_{\perp}^{(1)}$ and $V_{\perp}^{(2)}$ denote the transverse parts of the transformed Coulomb potentials of the target and projectile nucleus

$$V_\perp^{(1)}(\boldsymbol{r}_\perp) = Z_1\alpha \ln r_\perp^2,$$
$$V_\perp^{(2)}(\boldsymbol{r}_\perp) = Z_2\alpha \ln(\boldsymbol{r}_\perp - \boldsymbol{b})^2,$$

with the impact parameter $\boldsymbol{b}$ of the colliding nuclei. Exact analytical solutions for the corresponding Dirac equation have been derived by Baltz and McLerran (1998) in terms of an integral representation involving free Dirac spinors. The resulting electron-scattering amplitude is

$$\begin{aligned}\mathcal{M}_{p',p} &= 2\pi \mathrm{i}\delta(p_+ - p'_+)\, F_2(\boldsymbol{p}'_\perp - \boldsymbol{p}_\perp)\, u^\dagger(p')\alpha_+ u(p)\\ &\quad + 2\pi \mathrm{i}\delta(p_- - p'_-)\, F_1(\boldsymbol{p}'_\perp - \boldsymbol{p}_\perp)\, u^\dagger(p')\alpha_- u(p)\\ &\quad + \mathrm{i}\int \frac{\mathrm{d}^2 k_\perp}{(2\pi)^2}\, F_2(\boldsymbol{k}_\perp - \boldsymbol{p}'_\perp)\, F_1(\boldsymbol{p}_\perp - \boldsymbol{k}_\perp)\\ &\quad \times \left(u^\dagger(p') \frac{-\boldsymbol{\alpha}_\perp \cdot \boldsymbol{k}_\perp + \beta m}{p_+ p'_- - k_\perp^2 - m^2 + \mathrm{i}\epsilon} \alpha_+ u(p) \right.\\ &\quad \left. + u^\dagger(p') \frac{-\boldsymbol{\alpha}_\perp \cdot (\boldsymbol{p}_\perp + \boldsymbol{p}'_\perp - \boldsymbol{k}_\perp) + \beta m}{p_- p'_+ - (\boldsymbol{p}_\perp + \boldsymbol{p}'_\perp - \boldsymbol{k}_\perp)^2 - m^2 + \mathrm{i}\epsilon} \alpha_- u(p) \right)\end{aligned}$$

with

$$F_{(1,2)}(\boldsymbol{q}) = \int \mathrm{d}^2 r_\perp\, \mathrm{e}^{\mathrm{i}\boldsymbol{q}\cdot\boldsymbol{r}_\perp} (\mathrm{e}^{-\mathrm{i}V_\perp^{(1,2)}(\boldsymbol{r}_\perp)} - 1).$$

The resulting amplitude coincides with the single- and double-scattering contributions to the Watson series of multiple scattering (Watson 1953), implying that alternating interactions of the lepton with both ions vanish.

The direct calculation of the pair-production process is mandatory. It turns out that an exact calculation of the pair-creation process is not possible because diagrams including higher-order interactions of the lepton pair with both ions cannot be entirely evaluated in closed form. One obvious reason for this is the fact that it has not yet been feasible to account for all possible classes of diagram.

General considerations permit the decomposition of the exact amplitude into leading and subleading terms (Ivanov *et al.* 1999). Approximate expressions can be derived for a selected set of leading contributions. For example, the type of diagram shown in Figure 1.11 has been found to give rise to a very large Coulomb correction (Eichmann *et al.* 2000a; Ivanov *et al.* 1999).

In logarithmic accuracy the above-mentioned class of diagrams may be neglected for the approximate calculation of pair production. Clearly, the leading contribution is the two-photon diagram. The next-to-leading contributions consist of all diagrams in which one of the ions only exchanges one photon and the respective other ion interacts in higher orders with the lepton pair. The resulting corrections to the lowest-order cross-section arise from the squared sum of these diagrams and the interference with the two-photon diagram. In the Weizsäcker–Williams approximation they can

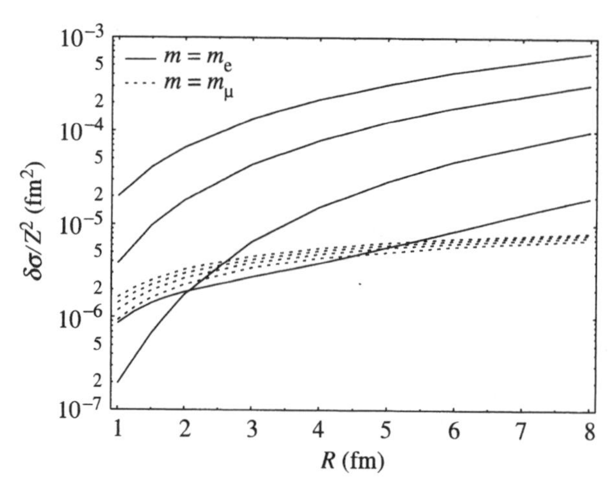

Figure 1.12 R-dependence of the nonCoulomb corrections for a homogeneously charged sphere. From top to bottom the nuclear charges are $Z = 10, 40, 60, 80$.

be expressed as the product of the photon distribution function of the ion interacting in first order and the higher-order corrections to the photoproduction of a lepton pair in the field of the ion interacting by multiphoton exchange.

An estimate of the higher-order effects on pair creation in ultrarelativistic heavy-ion collisions thus requires the investigation of high-energy photoproduction in the field of heavy ions (Eichmann 2000; Eichmann *et al.* 2000b). In the case of point-like nuclei these so-called Coulomb corrections were first calculated by Davis *et al.* (1954) employing Furry–Sommerfeld–Maue wave functions for the created leptons. Consideration of the extended nuclear-charge distribution leads to a qualitatively different behaviour of the production of light (e^+e^-) and heavy ($\mu^+\mu^-$), ($\tau^+\tau^-$) leptons.

The effect of finite nuclear extension on the total cross-section σ may be described in terms of a nonCoulomb correction $\delta\sigma$ to the point-charge cross-section σ^{point}, i.e.

$$\begin{aligned}\delta\sigma &= \sigma^{\text{point}} - \sigma \\ &= \frac{8\alpha m^2}{\pi}\int_0^{\pi} \mathrm{d}\varphi \int_0^1 \mathrm{d}r\, r \int_0^{\infty} \mathrm{d}x\, x^3 \\ &\quad \times (K_0^2(mx\xi) + \tfrac{2}{3}K_1^2(mx\xi))\,\mathrm{Re}(\mathrm{e}^{\mathrm{i}(\chi(xr)-\chi(x))} - \mathrm{e}^{2\mathrm{i}\nu\ln r})\end{aligned}$$

with $\xi = \sqrt{1 - 2r\cos\varphi + r^2}$. The eikonal

$$\chi(r) = 2Z\alpha \int \mathrm{d}^3r'\,\rho(r')\ln(|\boldsymbol{r}_\perp - \boldsymbol{r}'_\perp|)$$

depends on the nuclear charge-density distribution ρ.

Electron–positron pair production is rather unaffected by the charge distribution. It is sufficiently well described by assuming point-like nuclei. The tiny effects caused

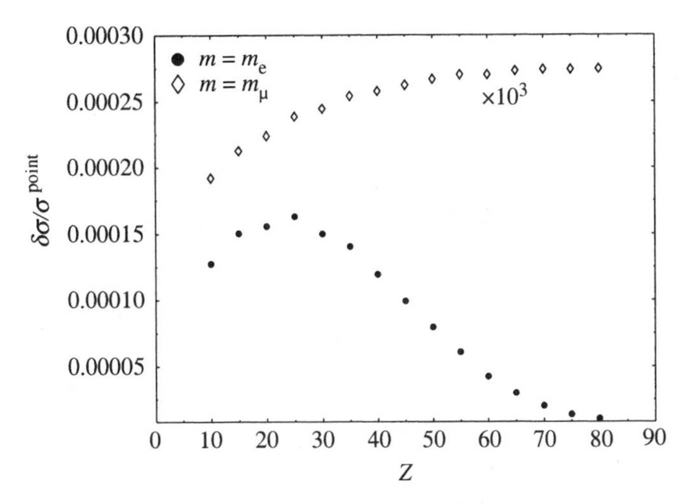

Figure 1.13 Corrections to the photoproduction cross-section relative to the point-charge value in the field of different nuclei. The nuclei are described by homogeneously charged spheres with $R = 1.12A^{1/3}$ fm, with A taken from the table of elements.

by the charge distribution may be assigned to an interesting coherence effect of multiple photon exchange which is similarly observed in the so-called eikonal form factor of the nucleus at small momentum transfers (Eichmann *et al.* 2000b). The influence of finite nuclear size on the nonCoulomb corrections is presented in Figure 1.12 for electron- and muon-pair production employing the homogeneously charged-sphere model. $\delta\sigma$ is found to increase with increasing nuclear radius R. However, the calculation of electron and muon pair production in the field of nuclei with different charge numbers Z shows that for electron pair production this rise is overcompensated by the rapid decrease of $\delta\sigma$ at large Z. For small nuclear charge- and mass-numbers the competing effects of the R- and the Z-dependence only partly cancel. Accordingly, for the photoproduction of electron pairs the relative corrections to the point-charge cross-section are smaller for heavy nuclei than for light nuclei. On the contrary, muon pair production shows the expected behaviour that with an increasing nuclear extension the relative corrections become larger (see Figure 1.13). With increasing lepton mass the importance of the nuclear-charge distribution increases whereas the influence of higher orders can be gradually neglected. For τ production it can be completely discarded.

The knowledge gained about high-energy photoproduction allows us to assess the higher-order corrections to lepton pair production in heavy-ion collisions. It is found that for the colliders RHIC (Au + Au) and LHC (Pb + Pb) the corrections to the total cross-section are of the order of -25% (RHIC) and -14% (LHC) for e^+e^-, -12.2%, i.e. -5% for $\mu^+\mu^-$ and -0.36%, i.e. 0.043% for $\tau^+\tau^-$ pair production (Eichmann *et al.* 2000b).

According to Baltz *et al.* (2001) the *inclusive* cross-section (summing over all numbers of pairs) can be formulated in terms of the retarded electron propagator and allows the use of the known exact solution of the Dirac equation (1.8) without the need to invoke crossing invariance.

1.4 Relativistic and QED Effects in Highly Charged Ions

Quantum electrodynamics, the relativistic quantum field theory of interacting charged particles and photons provides to an extremely high precision the theory of atoms (Kinoshita 1990; Labzowsky *et al.* 1993; Mohr *et al.* 1998). The comparison of theory and experiment for energy levels of bound electrons allows for a critical tests of quantum electrodynamics in external fields. In light systems, such as the hydrogen atom, QED effects are extremely important in connection with the determination of fundamental constants, as, for example, the fine-structure constant α or the electron mass (Mohr 1996; Niering *et al.* 2000). Complementary to that, the investigation of QED corrections in highly charged ions allows for a determination of the range of validity of QED in strong external fields.

QED can be considered to be one of the most precisely tested theories in physics at present. It provides an extremely accurate description of systems such as hydrogen and helium atoms, as well as for bound-leptonic systems, for example, positronium and muonium. Remarkable agreement between theory and experiment has been achieved with respect to the determination of the hyperfine structure and the Lamb shift. The same holds true for the electronic and muonic g-factors. The free-electron g-factor is determined at present as

$$g_{\text{free}}^{\text{exp}} = 2 + 2 \times 1\,159\,652\,188.4(4.3) \times 10^{-12},$$
$$g_{\text{free}}^{\text{th}} = 2 + 2 \times 1\,159\,652\,216.0(1.2)(67.8) \times 10^{-12},$$

where we refer to Dyck *et al.* (1987) and Hughes and Kinoshita (1999) for the experimental values and theoretical predictions, respectively. The second error indicated in the theoretical prediction reflects already the uncertainty of the value of α employed in this calculation (Jeffrey *et al.* 1997). Nowadays, similar precision is obtained in systems like positronium or for the Lamb shift in hydrogen, where the accuracy of theoretical predictions is limited by the insufficient knowledge of nuclear parameters (Karshenboim 1999; Pachucki 1999). The hyperfine-structure splitting of the ground state in hydrogen represents another impressive example of a quantity in nature that is most precisely known. Simultaneously, there are major theoretical difficulties associated with it. Measurement and calculation are conventionally not even presented in a comparable manner due to effects ascribed to the internal structure of the proton (Bodwin and Yennie 1988; Kinoshita 1990). The small deviation from the idealized point-dipole magnetic field still cannot be described theoretically from first principles. The current numbers for the transition frequency of the ground-state hyperfine-structure splitting in hydrogen obtained from experiment (Essen *et al.* 1971; Hellwig *et al.* 1970), respectively, provided by theory (Kinoshita 1990) are

$$\nu_{\text{HFS}}^{\text{exp}} = 1420.405\,751\,766\,7(9)\ \text{MHz},$$
$$\nu_{\text{HFS}}^{\text{th}} = 1420.451\,99(10)\ \text{MHz} + \text{nuclear-structure effects},$$

where the nuclear-structure effects include all contributions of the proton, from finite size and mass to form factors and internal structure. Most of the discrepancy between both numbers is reduced if the finite size of the proton is taken into account (Zemach 1956), but the theoretical precision does not increase. Quantum electrodynamical effects are included in the theoretical value up to a level of accuracy at which effects due to the internal proton structure become important as well.

Basically, QED is formulated in terms of a perturbation expansion of the S matrix, taking the fine-structure constant $\alpha \approx 1/137.036$ as an expansion parameter. The individual terms of the perturbation expansion are usually represented by Feynman diagrams. For light atomic systems the standard theoretical approach treats the nuclear Coulomb field perturbatively as well, where the (effective) nuclear coupling constant $Z\alpha$ is taken as an additional expansion parameter. Z denotes the nuclear charge number. However, this approach becomes inadequate for heavy systems in which the parameter $Z\alpha$ approaches unity. For example, in highly charged ions like uranium, the effective coupling constant is no longer a small parameter but amounts to $Z\alpha \approx 0.6$. Accordingly, a proper theoretical description of high-Z few-electron atoms should account for all orders in $Z\alpha$ simultaneously. This underlines that, from a theoretical point of view, QED of heavy atoms is essentially different from that for light systems and thus requires new calculational techniques and renormalization approaches. Also the experimental situation looks significantly different. The energy of the Ly-α transition in hydrogen amounts to 10.2 eV (about 1200 Å), while the same transition in hydrogen-like uranium belongs to the X-ray region with an energy of about 98 keV. As a result, solid-state X-ray detectors should be employed, which severely complicates any high-precision measurement of the 1s-Lamb shift in hydrogen-like high-Z ions.

Highly charged ions provide the strongest electromagnetic fields accessible to experimental investigation. Heavy ions in an arbitrary charge state or even bare heavy nuclei can be prepared. Hydrogen-like ions are obtained by stripping all but one electron from a heavy atom, for example, lead or uranium. The expectation value of the electric field strength in these systems is depicted in Figure 1.14.

The field strength at the nuclear surface may be even higher. For example, at the surface of a uranium nucleus it amounts to about $|\boldsymbol{E}| \cong 2 \times 10^{19}\ \text{V cm}^{-1}$.

Highly charged ions do not only provide a strong electric but also a strong magnetic field. In Figure 1.15 the expectation value for the magnetic field strength is given for hydrogen-like ions over the whole range of Z. It ranges from about 10^{-1} T for hydrogen to several times 10^5 T for the heaviest hydrogen-like ions accessible to experiment. For each nucleus characterized by a charge number Z and a mass number A, the odd-A isotope with the highest natural abundance or longest lifetime was chosen. Since the magnetic interaction involves powers of $1/c$, even this enormous

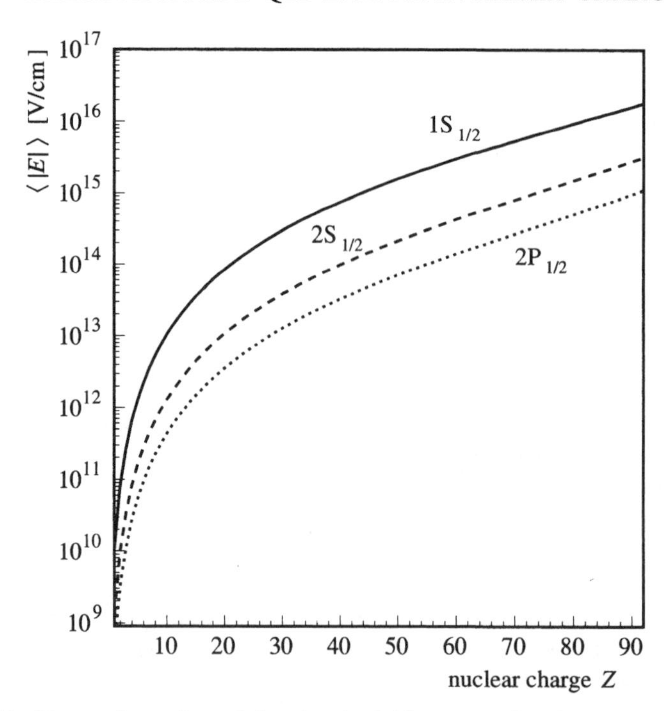

Figure 1.14 Expectation value of the electric field strength for the lowest-lying states of a hydrogen-like atom in the range $Z = 1$–92. Electron wave functions for extended nuclear-charge distributions are employed.

field strength leads to only a small influence on the atomic energy levels. As a result of the interaction of an electron with this magnetic field, a level splitting occurs corresponding to the possible values of the total angular momentum $\boldsymbol{F} = \boldsymbol{J} + \boldsymbol{I}$ of the ion, where $\boldsymbol{J}$ is the total electronic angular momentum and $\boldsymbol{I}$ denotes the total nuclear angular momentum. Only the total angular momentum $\boldsymbol{F}$ is an observable. The resulting hyperfine-structure splitting and its value can be determined quite accurately by spectroscopic means.

Evidently, the question arises of whether in such strong electric and magnetic fields 'usual' atomic physics—well established for hydrogen atoms, where the fields probed by the electron are up to six orders of magnitudes smaller—will still be valid or not. The determination of the range of validity of QED in strong external fields is also very promising for the detection of new physics beyond QED in such heavy atomic systems. During the last decade, great progress has been made in experimental investigations of heavy few-electron ions. The ground-state Lamb shift in hydrogen-like uranium was measured with an accuracy of a few per cent (Beyer 1995; Stöhlker *et al.* 2000), and further progress towards increasing the accuracy by an order of magnitude is anticipated (Stöhlker *et al.* 2000). Experimental data are also available for various hydrogen-like ions (Beyer *et al.* 1994; Mokler *et al.* 1995; Stöhlker *et al.* 1992). Despite inherent problems encountered with the renormalization of the bound-state

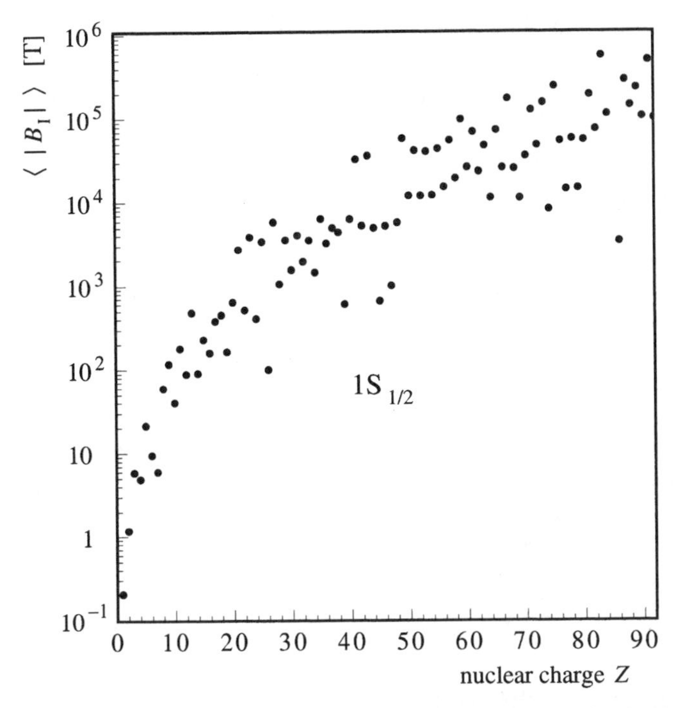

Figure 1.15 Expectation value of the nuclear magnetic field strength for the $1s_{1/2}$ state of a hydrogen-like atom in the range $Z = 1$–92. The values were calculated employing wave functions for extended nuclear-charge distributions (Beier 2000).

QED, its predictions can be tested to very high precision, e.g. via measurements of the Lamb shift of electron levels in highly charged ions. Dominant corrections to the energy spectrum are due to finite nuclear size and due to QED effects: self-energy and vacuum polarization of order α. Both radiative corrections have to be evaluated to all orders in $Z\alpha$ in the interaction with the external Coulomb potential to achieve agreement with Lamb-shift data measured with the relative precision of about 10^{-4} for hydrogen-like systems. Aiming at the utmost experimental precision, it is essential to determine the level of accuracy at which we leave the framework of pure QED. The natural limitation for testing QED is set by nuclear polarization effects and by the uncertainties of the nuclear parameter. In heavy systems, nuclear structure becomes nonnegligible at the level of a relative precision of about 10^{-6} (Nefiodov *et al.* 1996; Plunien and Soff 1995; Plunien *et al.* 1991). To provide predictions for the Lamb shift taking into account this ultimate standard requires the exact evaluation of all QED-radiative corrections of order α^2 (Beier *et al.* 1997a).

An even higher precision is achieved in experiments with highly charged lithium-like ions. The splitting between $2p_{1/2}$ and 2s levels in lithium-like uranium was determined experimentally (Schweppe *et al.* 1991) to be 280.59 eV with an uncertainty of only 0.09 eV. To a large extent, this result has initiated calculations of QED effects of second order in α as a goal for theorists, which is still under progress at present. Accurate experimental data are available at present for $2p_{1/2}$–2s and $2p_{3/2}$–2s transitions of several high-Z lithium-like ions (Beiersdorfer *et al.* 1998; Bosselmann *et al.* 1999; Feili *et al.* 2000; Staude *et al.* 1998). In helium-like ions, the two-electron contribution to the ground-state energy has been measured directly by comparing ionization energies of helium- and hydrogen-like ions (Marrs *et al.* 1995; Stöhlker *et al.* 1996). This experiment is of particular importance since the corresponding contributions due to the interelectronic interaction are calculated completely from QED up to second order in α (Persson *et al.* 1996b; Yerokhin *et al.* 1997a). For an overview of the present situation in heavy helium-like ions, we refer to the recent review (Shabaev *et al.* 2000a) and to a previous paper (Yerokhin *et al.* 1997a), where the most accurate theoretical prediction for these two-electron contributions to the ground-state energy of helium-like ions have been obtained.

At present, the hyperfine splitting in heavy hydrogen-like ions can also be measured with excellent precision (Klaft *et al.* 1994; López-Urrutia *et al.* 1996, 1998; Seelig *et al.* 1998). Investigations of the g-factor of a bound electron look very promising as well for testing higher-order QED effects, although its direct measurement has so far been accomplished only for light systems up to carbon (Häffner *et al.* 2000; Hermanspahn *et al.* 2000).

1.4.1 Relativistic description of few-electron systems

Atomic electrons interacting via the exchange of photons are bound in the electromagnetic field generated by the atomic nucleus. Because of the high nuclear mass and high nuclear charge number, the electromagnetic field generated by the nucleus can be approximated as a classical external field in which the fermion field and the free-photon field are quantized (bound-state QED). Although the fundamental interaction between electrons and photons is known, the influence of nuclear properties, which must be taken into account at a certain level of accuracy, in addition involves the solution of the nuclear many-body problem, which cannot be accomplished from first principles. This has far-reaching consequences even in one-electron systems, for which effects due to the electron–electron interaction are negligible. Although it might be possible to evaluate pure QED corrections to atomic spectra up to any desired accuracy, we have to include nuclear effects, which will inevitably set inherent limitations for tests of QED due to the uncertainties of nuclear parameters. Conversely, we may be able to determine such parameters from high-precision measurements of the atomic structure.

The relevant part of the action may be specified as

$$S = \int \mathrm{d}^4x \,\{\tfrac{1}{2}[\bar{\psi}(x), (\mathrm{i}\gamma^{\mu}\partial_{\mu} - m)\psi(x)] + \tfrac{1}{2}\{\mathcal{J}^{\mu}(x), \mathcal{A}_{\mu}(x)\} \\ - \tfrac{1}{8}\{\mathcal{F}^{\mu\nu}(x), \mathcal{F}_{\mu\nu}(x)\} + \mathcal{L}_{\mathrm{nuc}} + \mathcal{L}_{\mathrm{add}} + \mathcal{L}_{\mathrm{counter}}\},$$

with the standard free-field Lagrangians of the Dirac field (ψ) and the total Maxwell field ($\mathcal{F}$), respectively. The total nuclear Lagrangian $\mathcal{L}_{\text{nuc}}$ and additional interactions $\mathcal{L}_{\text{add}}$ beyond QED will not be specified explicitly. The counter-term Lagrangian $\mathcal{L}_{\text{counter}}$ is required for renormalization. The interaction term between the total Maxwell field $\mathcal{A}_\mu$ and the total electromagnetic source current $\mathcal{J}^\mu$ is written in a properly symmetrized form. The total source current appears as a sum of the Dirac current j^μ_ψ and of the nuclear source current j^μ_{nuc}, i.e.

$$\mathcal{J}^\mu(x) = j^\mu_{\text{nuc}} + j^\mu_\psi(x) = j^\mu_{\text{ext}}(x) + j^\mu_{\text{fluc}}(x) + j^\mu_\psi(x)$$

with

$$j^\mu_\psi(x) = -\tfrac{1}{2}e[\bar{\psi}(x), \gamma^\mu \psi(x)].$$

The nuclear current consists of a classical, external part $\langle j^\mu_{\text{nuc}}\rangle \equiv j^\mu_{\text{ext}}$ describing the nucleus in its ground state and a second quantized part j^μ_{fluc} describing internal nuclear degrees of freedom. Specification of this fluctuating current employs nuclear models. The action principle yields equations of motion for the coupled Dirac–Maxwell fields ψ and $\mathcal{A}^\mu$:

$$(\mathrm{i}\gamma^\mu \partial_\mu - m)\psi(x) - \tfrac{1}{2}e\gamma^\mu\{\mathcal{A}_\mu(x), \psi(x)\} = 0, \tag{1.9a}$$

$$\partial_\mu \mathcal{F}^{\mu\nu} = (\delta^\nu_\mu \Box - \partial^\nu \partial_\mu)\mathcal{A}^\mu = \mathcal{J}^\nu. \tag{1.9b}$$

Maxwell's equations (1.9) can be formally integrated with the aid of the free-photon propagator $D_{\mu\nu}$. According to the total current introduced above, the resulting total radiation field $\mathcal{A}^\mu$ may be decomposed as

$$\mathcal{A}^\mu(x) = A^\mu(x) + A^\mu_{\text{ext}}(x) + A^\mu_{\text{fluc}}(x) + A^\mu_\psi(x),$$

the sum of the classical (external) part A^μ_{ext} generated by the nucleus in its ground state and a total, second quantized part, which is given as the sum of the free radiation A^μ, the field A^μ_ψ due to the presence of electrons and the fluctuating field A^μ_{fluc} generated by the nuclear transition current j^μ_{fluc}. Going beyond the external-field approximation we may keep as the total radiation field

$$A^\mu_{\text{rad}} = A^\mu(x) + A^\mu_{\text{fluc}}(x).$$

The total Hamiltonian describing the total interacting many-body problem—*Dirac particles + radiation field + nucleus*—may be obtained from the T^{00}-component of the energy–momentum tensor. The part of the Hamiltonian relevant for the relativistic description of the atomic many-body problem in the presence of the external electromagnetic field of the nucleus including radiative corrections and possible interactions with internal nuclear degrees of freedom is given by

$$\begin{aligned} H_{\mathrm{QED}} &= \frac{1}{2}\int \mathrm{d}^3x\, [\psi^\dagger(x), h_{\mathrm{ext}}(x)\psi(x)] \\ &\quad + \frac{1}{2}\int \mathrm{d}^3x\, \mathrm{d}^4x'\, D_{\mu\nu}(x-x')\{j_\psi^\mu(x), j_\psi^\nu(x')\} \\ &\quad + \frac{1}{2}\int \mathrm{d}^3x\, \{j_\psi^\mu(x), A_\mu^{\mathrm{rad}}(x)\} + H_{\mathrm{counter}} \\ &= H_{\mathrm{ext}} + H_{\mathrm{e\text{-}e}} + H_{\mathrm{rad}} \end{aligned}$$

together with the one-particle Dirac Hamiltonian

$$h_{\mathrm{ext}}(x) = [-\mathrm{i}\boldsymbol{\alpha}\cdot(\boldsymbol{\nabla} + \mathrm{i}e\boldsymbol{A}_{\mathrm{ext}}(x)) + V_{\mathrm{ext}}(x) + \beta m].$$

Here $\alpha^\mu \equiv \gamma^\mu\gamma^0$ and $\gamma^0 \equiv \beta$ denote the Dirac matrices. The first term H_{ext} defines the dominant external field problem. The interaction with the classical electromagnetic field generated by the extended nucleus in its ground state has to be treated to all orders in the effective coupling $Z\alpha$. Here $\alpha = e^2$ denotes the fine-structure constant. Neglecting recoil effects due to the finite nuclear mass the external field may be considered as static. In the case of even–even nuclei and in the infinite-nuclear-mass limit the external field problem treats the motion of Dirac electrons in a static external Coulomb potential V_{ext} only. In odd-A nuclei a magnetic interaction potential $\boldsymbol{A}_{\mathrm{ext}}$ has also to be taken into account. The second term $H_{\mathrm{e\text{-}e}}$ accounts for the electron–electron interaction self-consistently up to order α. Finally, the term H_{rad} describes the interaction between bound electrons and the vacuum of the total radiation field. This gives rise to various QED-radiative corrections (A_μ) and nuclear (recoil and polarization) effects (A_μ^{fluc}), which can formally be treated at the same level. All effective radiative effects (A_μ^{rad}) are evaluated perturbatively in powers of the coupling constant α (and m/M for nuclear effects, respectively). The Hamiltonian also contains a part H_{counter} which denotes all necessary counter terms required for the renormalization of mass and charge divergences. It has to be incorporated into the radiative part H_{rad}.

A proper definition of (quasi-)particle-creation and (quasi-)particle-annihilation operators $\tilde{a}_n^\dagger$ and $\tilde{a}_n$ is provided by diagonalization of the (time-independent) unperturbed part $H_0 = H_{\mathrm{ext}} + H_{\mathrm{e-e}}$ of the total Hamiltonian. After the iteration is performed (e.g. on the Dirac–Fock level) the latter may be cast into the form

$$H_0 = \frac{1}{2}\int \mathrm{d}^3r\, [\psi^\dagger(x), h_{\mathrm{eff}}(\boldsymbol{r})\psi(x)],$$

where h_{eff} denotes an effective one-particle Hamiltonian. The Dirac-field operator ψ may be expanded in terms of noninteracting (quasi-)particle annihilation and creation operators

$$\psi(x) = \sum_{n>\mathrm{F}} \tilde{a}_n\varphi_n(x) + \sum_{n<\mathrm{F}} \tilde{b}_n^\dagger\varphi_n(x), \qquad \varphi_n(x) = \varphi_n(\boldsymbol{r})\,\mathrm{e}^{-\mathrm{i}\mathcal{E}_n t}.$$

F denotes the Fermi surface fixed at the Fermi energy E_F located above the highest occupied electron state. The operator $\tilde{a}_n$ annihilates an electron described by the spinor φ_n and the energy eigenvalue $\mathcal{E}_n > E_F$, and similarly $\tilde{b}_n^\dagger$ creates an electron-hole in the state φ_n with the energy eigenvalue $\mathcal{E}_n < E_F$. In the case of bare nuclei, the QED vacuum state is defined with respect to the Fermi surface F_0 with a Fermi energy E_{F_0} located above $E = -1$ and below the lowest (unoccupied) bound state. The operator $\tilde{b}_n^\dagger = \tilde{a}_n$ with $\mathcal{E}_n < E_{F_0}$ creates a positron in the state φ_n with energy $(-\mathcal{E}_n)$. Over the range Z of the known periodic system all electron bound states occur at level energies $\mathcal{E}_n > 0$, so we may choose $E_{F_0} = 0$.

Although a number of different approaches exist, the general solution to the problem of setting up a relativistic Hamiltonian from QED describing many-electron systems is not known. While the spectrum of the nonrelativistic many-body Schrödinger Hamiltonian is bounded from below, fundamental difficulties arise due to the negative energy continuum states. Most of the existing methods may be considered as prescriptions to avoid these difficulties in one way or another. A pragmatic approximation consists of the introduction of Hamiltonians, where the negative-energy states are projected out (so-called no-pair Hamiltonians). The commonly accepted point of view is that QED effects associated with the negative Dirac sea may be treated perturbatively as corrections to the many-electron problem. Accordingly, most of the relativistic many-body calculations (see, for example, Sucher 1980) take the no-pair Hamiltonian as a starting point. It may be written in a second quantized form as

$$\begin{aligned} H &= H_{\text{ext}} + V + B \\ &= \sum_i E_i a_i^\dagger a_i + \frac{1}{2} \sum_{ijkl} v_{ijkl} a_i^\dagger a_j^\dagger a_k a_l + \frac{1}{2} \sum_{ijkl} b_{ijkl} a_i^\dagger a_j^\dagger a_k a_l. \end{aligned}$$

The creation/annihilation operators $a_i^\dagger / a_i$ denote the one-particle operators which diagonalize the Hamiltonian H_{ext}. The summation indices i, j, k, l denote the usual set of one-electron quantum numbers and run over positive-energy states only. The quantities v_{ijkl} are two-electron Coulomb matrix elements and the quantities b_{ijkl} denote two-electron Breit matrix elements, respectively. We specify their static limit (neglecting any frequency dependence):

$$v_{ijkl} = \left\langle \varphi_i \varphi_j \left| \frac{1}{r_{12}} \right| \varphi_l \varphi_k \right\rangle$$

and

$$b_{ijkl} = -\left\langle \varphi_i \varphi_j \left| \frac{\boldsymbol{\alpha}_1 \cdot \boldsymbol{\alpha}_2 + \boldsymbol{\alpha}_1 \cdot \hat{\boldsymbol{r}}_1 \, \boldsymbol{\alpha}_2 \cdot \hat{\boldsymbol{r}}_2}{2r_{12}} \right| \varphi_l \varphi_k \right\rangle,$$

where $r_{12} = |\boldsymbol{r}_1 - \boldsymbol{r}_2|$ and $\hat{\boldsymbol{r}} = \boldsymbol{r}/r$. It remains a challenge for the theory to examine the possibilities of merging QED and relativistic many-body methods from the very beginning. This subject is still under investigation (Lindgren 2000).

1.4.2 Relativistic model Hamiltonians for many-electron systems

As mentioned above, QED is considered to be the theory describing low-energy processes between charged particles, i.e. electrons and ions. There is strong experimental evidence that it provides a relativistic description most relevant for chemistry and atomic physics. Nevertheless, the impressive success in describing fundamental physics contrasts with a poor mathematical understanding of QED. Accordingly, it is worthwhile trying to close this gap in a mathematically rigorous manner by investigating relativistic models which could serve as testing cases. These should be simple enough to be treated rigorously but still contain physically essential and generic features of the real system under consideration.

Recent investigations along this line have dealt with models, for example, the naive relativistic Hamiltonians (Chandrasekhar), no-pair Hamiltonians (see, for example, Sucher 1980) and with models of QED including the electric interaction but disregarding the magnetic one (interacting electron–positron field). One basic feature of all these models should be the boundedness of the energy per particle (stability of matter). Without this basic property, no further detailed chemical computations are worthwhile. That this is indeed a suitable question has been pointed out and solved partly by Lieb and Yau (1988) in their basic work on relativistic stability. Recent activities have continued along this path by investigating (pre-)QED models (Bach *et al.* 1998, 1999; Chaix and Iracane 1989; Chaix *et al.* 1989; Evans *et al.* 1996).

To start with, consider systems consisting of N dynamical electrons and positrons and K fixed nuclei with Coulomb interactions between all pairs of particles. The clamped-nuclei approximation (the Born–Oppenheimer approximation) may be legitimate because of the huge difference in mass between electrons and nuclei. Stability of matter means that the energy of such a model system is bounded from below by a negative constant times the number of particles: $E \geqslant -C(N+K)$. Such a condition is necessary for some basic physical properties such as the existence of the thermodynamical limit.

Relativistic N-particle model Hamiltonians

Recent investigations (Lieb *et al.* 1996) have concentrated on, as a first task, optimizing the stability result for the Chandrasekhar operator

$$\begin{aligned} H &= T + V_{\mathrm{C}} \\ &= \sum_{i=1}^{N} \sqrt{-\hbar^2 c^2 \Delta + m^2 c^4} \\ &\quad + \sum_{\substack{i,j=1 \\ i<j}}^{N} \frac{e^2}{|\boldsymbol{r}_i - \boldsymbol{r}_j|} + \sum_{\substack{i,j=1 \\ i<j}}^{K} \frac{e^2 Z_i Z_j}{|\boldsymbol{R}_i - \boldsymbol{R}_j|} - \sum_{i=1}^{K} \sum_{j=1}^{N} \frac{e^2 Z_j}{|\boldsymbol{r}_i - \boldsymbol{R}_j|}. \end{aligned}$$

This operator defines a pseudorelativistic model which includes neither positrons nor the spin of the electrons. These authors have improved on some of results derived earlier (Lieb and Yau 1988), notably in the presence of magnetic fields. In the physical case, $\alpha = 1/137$, they proved stability for all elements with nuclear charge numbers Z less or equal to 59. The short proof presented employed several inequalities also used in earlier work. One of them concerns the estimate of the localization error for the kinetic energy (Lieb and Yau 1988). The size of the error is determined by the fourth power of the L^4-norm of some function Y. It is expected that any improvement of this estimate could increase the critical Z-value considerably. With numerical computations, Lieb *et al.* (1996) succeeded in demonstrating the stability up to a critical charge number $Z = 60$. Relativistic no-pair Hamiltonians have also been considered. The stability of the Brown–Ravenhall operator—neglecting magnetic field components—has been investigated by Hoever and Siedentop (1999), while in Lieb *et al.* (1997a,b) the magnetic field generated by the electrons has been included.

Electron–positron field

The mathematical problem associated with the Dirac Hamiltonian, i.e. the starting point of the relativistic theory of atoms, can be phrased in simple terms. The electron–positron field can have states of arbitrarily negative energy. As a general feature of the Dirac spectrum this instability occurs even in the case of extended nuclei and even in the absence of any nucleus (free Dirac spectrum), the energy is not bounded from below. This gives rise to the necessity of renormalization and well-established renormalization schemes have been around for many decades. Despite their successful applications in physics, we may ask instead whether there exist states that allow for positivity of the energy.

A rigorous mathematical model for the relativistic electron–positron field in the Hartree–Fock approximation has been recently proposed (Bach *et al.* 1999). It describes electrons and positrons with the Coulomb interaction in second quantization in an external field using generalized Hartree–Fock states. It is based on the standard QED Hamiltonian neglecting the magnetic interaction $\boldsymbol{A} = \boldsymbol{0}$ and is motivated by a physical treatment of this model (Chaix and Iracane 1989; Chaix *et al.* 1989).

Being a true relativistic model, the one-particle energy is given by the Dirac Hamiltonian $H = \boldsymbol{\alpha} \cdot (\boldsymbol{p} + e\boldsymbol{A}) + m\beta + V$ acting in the Hilbert space $\mathfrak{H} := L^2(\mathbb{R}^3) \otimes \mathbb{C}^4$. The Dirac operator includes both electrons and positrons. In order to second quantize the field, we have to split $\mathfrak{H}$ into the electron subspace $\mathfrak{H}_+$ and the positron subspace $\mathfrak{H}_-$. There is no unique way of doing this. For the proposed model (Bach *et al.* 1999), however, there exists a distinct way of doing so. As shown rigorously in Bach *et al.* (1999), the highest ground-state energy is indeed obtained when choosing the electron subspace to be the positive spectral subspace of H. For a related model this has been assumed by Mittleman (1981) on physical grounds. The corresponding orthogonal projections may be denoted here by P_+ and P_-.

Instead of investigating the QED energy in generalized Hartree–Fock states in the electron–positron Fock space, it turns out that it is sufficient to consider charge-density matrices $\gamma : \mathfrak{H} \to \mathfrak{H}$ and their energy (Hundertmark *et al.* 2000). The charge-density matrices have the properties

$$\gamma \text{ self-adjoint}, \qquad \operatorname{tr}\gamma < \infty, \qquad -P_- \leqslant \gamma \leqslant P_+; \tag{1.10}$$

the number of particles is $N(\gamma) := \operatorname{tr}(P_+ - P_-)\gamma$ and the charge is $Q(\gamma) := \operatorname{tr}\gamma$. Their energy in the model is given by the Hartree–Fock functional

$$\mathcal{E}(\gamma) = \operatorname{tr}(H, \gamma) + \tfrac{1}{2}\alpha \int \mathrm{d}^3r\,\mathrm{d}^3r'\,\frac{\rho_\gamma(\boldsymbol{r})\rho_\gamma(\boldsymbol{r}')}{|\boldsymbol{r}-\boldsymbol{r}'|} - \tfrac{1}{2}\alpha \int \mathrm{d}^3r\,\mathrm{d}^3r'\,\frac{|\gamma(\boldsymbol{r},\boldsymbol{r}')|^2}{|\boldsymbol{r}-\boldsymbol{r}'|},$$

where $\rho_\gamma(\boldsymbol{r})$ denotes the charge density at the point $\boldsymbol{r}$ and $\gamma(\boldsymbol{r}, \boldsymbol{r}')$ forms an integral kernel of γ.

According to Chaix *et al.* (1989), in the absence of external potentials ($|\boldsymbol{A}| = V = 0$) it has been argued that for stability we need $\alpha < 2/\pi$ and for instability we need $\alpha > \pi/\log 4$. The stability result has been proved by Bach *et al.* (1999). To demonstrate the instability result a translationally invariant density matrix γ_{C} based on a rotation in momentum space has been employed (Chaix *et al.* 1989). But since the model is defined on $\mathbb{R}^3$, such a density matrix is either zero or has an infinite number of particles, it has either zero energy or infinite energy. This is because the state represented by the density matrix is the same in every unit volume. Therefore, the arguments presented by Chaix *et al.* (1989) could just show the instability within a unit volume.

This investigation has been improved recently (Hundertmark *et al.* 2000) by finding a way to implement asymptotically this density matrix in $\mathbb{R}^3$. The basic idea was to put γ_{C} in a box and let its radius r tend to infinity. But besides the technical difficulties with estimating the energy of such a density matrix, there is the fundamental problem that for every r the inequality (1.10) has to be fulfilled. This implies that electrons and positrons may not mix. A possibility for overcoming this problem is to introduce a cut-off in the coordinates, where the electron subspace consists of the first two spin components and the positron subspace of the second two. As a result, the correct angle for the rotation, which can be guessed using a method of Evans *et al.* (1996), has been derived yielding the sharp result: instability occurs for $\alpha > 4/\pi$.

The stability of atoms ($V = -Z\alpha|\boldsymbol{r}|^{-1}$) including an external magnetic field has been demonstrated for $Z \leqslant 68$ (Bach *et al.* 1998). In a following paper (Bach *et al.* 1999) the stability of matter has finally been proved. For the atomic problem an inequality for the moduli of Dirac operators,

$$|-\mathrm{i}\boldsymbol{\alpha}\cdot\nabla + m\beta - Z\alpha|\boldsymbol{r}|^{-1}| \geqslant d|-\mathrm{i}\boldsymbol{\alpha}\cdot\nabla|$$

has been shown. Although the proof is very short and elegant, the resulting numbers turned out to be unsatisfactory.

In a recent work (Brummelhuis *et al.* 2001) the stability result has been improved up to charge numbers $Z \leqslant 117$, which covers the range of all known elements. Again the proof starts by taking the square of the inequality, because the modulus of the Dirac operator with the Coulomb potential is not easy to handle. However, instead of using the triangle inequality, the dilation homogeneity of the massless ($m = 0$) Dirac operator was employed to reduce the problem to a related one for multiplication operators. With this method the sharp constant for the squared inequality can actually be shown. The massive case reduces to the massless one by using the positivity of the ground-state energy to control the mass terms.

As expected the lowest energy of these models is the generalized Hartree–Fock vacuum state. To describe atoms, we need to restrict the allowed charge or particle number. One way of changing that is to subtract a part of the rest energy $mc^2N(\gamma)$ of the charge-density matrix γ from the total energy. For the free system with $Z = 0$ the vacuum turns out to still be the ground state if 0.999 983 times the rest mass is subtracted. On the other hand, subtracting the full rest mass or more, the model does not possess a ground state. To provide a similar proof for $Z > 0$ would be more difficult, because to compute the number of particles of a state involves knowledge of the electron and positron subspaces, which are much more complicated in that case (see Röhrl 2000). Some of the problems associated with the charge and mass renormalization have been recently addressed (Lieb and Siedentop 2000).

1.4.3 Bound-state QED

Bound-state QED provides a proper and practicable description of few-electron systems. Both QED-radiative corrections and electron–electron interactions may be treated perturbatively with respect to the coupling $\alpha = e^2$, counting the number of virtual photons involved, while the interaction with the external nuclear fields is included to all orders in $Z\alpha$.

The QED radiative effects are treated perturbatively by the inclusion of H_{rad}. Accordingly, the unperturbed Hamiltonian H_0 reduces to the external-field problem (normal ordering is indicated by : :)

$$H_0 \equiv H_{\text{ext}} = \frac{1}{2}\int \mathrm{d}^3r \;:[\psi^\dagger(x), h_{\text{ext}}(\boldsymbol{r})\psi(x)]:. \tag{1.11}$$

Even–even nuclei may be described by a spherically symmetric Coulomb potential V_{ext} of an extended nucleus with charge number Z. Pure QED effects due to the interaction with the free radiation field are carried by the interaction Hamiltonian

$$H_{\text{I}}(t) = -\tfrac{1}{2}e\int \mathrm{d}^3r\,[\bar{\psi}(x), A\!\!\!/_{\mu}(\boldsymbol{r})\psi(x)] - \tfrac{1}{2}\delta m\int \mathrm{d}^3r\,[\bar{\psi}(x), \psi(x)]. \tag{1.12}$$

The field operator ψ is now expanded into the set of solutions ϕ_i of the corresponding one-particle Dirac equation $h_{\text{ext}}\phi_i = E_i\phi_i$ according to

$$\psi(x) = \sum_{E_i>0} a_i\,\phi_i(x) + \sum_{E_i<0} b_i^\dagger\phi_i(x), \tag{1.13}$$

with

$$\phi_i(x) = \phi_i(\boldsymbol{r})\,\mathrm{e}^{-\mathrm{i}E_it}.$$

The subscript i labels the principle quantum number and angular momentum quantum numbers $(njlm)$. Here a_i and $b_i^\dagger$ denote electron-annihilation and positron-creation operators, respectively, defined via diagonalization of the unperturbed Hamiltonian (1.11)

$$H_0 = \sum_{E_i>0} E_i a_i^\dagger a_i + \sum_{E_i<0} E_i b_i^\dagger b_i .$$

Unperturbed N-electron state vectors are generated according to

$$|N\rangle = \sum_{i_1,\ldots,i_N} c_{i_1,\ldots,i_N} a_{i_1}^\dagger \cdots a_{i_N}^\dagger |0\rangle .$$

The main application of bound-state QED is the evaluation of energy levels of few-electron atoms. The energy levels appear as a series of even powers of the coupling with the radiation field since only virtual photons, each of which enters with two powers of the coupling constant (e^2), are involved

$$E = E^{(0)} + E^{(2)} + E^{(4)} + \cdots .$$

In this formulation the zeroth-order level energy is just the Dirac eigenvalue E_{nj} summed over the electrons in the unperturbed state. In particular, $E_{\mathrm{N}}^{(0)} = \langle N|H_0|N\rangle$, where the unperturbed states and the action of H_0 are explicitly given by

$$|nljm\rangle = a_{nljm}^\dagger |0\rangle ,$$
$$H_0|nljm\rangle = E_{nj}|nljm\rangle$$

for one-electron states, and by

$$|nljn'l'j'JM\rangle = \sum_{m,m'} C_{jmj'm'}^{JM} a_{nljm}^\dagger a_{n'l'j'm'}^\dagger |0\rangle ,$$
$$H_0|nljn'l'j'JM\rangle = (E_{nj} + E_{n'j'})|nljn'l'j'JM\rangle$$

for two-electron states. Two-electron states are summed with vector addition coefficient weights to generate states of definite total angular momentum.

The particular interaction picture defined by Equations (1.11)–(1.13) treating the interaction with the external nuclear Coulomb potential exact to all orders in $Z\alpha$ is called the Furry picture. The derivation of the general formula for the energy shift caused by the interaction (1.12) utilizes the concept of adiabatic switching, i.e. $H_{\mathrm{I}}(t) \rightarrow \lambda H_{\mathrm{I}}^{\epsilon}(t) = \lambda \mathrm{e}^{-\epsilon|t|} H_{\mathrm{I}}(t)$. The exponential factor renders the integrations over time involved in S matrix elements or energy shifts finite. Restoring the original interaction in the limit $\epsilon \rightarrow 0$ leads to energy conservation (δ functions) at each vertex. The parameter λ is introduced to trace back the order in the coupling constant e. This allows us to collect singularities arising from separate terms (same order in λ) of the energy shift expression which cancel if the appropriate diagrams are evaluated simultaneously. The S operator may be expanded in powers ν of the coupling e:

$$S_{\epsilon,\lambda} = \sum_{\nu=0}^{\infty} \frac{(-\mathrm{i}\lambda)^{\nu}}{\nu!} \int \mathrm{d}t_1 \cdots \int \mathrm{d}t_{\nu}\, T[H_{\mathrm{I}}^{\epsilon}(t_1)\cdots H_{\mathrm{I}}^{\epsilon}(t_{\nu})].$$

According to Gell-Mann and Low (1951) and Sucher (1957) the energy shift of an unperturbed N-electron state $|N\rangle$ can be derived from the S matrix element (up to an irrelevant, state-independent constant)

$$\Delta E_{\mathrm{N}} = \lim_{\substack{\epsilon\to 0\\ \lambda\to 1}} \frac{\mathrm{i}\epsilon}{2} \frac{(\partial/\partial\lambda)\langle N|S_{\epsilon,\lambda}|N\rangle_{\mathrm{c}}}{\langle N|S_{\epsilon,\lambda}|N\rangle_{\mathrm{c}}}.$$

The subscript 'c' indicates that only connected diagrams have to be included because the contributions due to disconnected graphs can be eliminated using the factorization property (Goldstone 1957)

$$\langle N|S_{\epsilon,\lambda}|N\rangle = \langle N|S_{\epsilon,\lambda}|N\rangle_{\mathrm{c}}\langle 0|S_{\epsilon,\lambda}|0\rangle_{\mathrm{c}}.$$

The Gell-Mann–Low–Sucher formula necessitates the evaluation of vacuum expectation values of time-ordered products involving multiple fermion and photon operators. Such products can be evaluated by means of Wick's theorem (Wick 1950), which decomposes time-ordered products into a sum of normal-ordered terms and complete contractions. To give an explicit example we specify the energy shift to lowest order in α for one-electron systems. Note that the derivation of formulae for the energy shift based on the Gell-Mann–Low–Sucher formula and in particular the renormalization becomes increasingly cumbersome if more than one electron is involved. An equivalent but technically far more convenient approach is provided by the two-time Green function approach (Shabaev 1990a,b, 1991), which has been widely used in recent calculations of QED effects in few-electron systems (see, for example, Artemyev *et al.* 1999; Shabaev 1993, 1994; Yerokhin *et al.* 1999).

The energy shift of a bound-electron state $|\phi_a\rangle$ is derived as a sum of completely contracted terms of the second-order S matrix element $S_{\epsilon,\lambda}^{(2)}$ taking into account the mass-renormalization counter term provided by $S_{\epsilon,\lambda}^{(1)}$ according to

$$\Delta E_a^{(2)} = \lim_{\epsilon\to 0} \tfrac{1}{2}\mathrm{i}\epsilon(2\langle\phi_a|S_{\epsilon,\lambda}^{(2)}|\phi_a\rangle_{\mathrm{c}} + \langle\phi_a|S_{\epsilon,\lambda}^{(1)}|\phi_a\rangle_{\mathrm{c}}).$$

Details of the derivation of general expressions for energy shifts at a given order can be found in Mohr *et al.* (1998). Contractions between pairs of fermion ψ or boson field operators A_{μ} lead to electron and photon propagator functions. The exact electron propagator in a static external field is homogeneous in time and appears as

$$\begin{aligned} \mathrm{i}S(x,x') &= \langle 0|T[\psi(x)\bar{\psi}(x')]|0\rangle \\ &= \int_{C_{\mathrm{F}}} \frac{\mathrm{d}E}{2\pi}\, \mathrm{e}^{-\mathrm{i}E(t-t')} G(\boldsymbol{r},\boldsymbol{r}',E)\,\gamma^0 \\ &= \int \frac{\mathrm{d}E}{2\pi}\, \mathrm{e}^{-\mathrm{i}E(t-t')} \sum_n \frac{\phi_n(\boldsymbol{r})\bar{\phi}_n(\boldsymbol{r}')}{E - E_n(1-\mathrm{i}\eta)}, \end{aligned}$$

where the temporal Fourier transform of the exact Green function G fulfils the Dirac equation in the external Coulomb field of the nucleus:

$$[E - (-\mathrm{i}\boldsymbol{\alpha} \cdot \nabla + V_{\mathrm{ext}}(\boldsymbol{r}) + \beta m)]\, G(\boldsymbol{r}, \boldsymbol{r}', E) = \delta(\boldsymbol{r} - \boldsymbol{r}').$$

This Green function is analytic in the complex energy plane except for the bound-state poles at E_n, with branch points at $|E| = 1$ and cuts along the real axis for $|E| > 1$. Bound states occur only at energies $E > 0$. The free-photon propagator appears as a time-ordered product of free-photon field operators (in Feynman gauge)

$$\begin{aligned} \mathrm{i}D_{\mu\nu}(x - x') &= \langle 0|T[A_\mu(x)A_\nu(x')]|0\rangle = g_{\mu\nu}\,\mathrm{i}\,D(x - x') \\ &= -g_{\mu\nu} \int_{C_{\mathrm{F}}} \frac{\mathrm{d}E}{2\pi}\, \mathrm{e}^{-\mathrm{i}E(t-t')}\, \frac{\mathrm{e}^{-b|\boldsymbol{r}-\boldsymbol{r}'|}}{|\boldsymbol{r} - \boldsymbol{r}'|}, \end{aligned}$$

with $b = -\mathrm{i}\sqrt{E^2 - \mathrm{i}\varepsilon}$. The propagation function D is analytic in the complex E-plane except for branch points and cuts at $\mathrm{Re}(E) > 0$. The analytic properties of both the electron propagator G and of the free-photon propagator $D_{\mu\nu}$ play a key role in the evaluation and renormalization of QED-radiative corrections.

As QED-radiative effects of order α, we identify the formal expression for the self-energy correction

$$\begin{aligned} \Delta E_a^{\mathrm{SE}} &= \mathrm{i}\alpha \int \mathrm{d}^3r\, \mathrm{d}^3r'\, \phi_a^\dagger(\boldsymbol{r}) \int_{C_{\mathrm{F}}} \frac{\mathrm{d}E}{2\pi}\, \alpha^\mu\, G(\boldsymbol{r}, \boldsymbol{r}', E)\, \alpha^\nu \\ &\quad \times D_{\mu\nu}(\boldsymbol{r}, \boldsymbol{r}', E_a - E)\, \phi_a(\boldsymbol{r}') - \delta m \int \mathrm{d}^3r\, \phi_a^\dagger(\boldsymbol{r})\beta\phi_a(\boldsymbol{r}) \\ &= \langle \phi_a|\Sigma^{\mathrm{SE}}(E)|\phi_a\rangle - \delta m\langle\phi_a|\beta|\phi_a\rangle, \end{aligned} \tag{1.14}$$

where the first-order self-energy operator Σ^{SE} has been introduced, and for the vacuum-polarization correction

$$\begin{aligned} \Delta E_a^{\mathrm{VP}} &= -\mathrm{i}\alpha \int \mathrm{d}^3r\, \mathrm{d}^3r'\, \phi_a^\dagger(\boldsymbol{r})\alpha^\mu\phi_a(\boldsymbol{r})D_{\mu\nu}(\boldsymbol{r}, \boldsymbol{r}', E = 0) \\ &\quad \times \int_{C_{\mathrm{F}}} \frac{\mathrm{d}E}{2\pi}\, \mathrm{Tr}[\alpha^\nu G(\boldsymbol{r}, \boldsymbol{r}', E)] \\ &= \langle\phi_a|\mathcal{U}^{\mathrm{VP}}|\phi_a\rangle, \end{aligned} \tag{1.15}$$

which appears as a first-order perturbation with respect to the vacuum-polarization potential $\mathcal{U}^{\mathrm{VP}}$. In Figure 1.16 the corresponding Feynman diagrams are depicted. The self-energy of the electron arises due to the emission and absorption of a virtual photon by the bound electron. The vacuum-polarization correction can be viewed as an additional interaction of the bound electron with virtual electron–positron pairs induced by the Coulomb potential of the nucleus which modifies the external field.

Bound-state QED thus treats the interaction with the external field exact to all orders in $Z\alpha$, while additional effects due to the interaction with the quantized, free radiation field are treated perturbatively to any desired order in α. The order of a

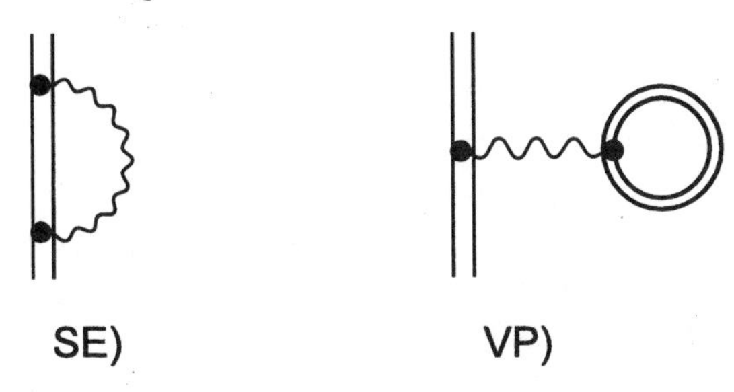

Figure 1.16 Feynman diagrams for the self-energy (SE) and the vacuum polarization (VP) of a bound electron of order α. The double lines indicate wave functions and propagators in the external Coulomb potential of the nucleus.

given (bound-state) QED diagram is given by the number of free internal photon lines involved.

1.4.4 Self-energy correction

In Figure 1.16 the self-energy correction to first order in α is depicted. We sketch the exact evaluation of the self-energy corrections for hydrogen-like systems placing particular emphasis on the covariant renormalization approach (Indelicato and Mohr 1992; Mohr 1974a,b; Mohr and Soff 1993) and on the noncovariant partial-wave renormalization (PWR) (Persson *et al.* 1993a; Quiney and Grant 1993). Both methods have been employed successfully for evaluating radiative corrections of first and second order in α (see, for example, Persson *et al.* 1993a,b, 1996a; Quiney and Grant 1994). All these calculations of the self-energy are based on angular-momentum decompositions of propagators and electron states. The unrenormalized self-energy is infinite, a method of subtracting off the infinite mass renormalization term must be employed which is suitable for a numerical calculation as well. This is accomplished by constructing counter terms that can be calculated analytically and that are also expanded in terms of angular momentum eigenfunctions so that a term-by-term subtraction can be performed. One aspect in which the various approaches differ is in the detailed form of the term subtracted.

Regularization and the singular part

The self-energy correction (1.14) includes an integration over intermediate energies which diverges for spatial arguments $r_{21} = |\boldsymbol{r}_2 - \boldsymbol{r}_1| \approx 0$. Accordingly, appropriate regularizations are required to make the integral finite and to isolate the infinite mass divergence. For instance, in coordinate space the Pauli–Villars regularization (Pauli and Villars 1949) is implemented by the replacement of the photon propagator in the integrand of Equation (1.14),

$$D(\boldsymbol{r}_1 - \boldsymbol{r}_2, E_a - E) \sim \frac{\mathrm{e}^{-br_{21}}}{r_{21}} \to \frac{\mathrm{e}^{-br_{21}}}{r_{21}} - \frac{\mathrm{e}^{-b'r_{21}}}{r_{21}},$$

with

$$b' = -\mathrm{i}\sqrt{(E_a - E)^2 - \Lambda^2 - \mathrm{i}\varepsilon} \quad \text{and} \quad b = -\mathrm{i}\sqrt{(E_a - E)^2 - \mathrm{i}\varepsilon},$$

respectively. Λ denotes the regularization parameter. At the same time, the mass renormalization term δm (free-electron self-energy) is replaced by the corresponding term calculated with the regularized photon propagator, i.e. $\delta m \to \delta m(\Lambda)$. The regulator-dependent expression for the self-energy then reads

$$\begin{aligned} E_a^{\mathrm{SE}}(\Lambda) = -\mathrm{i}\alpha \int \mathrm{d}^3 r_2\, \mathrm{d}^3 r_1\, \phi_a^\dagger(\boldsymbol{r}_2)\alpha_\mu \int_{C_\mathrm{F}} \frac{\mathrm{d}E}{2\pi}\, G(\boldsymbol{r}_2, \boldsymbol{r}_1, E)\alpha^\mu \phi_a(\boldsymbol{r}_1) \\ \times \left[\frac{\mathrm{e}^{-br_{21}}}{r_{21}} - \frac{\mathrm{e}^{-b'r_{21}}}{r_{21}}\right] - \delta m(\Lambda) \int \mathrm{d}^3 x\, \phi_a^\dagger(\boldsymbol{r})\beta\phi_a(\boldsymbol{r}). \end{aligned}$$

It yields the finite physical energy shift ΔE_a^{SE} in the limit $\Lambda \to \infty$. Singular terms arise from the high-energy region of the integration over E. They can be isolated in the first few terms in the expansion of the electron propagator

$$G(\boldsymbol{r}_2, \boldsymbol{r}_1, E) = \langle \boldsymbol{r}_2 | G(E) | \boldsymbol{r}_1 \rangle$$

in powers of the external potential. The resolvent operator

$$G(E) = (h_0 + V_{\mathrm{ext}} - E)^{-1}$$

may be represented in the form

$$\begin{aligned} G(E) &= F(E) + F(E)\, V_{\mathrm{ext}}\, F(E) + F(E)\, V_{\mathrm{ext}}\, G(E)\, V_{\mathrm{ext}}\, F(E) \\ &= G_{\mathrm{as}}(E) + [G(E) - G_{\mathrm{as}}(E)], \end{aligned} \tag{1.16}$$

together with the free resolvent operator $F(E) = (h_0 - E)^{-1}$. The term in square brackets contains the leading divergent terms in the asymptotic expansion in $|E|^{-1}$. The asymptotic term $G_{\mathrm{as}}(E)$ can be derived from the high-energy region by extracting the leading terms in a form amenable to analytic calculation and evaluating the remainder $(G(E) - G_{\mathrm{as}}(E))$ numerically. The various methods of calculation can be characterized in part by the choice of $G_{\mathrm{as}}(E)$, which is made from variations of the expansion of $G(E)$ in powers of V_{ext}. The first method of numerically calculating the QED self-energy was proposed in Brown *et al.* (1959), which was subsequently corrected and implemented (Desiderio and Johnson 1971). Different variants of self-energy calculations have been developed by several authors (Blundell and Snyderman 1991; Indelicato and Mohr 1992; Mohr 1974a,b; Mohr and Soff 1993). Mohr's method deals with the choice

$$G_{\mathrm{as}}(E) = \frac{1}{h_0 - E} - V_{\mathrm{ext}} \left.\frac{1}{(h_0 - E)^2}\right|_{\mathrm{D}} \equiv F(E) - V_{\mathrm{ext}}\, [F(E)]^2|_{\mathrm{D}},$$

where the off-diagonal terms that connect the large to the small components of the electron wave function are omitted (indicated by the notation $|_{\mathrm{D}}$) in this term.

As a final result we obtain the exact self-energy of order α but to all orders in $Z\alpha$ for bound ns states with principal quantum number n. We may write

$$\Delta E_n^{\mathrm{SE}} = \frac{\alpha}{\pi}\frac{(Z\alpha)^4}{n^3} F^{\mathrm{SE}}(Z\alpha, R). \tag{1.17}$$

The most accurate calculations of the SE correction were carried out in Mohr (1974a, 1992) and in Indelicato and Mohr (1998) for the point nucleus, and in Mohr and Soff (1993) for the extended nucleus. For heavy systems ($Z > 50$) the dependence of the self-energy correction F^{SE} on the nuclear radius R also Ahas to be taken into account (Soff 1993).

1.4.5 Vacuum polarization

In Figure 1.16 the Feynman diagram corresponding to the vacuum-polarization correction of order α is depicted. Here we briefly discuss the method to evaluate this QED correction to all orders in $Z\alpha$ involving exact electron orbitals and propagators in external Coulomb potentials.

According to Equation (1.15) the exact first-order vacuum polarization potential induced by the static external Coulomb field of a nucleus reads (in the Feynman gauge)

$$\begin{aligned}\mathcal{U}^{\mathrm{VP}}(\boldsymbol{r}) &= -\mathrm{i}\alpha \int \mathrm{d}^3 r_1 \frac{1}{|\boldsymbol{r}-\boldsymbol{r}_1|}\int_{C_{\mathrm{F}}}\frac{\mathrm{d}E}{2\pi}\,\mathrm{Tr}[G(\boldsymbol{r},\boldsymbol{r}',E)] \\ &= e\int \mathrm{d}^3 r_1 \frac{\rho^{\mathrm{VP}}(\boldsymbol{r}_1)}{|\boldsymbol{r}-\boldsymbol{r}_1|},\end{aligned}$$

where the total induced vacuum polarization charge density ρ^{VP} follows as an energy integral over the trace of the exact electron propagator G along the contour C_{F}. Utilizing the integral equation (1.16) for the Green function G, we obtain the vacuum polarization density ρ^{VP} as a sum of the Uehling part (one-potential term)

$$\rho^{\mathrm{VP}}_{\mathrm{Ueh}}(\boldsymbol{r}) = -\mathrm{i}e\int_{C_{\mathrm{F}}}\frac{\mathrm{d}E}{2\pi}\int \mathrm{d}^3 r'\,\mathrm{Tr}[F(\boldsymbol{r},\boldsymbol{r}',E)V_{\mathrm{ext}}(\boldsymbol{r}')F(\boldsymbol{r}',\boldsymbol{r},E)]$$

and of the Wichmann–Kroll part summarizing all multiple-potential terms of higher orders in $Z\alpha$

$$\begin{aligned}\rho^{\mathrm{VP}}_{\mathrm{WK}}(\boldsymbol{r}) = -\mathrm{i}e\int_{C_{\mathrm{F}}}\frac{\mathrm{d}E}{2\pi}\bigg\{&\mathrm{Tr}[G(\boldsymbol{r},\boldsymbol{r},E)] \\ &-\int \mathrm{d}^3 r'\,\mathrm{Tr}[F(\boldsymbol{r},\boldsymbol{r}',E)V_{\mathrm{ext}}(\boldsymbol{r}')F(\boldsymbol{r}',\boldsymbol{r},E)]\bigg\}.\end{aligned} \tag{1.18}$$

The zero-potential term $\sim (Z\alpha)^0$, as well as all terms of even power $\sim (Z\alpha)^{2n}$, vanish by virtue of Furry's theorem (Furry 1951).

Uehling potential

The Uehling potential represents the dominant vacuum-polarization correction of order $\alpha(Z\alpha)$ to the one-photon exchange potential between the nuclear-charge distribution and the bound electron. Finally, we derive the analytically known renormalized polarization function in terms of an integral representation. For spherically symmetric external charge distributions we obtain the renormalized Uehling potential (Klarsfeld 1977):

$$\left.\begin{aligned} \mathcal{U}^{\mathrm{VP}}_{\mathrm{Ueh}}(r) &= -\frac{2e}{3\pi}\frac{Z\alpha}{r}\int_0^\infty \mathrm{d}r'\,r'\,\rho_{\mathrm{ext}}(r')\,[\chi_2(2|r-r'|)-\chi_2(2(r+r'))],\\ \chi_n(z) &= \int_1^\infty \mathrm{d}t\left(1-\frac{1}{t^2}\right)^{1/2}\left(1+\frac{1}{2t^2}\right)\frac{\mathrm{e}^{-zt^2}}{t^n}. \end{aligned}\right\} \tag{1.19}$$

Wichmann–Kroll contribution

The representation (1.18) implies a subtraction scheme for calculating the finite part of the Wichmann–Kroll potential and the vacuum polarization charge density $\rho^{\mathrm{VP}}_{\mathrm{WK}}$. It was first considered by Wichmann and Kroll (1956). A detailed discussion of the evaluation of this contribution for high-Z nuclei of finite extent is presented in Soff and Mohr (1988) and Soff (1989). A special application of the computed vacuum polarization potential to muonic atoms has been presented in Schmidt *et al.* (1989).

The numerical approach developed in Soff and Mohr (1988) utilizes the decomposition of the exact Green function into partial waves and derives the radial vacuum polarization charge density after the integration contour has been Wick-rotated

$$\rho^{\mathrm{VP}}(r) = -\mathrm{i}e\int\frac{\mathrm{d}u}{2\pi}\,|\kappa|\sum_{|\kappa|=1}^{\infty}\mathrm{Re}\left\{\sum_{i=1}^{2}G^{ii}_{\kappa}(r,r',\mathrm{i}u)\right\}.$$

This formal expression contains the divergent Uehling term of order $\alpha(Z\alpha)$, while all higher-order terms ($\alpha(Z\alpha)^n$ with $n \geqslant 3$) are finite. A successful renormalization scheme for the total vacuum polarization charge density is therefore provided by subtracting off the corresponding partial-wave decomposition of the Uehling contribution. The effect of the Uehling potential is then calculated separately based on the analytic expression (1.19). Denoting the partial wave components of the free Green function by F^{ij}_{κ}, the final expression for the vacuum polarization charge density of order $\alpha(Z\alpha)^n$ with $n \geqslant 3$ reads

$$\begin{aligned} \rho_{\mathrm{WK}}(r) &= \frac{e}{2}\int_0^\infty\frac{\mathrm{d}u}{2\pi}\sum_{\kappa=\pm1}^{\pm\infty}|\kappa|\\ &\quad\times\mathrm{Re}\left\{\sum_{i=1}^{2}G^{ii}_{\kappa}(r,r,\mathrm{i}u)+\int_0^\infty\mathrm{d}r'\,r'^2\,V_{\mathrm{ext}}(r')\sum_{m,n=1}^{2}[F^{mn}_{\kappa}(r,r',\mathrm{i}u)]^2\right\}. \end{aligned} \tag{1.20}$$

Given the Wichmann–Kroll density we can calculate first the contribution to the vacuum polarization potential and then the corresponding energy shift. The energy correction associated with the Wichmann–Kroll potential caused by the density (1.20) is usually expressed in terms of a function $H_{\rm WK}$. Again for bound ns states we may write similarly to Equation (1.17)

$$\Delta E_a^{\rm VP}({\rm WK}) = \frac{\alpha}{\pi}\frac{(Z\alpha)^4}{n^3} H^{\rm WK}(Z\alpha, R).$$

For extended nuclei the remaining Wichmann–Kroll contribution was first calculated in Soff and Mohr (1988) and Soff (1989), while corresponding results for the point nucleus can be found in Manakov *et al.* (1989). The renormalization scheme for the Wichmann–Kroll part was also used later (Persson *et al.* 1993b) to generate more precise results. Finite nuclear size effects have also been taken into account, and the results are in fair agreement with the corresponding data given by Soff and Mohr (1988). The calculations performed in Persson *et al.* (1993b) and Soff and Mohr (1988) underline the importance of including finite-size effects in any state-of-the-art calculations of the vacuum-polarization correction for high-Z systems.

1.4.6 Lamb-shift calculations for highly charged ions

The ground-state Lamb shift in heavy hydrogen-like atoms

The term *Lamb shift* of a single atomic level usually refers to the difference between the Dirac energy for point-like nuclei and its observable value shifted by nuclear and QED effects. Nuclear effects include energy shifts due to static nuclear properties such as the size and shape of the nuclear charge density distribution and due to nuclear dynamics, i.e. recoil correction and nuclear polarization. To a zeroth approximation, the energy levels of a hydrogen-like atom are determined by the Dirac equation. For point-like nuclei the eigenvalues of the Dirac equation can be found analytically. In the case of extended nuclei, this equation can be solved either numerically or by means of successive analytical approximation (see Rose 1961; Shabaev 1993).

Besides nuclear finite-size effects, the leading corrections to the Dirac energy levels originate from QED-radiative corrections of order α, i.e. the first-order self-energy (SE) and vacuum polarization (VP) (see Figure 1.16). We refer to the tabulation provided in Johnson and Soff (1985). At present, the calculation of the first-order QED corrections can be considered as well established. Extensive tabulations for the Wichmann–Kroll part of the vacuum polarization are presented in Beier *et al.* (1997b) and for the self-energy correction in Beier *et al.* (1998). At the present level of experimental precision all available 1s-Lamb-shift data for hydrogen-like ions can be explained on the basis of these dominant QED corrections and finite nuclear size effects. However, the accuracy of about 1 eV, which is aimed for in the experiments under preparation at the Gesellschaft für Schwerionenforschung mbH (GSI) (Stöhlker *et al.* 2000), demands that theory provide a calculation of the whole set of

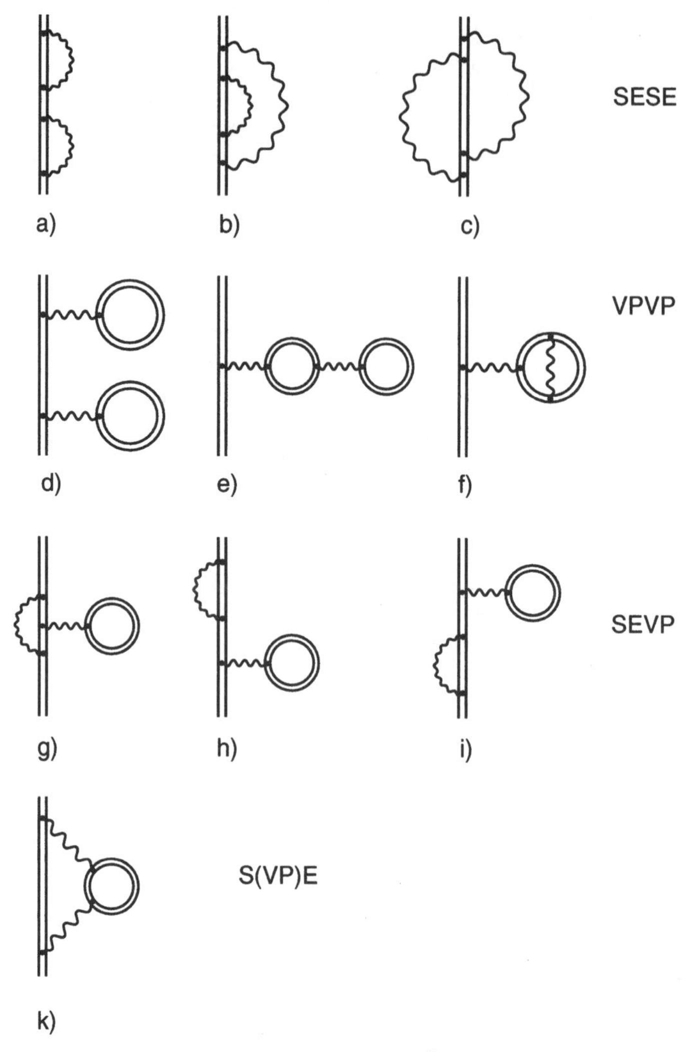

Figure 1.17 QED corrections of order α^2 in hydrogen-like ions.

QED corrections of order α^2. The corresponding Feynman diagrams are depicted in Figure 1.17.

The set of second-order QED corrections can be divided into separately gauge-invariant subsets: the two-photon self-energy diagrams SESE (a)–(c), the loop-after-loop vacuum polarization VPVP (d), the two-loop vacuum polarization VPVP (e) together with the diagram VPVP (f), the mixed self-energy vacuum polarization SEVP (g)–(i) and finally the effective self-energy diagram S(VP)E (k). Most of these corrections have been calculated by employing various modifications of the methods developed for the evaluation of the first-order self-energy and vacuum-polarization corrections. By this means the contributions of all diagrams (d)–(k) have been evaluated by a number of authors. However, we should note that the corrections VPVP (f) and S(VP)E (k) are at present calculated only in the Uehling approximation, which should be considered only as an estimate of the real contribution.

Renormalization schemes for the two-photon self-energy were first developed by Labzowsky and Mitrushenkov (1996) and later by Lindgren *et al.* (1998), where the Feynman gauge has been employed. Within this gauge choice the irreducible contribution of the loop-after-loop diagram SESE (a) can be separately calculated. The first calculations were presented in Mitrushenkov *et al.* (1995). The loop-after-loop vacuum-polarization correction VPVP (d) has been considered (Persson *et al.* 1993b) and tabulated (Beier *et al.* 1997b). The Källén–Sabry contributions, i.e. the VPVP (e,f) corrections in the Uehling approximation, have been investigated by Beier and Soff (1988) for point-like and Schneider *et al.* (1993a) for extended nuclei. A renormalization scheme for deducing the energy shifts due to higher orders in $Z\alpha$ of the two-loop vacuum polarization VPVP (e) diagram and corresponding numerical results have been delivered in Plunien *et al.* (1998). It allows us to assign separately a value for the Uehling part of the VPVP (f) diagram. The mixed self-energy vacuum polarization SEVP (g)–(i) was first evaluated in Lindgren *et al.* (1993) taking into account only the Uehling part of the vacuum-polarization potential, and later in Persson *et al.* (1996a). Until now the effective self-energy S(VP)E (k) has been calculated only within the Uehling approximation (Mallampalli and Sapirstein 1996; Persson *et al.* 1996a). While a considerable effort has been made during the last decade in evaluating the α^2-corrections (a) and (d)–(k), the calculation of the remaining two-photon self-energy diagrams SESE (b) and SESE (c) together with the reducible part of SESE (a) (employing the Feynman gauge) turns out to be a far more difficult task.

The first attempt to perform this was carried out by Mallampalli and Sapirstein (1998), where a specific part for this correction was evaluated. However, since there is nothing to suggest that this part represents a dominant contribution, this calculation still does not provide any information about the actual value of this correction. Combined with the results of a previous investigation (Mallampalli and Sapirstein 1998), a direct evaluation of the two-loop self-energy contributions SESE (b,c) to the 1s-Lamb shift in hydrogen-like uranium has been reported recently (Yerokhin and Shabaev 2001). These calculations employ a covariant renormalization scheme based on the potential expansion of electron propagators. All the corrections discussed so far are treated in the external-field approximation. This means that the nucleus is considered only as a source of an external Coulomb field. Going beyond this approximation requires as a first step the evaluation of nuclear-recoil corrections. In contrast to nonrelativistic quantum mechanics, where the recoil effect is accounted for simply by the reduced mass, a full relativistic theory of the nuclear recoil has to be formulated within the QED framework. Investigations reveal that the nonrelativistic treatment of the nuclear recoil is incorrect by more than 50% for heavy hydrogen-like systems like uranium. A formula for the recoil effect in a hydrogen-like ion to first order in m/M (the mass ratio of the electron and the nucleus) that accounts for the complete

Table 1.3 Lamb-shift contribution for the 1s ground state of $^{238}U^{91+}$ and $^{208}Pb^{81+}$ (in eV), including the full nuclear-structure corrections.

Corrections (in eV):	$^{238}U^{91+}$		$^{208}Pb^{81+}$	
Finite nuclear size	198.82	±0.10	67.25	±0.02
Self-energy (order α)	355.05		226.33	
Vacuum polarization (order α)	−88.60		−48.41	
SESE (a) (irred.)	−0.97		−0.51	
SESE (a) (red.) (b) (c)	−0.90	±0.10	±0.00[a]	
VPVP (d)	−0.22		−0.09	
VPVP (e)	−0.15		−0.07	
VPVP (f) (Uehling approx.)	−0.60	±0.10	−0.34	
SEVP (g)–(i)	1.12		0.53	
S(VP)E (k) (Uehling approx.)	0.13		0.07	
Nuclear recoil	0.46		0.37	
Nuclear polarization	−0.20	±0.10	0.00	
Lamb shift (theory)	463.94	±0.40	245.13	±0.02
Lamb shift (experiment)	468	±13	290	±75

[a]A result for the SESE (b,c) contribution for lead is not yet available.

$Z\alpha$-dependence was first derived in Shabaev (1985). A detailed description of the method can be found in Shabaev (1998).

The numerical evaluation of nuclear-recoil effects to all orders in $Z\alpha$ has been carried out by Artemyev *et al.* (1995) and Shabaev *et al.* (1998a) for point-like and for extended nuclei, respectively. A second step beyond the external-field approximation accounts for the internal nuclear dynamics, i.e. the polarizability of the nucleus. The nuclear-polarization correction was derived and evaluated in the framework of the effective photon propagator in Plunien *et al.* (1991), Plunien and Soff (1995) and Nefiodov *et al.* (1996). For a detailed discussion we also refer to Mohr *et al.* (1998). The individual contributions to the 1s-ground-state Lamb shift in $^{238}U^{91+}$ and $^{208}Pb^{81+}$ are presented in Table 1.3. The finite nuclear size correction is calculated for a Fermi distribution with a root-mean-square (RMS) radius $\langle r^2\rangle^{1/2} = 5.860(2)$ fm for uranium and $\langle r^2\rangle^{1/2} = 5.505(1)$ fm for lead (Zumbro *et al.* 1984). The indicated uncertainties of 0.10 eV for uranium and 0.02 eV for lead reflect the uncertainty in the RMS radii. We also note that an uncertainty of 0.38 eV for uranium and 0.14 eV for lead may be ascribed to the nuclear size effect when calculating the difference between the corrections using a Fermi model or a homogeneously charged sphere distribution with the same RMS radius (see Beier *et al.* 1997a, for details).

To obtain the total binding energies, the (point-nucleus) Dirac eigenvalues of −132 279.92(1) eV for uranium and −101 581.37(1) eV for lead should be added to the Lamb-shift contribution quoted in Table 1.3. We may note, for example, that in the case of uranium an error of 0.01 eV of the Dirac binding energy even results from the uncertainty of the Rydberg constant (Mohr and Taylor 2000). As can be seen from the table, the present level of experimental precision provides a test of QED effects to first order in α at the level of 5%.

$2p_{1/2}$–2s Lamb shift in lithium-like uranium

In heavy lithium-like ions, in addition to the one-electron QED and nuclear corrections discussed above, additional contributions due to the electron–electron interaction have to be taken into account. Traditionally, energy levels in high-Z lithium-like ions have been calculated employing the relativistic many-body perturbation theory (RMBPT) (Blundell 1993; Kim *et al.* 1991; Ynnerman *et al.* 1994), the multiconfiguration Dirac–Fock method (Indelicato and Desclaux 1990) and the relativistic configuration-interaction method (Chen *et al.* 1995; Cheng *et al.* 2000). These methods account for the correlation and the relativistic corrections but do not provide a rigorous treatment of the QED effects. This fact does not create any problems in first order in α, where the correlation effects are naturally separated from the radiative corrections, and the first-order QED corrections can be simply added to the many-body results. However, a calculation valid up to order α^2 must be carried out rigorously starting from first principles of QED. To some extent, the electron–electron interaction can be accounted for by considering the first-order QED corrections for an electron in an effective field, that is, the sum of the nuclear potential and an additional spherically symmetric potential generated by other electrons. We refer to this approach as the *effective-potential* approximation. By effective potential we mean here the direct part of the electron–electron interaction,

$$V_{\mathrm{scr}}(r_1) = \alpha \sum_i \int_0^{\infty} \mathrm{d}r_2\, r_2^2 \frac{1}{r_>}[g_i^2(r_2) + f_i^2(r_2)],$$

where $g_i(r)$ and $f_i(r)$ are the upper and the lower radial components of the wave function, and $r_> = \max(r_1, r_2)$.

In that method, a part of the second-order effects is included in the first-order QED corrections (Blundell 1993; Chen *et al.* 1995; Cheng *et al.* 1991, 2000; Indelicato 1991).

Ab initio QED calculations for heavy few-electron atoms are generally performed by perturbation theory. In recent research (Yerokhin *et al.* 2000, 2001), in the zeroth approximation the electrons interact only with the Coulomb field of the nucleus. To zeroth order the binding energy is given by the sum of one-electron binding energies. The interelectronic interaction and the radiative corrections are accounted for by perturbation theory in the parameters $1/Z$ and α, respectively. Since $1/Z \sim \alpha$ for very-high-Z ions, for simplicity we can classify all corrections by the parameter α.

The correction corresponding to the interelectronic interaction of order α (one-photon exchange) can easily be evaluated numerically. While the QED corrections of first order in α can be considered as well established at present (see, for example,

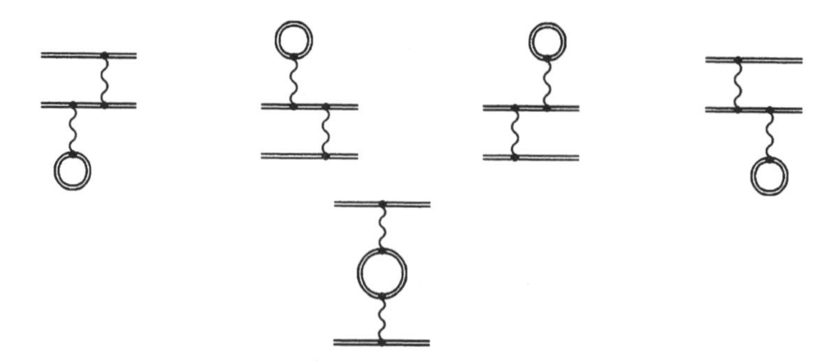

Figure 1.18 Vacuum-polarization screening diagrams.

Shabaev *et al.* 2000a, for a recent review), the complete second-order calculation is not yet finished.

The two-electron QED corrections of order α^2 for the ground state of helium-like ions with $Z > 20$ have been calculated completely. The corresponding two-photon exchange correction was calculated in Blundell *et al.* (1993) and Lindgren *et al.* (1995). For helium-like systems the self-energy and vacuum-polarization screening diagrams have been evaluated (Persson *et al.* 1996b; Yerokhin *et al.* 1997a). The related calculations for excited states of helium-like ions were performed for the vacuum-polarization screening diagrams in Artemyev *et al.* (2000) and, in the case of nonmixed states, for the two-photon exchange diagrams in Mohr and Sapristein (2000). As a result of several years of effort, the complete evaluation of all two-electron QED corrections of order α^2 for the $2p_{1/2}$–2s transition in heavy lithium-like ions has recently been reported (Yerokhin *et al.* 2000, 2001).

The set of α^2-corrections can be divided into three separately gauge-invariant subsets: the vacuum-polarization screening contribution (Figure 1.18), the self-energy screening correction (Figure 1.19), and the two-photon exchange diagrams (Figure 1.20). We refer to the diagram in Figure 1.20(c) as the *three-electron* contribution and to the diagrams in parts (a) and (b) of Figure 1.20 as the *ladder* and *crossed* contributions, respectively.

For the $2p_{1/2}$–2s transition in lithium-like high-Z ions, the gauge-invariant sets of the screened vacuum-polarization corrections and of the screened-self-energy have been evaluated recently by Artemyev *et al.* (1999) and Yerokhin *et al.* (1999), respectively. Finally, the complete two-electron contributions have been presented recently, even beyond the Breit level (Yerokhin *et al.* 2000, 2001).

Most of these diagrams contain two intermediate electron propagators and, therefore, double summations over the whole spectrum of the Dirac equation in the external nuclear field. This makes their computation numerically intensive. Both the self-energy and vacuum-polarization screening corrections are ultraviolet divergent and require renormalization to yield a finite result.

The numerical evaluation of the two-photon exchange corrections has been carried out in a way similar to that for the $(1s)^2$ state in Blundell *et al.* (1993). The summa-

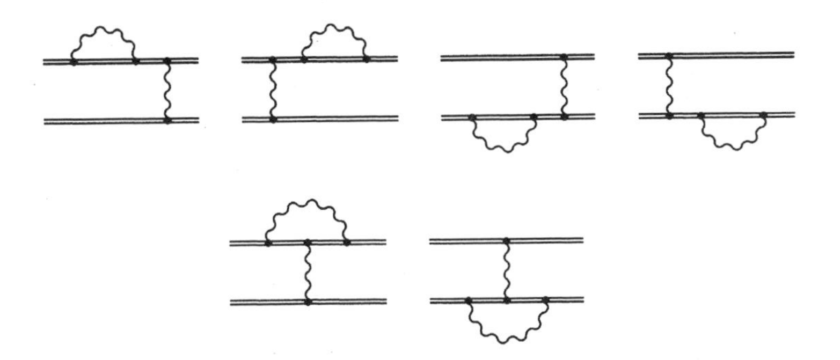

Figure 1.19 Self-energy screening diagrams.

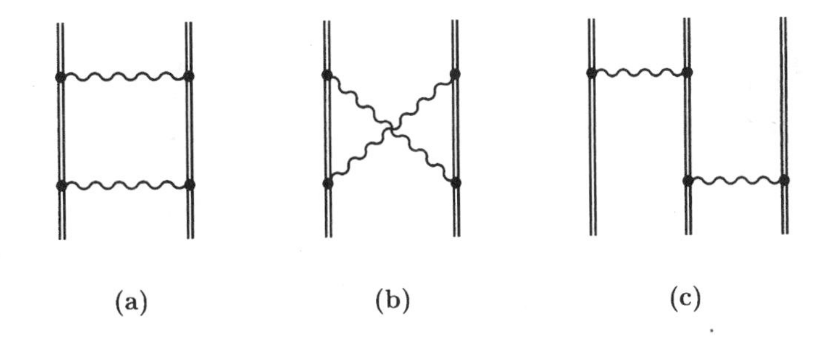

Figure 1.20 Two-photon exchange diagrams.

tion over the whole spectrum of the intermediate states was performed by using the B-spline approach for the Dirac equation (Johnson *et al.* 1988). The finite size of the nucleus is taken into account by using a homogeneously charged sphere distribution of the nuclear charge with RMS radii taken from Yerokhin *et al.* (1999). Calculations performed both in the Feynman and in the Coulomb gauge exhibit an excellent agreement. The direct and the exchange parts are found to be separately gauge invariant on the level of numerical accuracy. Their contribution has been accounted for within the Breit approximation (Shabaev *et al.* 2000b). Due to the restrictions of the Breit approximation the uncertainty assigned to this contribution may be about 50%. A complete overview of the present status of QED calculations of the interelectron interaction in two- and three-electron ions was recently given by Andreev *et al.* (2001).

In Table 1.4 the individual contributions to the $2p_{1/2}$–2s transition energy in lithium-like uranium are compiled. The one-electron QED-corrections are included as well. The predicted value of 280.46(9) $\pm$ 0.20 eV for the total Lamb shift is in agreement with the related experimental result of 280.59(10) eV (Schweppe *et al.* 1991).

The error of 0.20 eV ascribed to the theoretical value originates from the second-order one-electron QED contribution, which is not yet completely established. The

Table 1.4 Various contributions to the $2p_{1/2}$–2s transition in $^{238}U^{89+}$ (in eV).

One-electron nuclear size		−33.35(6)
One-photon exchange		368.83
First-order	SE	−55.87
	VP	12.94
Two-photon exchange	Breit approximation	−13.54
	beyond Breit approximation	0.17
Two-electron	SE	1.52
	VP	−0.36
⩾3 photon exchange		0.16(7)
Nuclear recoil		−0.07
Nuclear polarization		0.03(1)
One-electron α^2 QED		±0.20
Total theory		280.46(9) ± 0.20
Experiment		280.59(9)

total two-photon exchange correction can be conveniently divided into a dominant part that can be evaluated within the Breit approximation, and a remainder which can be interpreted as a contribution of the second-order QED effects. For the exact definition of the two-photon exchange contribution within the Breit approximation we refer to (Yerokhin *et al.* 2000, 2001). As can be read off from the table, the first-order QED corrections contribute with −42.93 eV, while the total second-order QED correction beyond the Breit approximation amounts to about 1.33 ± 0.20 eV. In summary, the present status in the prediction of the Lamb shift in lithium-like uranium provides a sensitive test for QED effects of first order in α on a level of accuracy of about 0.5%, and of QED corrections of order α^2 on a level of accuracy of about 15%. Thus, for the first time, this opens up the possibility of probing second-order QED effects in a strong Coulomb field by a comparison between theory and experiment. However, we should be aware of the fact that a reliable estimate of the uncertainty of the predicted value can be achieved only after the calculation of the second-order one-electron QED corrections have been established.

1.4.7 Hyperfine structure and bound-electron g-factor

Basic formulae

Magnetic interactions between bound electrons and external magnetic fields are described by the additional interaction Hamiltonian

$$H_{\text{int}} = -\boldsymbol{\mu} \cdot \boldsymbol{B} = e\,\boldsymbol{\alpha} \cdot \boldsymbol{A},$$

where $\boldsymbol{\mu}$ is the magnetic moment operator, $\boldsymbol{B}$ denotes the vector of the magnetic field and $\boldsymbol{A}$ denotes the corresponding vector potential. The two major effects involving magnetic fields are the Zeeman effect, where the magnetic moment of the electron $\boldsymbol{\mu}_J$ interacts with a homogeneous external magnetic field $\boldsymbol{B}_{\text{ext}}$ (respectively $\boldsymbol{A}_{\text{ext}}$), and the hyperfine-structure splitting, where the magnetic moment of the nucleus $\boldsymbol{\mu}_I$ interacts with the magnetic field $\boldsymbol{B}_{\text{hfs}}$ (respectively $\boldsymbol{A}_{\text{hfs}}$), caused by the motion of the electron. For these two different kinds of perturbation the potentials are given by

$$\boldsymbol{A}_{\text{ext}}(\boldsymbol{r}) = \tfrac{1}{2}\boldsymbol{B}_{\text{ext}} \times \boldsymbol{r},$$
$$\boldsymbol{A}_{\text{hfs}}(\boldsymbol{r}) = \frac{\boldsymbol{\mu}_I \times \boldsymbol{r}}{4\pi r^3}.$$

The operators of the magnetic moments of electron and of the nucleus read

$$\boldsymbol{\mu}_J = -g_J \mu_{\text{B}} \frac{\boldsymbol{J}}{\hbar},$$
$$\boldsymbol{\mu}_I = g_I \mu_{\text{N}} \frac{\boldsymbol{I}}{\hbar},$$

where $\boldsymbol{J}$ now denotes the total angular momentum of the electron and $\boldsymbol{I}$ denotes the angular momentum of the nucleus. The quantities g_J and g_I denote the g-factor of the bound electron and of the nucleus, respectively. The latter accounts for the complex structure of the nucleus. The magnetic moments are expressed in units of the Bohr magneton for an electron

$$\mu_{\text{B}} = \frac{e\hbar}{2m_{\text{e}}} \quad (\approx 0.579 \times 10^{-4}\ \text{eV T}^{-1})$$

and the nuclear magneton μ_{N}

$$\mu_{\text{N}} = \frac{e\hbar}{2m_{\text{p}}} \quad (\approx 3.152 \times 10^{-8}\ \text{eV T}^{-1}).$$

For a free electron, the Dirac theory yields a value $g_{\text{free}} = 2$. Corrections to the free-electron g-factor are essentially due to the interaction with the free radiation field. The determination of the g-factor of a bound electron interacting with an additional, homogeneous magnetic field $\boldsymbol{B}_{\text{ext}}$ via measurements of the induced energy shift on atomic levels is one of the basic experiments for testing QED. The deviation $(g - 2)$ can be measured rather precisely.

The generic magnetic interaction gives rise to a level energy shift

$$\Delta E_{nFM}^{\text{mag}} = \langle \Phi_{nFM} | e\boldsymbol{\alpha} \cdot \boldsymbol{A} | \Phi_{nFM} \rangle.$$

The matrix element has to be evaluated with states

$$|\Phi_{nFM}\rangle = \sum_{m_I m_j} C_{Im_I jm_j}^{FM} |\phi_{njm_j}\rangle |\xi_{Im_I}\rangle,$$

where the angular momentum of the electron and the nucleus are coupled to total angular momentum quantum numbers F and M, respectively. Due to the presence of a Zeemann-type interaction, the generic energy shift of a bound electron characterized by quantum numbers n and κ in hydrogen-like ions reads

$$\Delta E_{n\kappa}^{\text{ext}} = m\frac{4\kappa}{4\kappa^2-1}\left(\int_0^\infty \mathrm{d}r\, r^3\, f_{n,\kappa}(r) g_{n,\kappa}(r)\right)\mu_{\mathrm{B}} B_{\text{ext}}, \tag{1.21}$$

which defines the g_J-factor according to

$$g_J = \frac{4\kappa}{4\kappa^2-1}\left(\int_0^\infty \mathrm{d}r\, r^3 f_{n,\kappa}(r) g_{n,\kappa}(r)\right).$$

Assuming, for example, a point-like atomic nucleus and taking into account the pure Coulomb potential, where the wave functions are analytically known, yields for the 1s state (Breit 1928)

$$g_{J,\text{point like}} = \tfrac{2}{3}(1 + 2\sqrt{1-(Z\alpha)^2}).$$

Similarly, the generic expression for the energy shift of a level induced via the hyperfine interaction is given by

$$\Delta E_{n\kappa}^{\text{HFS}} = \frac{e}{4\pi} g_I\, \mu_{\mathrm{N}}\, \frac{4\kappa}{4\kappa^2-1}[F(F+1) - I(I+1) - j(j+1)] \times \int_0^\infty f_{n,\kappa}(r)\, g_{n,\kappa}(r)\, \mathrm{d}r. \tag{1.22}$$

Beside finite nuclear size effects that are carried essentially by the bound-electron wave functions $g_{n\kappa}$ and $f_{n\kappa}$, QED corrections will contribute to observable modifications of the generic level shifts, Equations (1.21) and (1.22), respectively. Carrying out the complete derivation and numerical evaluations based on the S matrix formalism is rather cumbersome. Alternatively, in the case of weak magnetic interactions all expressions can also be obtained by considering the perturbations of the electron wave function ϕ_a, the corresponding binding energy E_a, and of the electron propagator (Indelicato 1991). Performing the following replacements in a given generic bound-state QED diagram of order α, i.e.

$$\begin{aligned}|\phi_a\rangle &\to |\phi_a\rangle + \sum_{\substack{n\\E_n\neq E_a}} \frac{|\phi_n\rangle\langle\phi_n|e\,\boldsymbol{\alpha}\cdot\boldsymbol{A}|\phi_a\rangle}{E_a - E_n} + \cdots \\ &= |\phi_a\rangle + |\delta\phi_a\rangle + \cdots,\end{aligned} \tag{1.23}$$

$$S_{\mathrm{F}}(\boldsymbol{x},\boldsymbol{y},E) \to S_{\mathrm{F}}(\boldsymbol{x},\boldsymbol{y},E) + \int \mathrm{d}^3z\, S_{\mathrm{F}}(\boldsymbol{x},\boldsymbol{z},E)\, e\,\boldsymbol{\alpha}\cdot\boldsymbol{A}\, S_{\mathrm{F}}(\boldsymbol{z},\boldsymbol{y},E) + \cdots, \tag{1.24}$$

$$\begin{aligned}E_a &\to E_a + \langle\phi_a|\, e\,\boldsymbol{\alpha}\cdot\boldsymbol{A}|\phi_a\rangle + \cdots \\ &= E_a + \Delta E_a^{\text{mag}} + \cdots\end{aligned}$$

will lead, for example, to a modification of the self-energy operator

$$\Sigma(E_a) \to \Sigma(E_a) + \Delta E_a^{\text{mag}} \frac{\partial}{\partial E} \Sigma(E)\bigg|_{E=E_a} + \cdots .$$

The dots indicate higher orders of perturbation theory and are not considered here. This method of deriving expressions has been employed frequently for magnetic perturbations as well as for the related electron–electron interaction (Persson *et al.* 1996b,c; Sunnergren *et al.* 1998).

Hyperfine-structure calculations

The theoretical description of QED corrections to the hyperfine-structure splitting (HFS) is practically the same as for the g-factor. In both cases radiative corrections to a magnetic perturbation are considered. The HFS in highly charged ions has been investigated experimentally until now at the GSI (Klaft *et al.* 1994; Seelig *et al.* 1998; Winter *et al.* 1999) as well as at the Lawrence Livermore National Laboratory (López-Urrutia *et al.* 1996, 1998). Corresponding to the rather high experimental precision achieved in such systems, theoretical investigations have been carried out in particular on the QED contributions of order α. As in the case of the Lamb shift, calculations that are nonperturbative in $Z\alpha$ are required. Nuclear-size corrections may be taken into account either by solving numerically the Dirac equation for the wave functions or by employing approximate analytical formulae. For hydrogen-like ions the HFS of the ground state can be written in the form

$$\Delta E = \tfrac{4}{3}\alpha(\alpha Z)^3 \frac{\mu_I}{\mu_{\mathrm{N}}} \frac{m}{m_{\mathrm{p}}} \frac{2I+1}{2I} mc^2 [A(Z\alpha)(1-\delta)(1-\epsilon) + \chi_{\mathrm{QED}}], \qquad (1.25)$$

where m is the electron mass, m_{p} is the proton mass, μ_I is the nuclear magnetic moment and μ_{N} denotes the nuclear magneton. The terms in the square brackets represent the corrections to the classical nonrelativistic hyperfine splitting: the relativistic factor $A(Z\alpha)$ for the $1s_{1/2}$ state is given by (Pyykkö *et al.* 1973)

$$A(Z\alpha) = \frac{1}{\gamma(2\gamma - 1)} \quad \text{with } \gamma = \sqrt{1 - (Z\alpha)^2},$$

the finite-size nuclear-charge distribution correction is δ, the finite-size nuclear-magnetization distribution correction is ϵ, and the QED corrections are denoted by χ_{QED}. The effect due to a deviation of the nuclear magnetization distribution from the point-dipole model is often termed the Bohr–Weisskopf effect. The first numerical investigations date back to Bohr (1951) and Bohr and Weisskopf (1950), where calculations based on the nuclear single-particle model were performed. Within this approach the extended magnetization distribution is due to a single-valence nucleon moving in the effective nuclear field generated by the core nucleus. A more elaborate version of this model has been employed to obtain numerical values for this effect in the range $Z = 49$–83 for hydrogen- and lithium-like ions (Shabaev 1994; Shabaev *et al.*

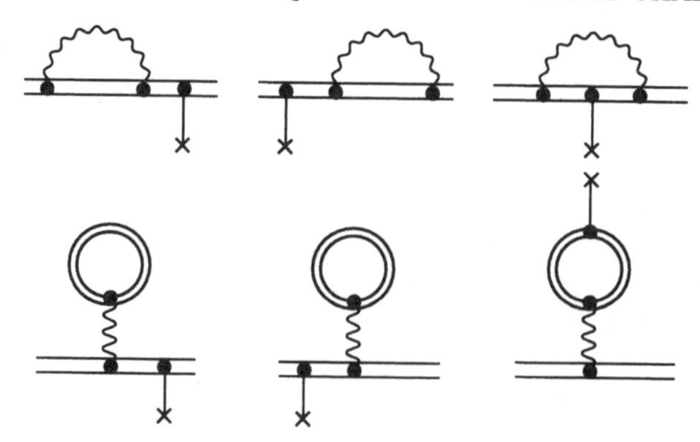

Figure 1.21 The self-energy and vacuum-polarization correction to a bound electron perturbed by an external magnetic field. The solid line terminated by a cross denotes the interaction with the external magnetic field.

1997, 1998c). For $^{209}Bi^{82+}$ a relativistic dynamic-proton model developed by Labzowsky *et al.* (1995, 1997) yields similar results. Here we refer to a number of other approaches that only try to model the magnetization distribution inside of the nucleus (Arima and Horie 1955; Finkbeiner *et al.* 1993; Noya *et al.* 1958; Schneider *et al.* 1993b; Tomaselli *et al.* 1995, 1997). A detailed discussion on the current difficulties in estimating the Bohr–Weisskopf effect is given in a recent review (Beier 2000).

QED radiative corrections of order α to a bound electron interacting with a perturbing magnetic field (depicted in Figure 1.21) have all been calculated independently during the last decade by several groups for the $1s_{1/2}$ and the $2s_{1/2}$ state (Blundell *et al.* 1997; Persson *et al.* 1996c; Schneider *et al.* 1994; Shabaev *et al.* 1997, 1998c, 2000b; Sunnergren *et al.* 1998; Yerokhin *et al.* 1997b). It turns out, however, that their magnitude is of similar size to the uncertainty of the Bohr–Weisskopf effect caused by the inherent model dependence of the nuclear magnetization distribution. Measuring the hyperfine-structure splitting in highly charged ions provides an alternative scenario for precision tests of bound-state QED. As was pointed out in Shabaev *et al.* (1998b) and Shabaev (1998), it is yet possible to combine measurements of the hyperfine-structure splitting in hydrogen-like and lithium-like ions of the same species. By extracting a value for the Bohr–Weisskopf effect from one experiment, its magnitude for the other charge state of the ion can be adjusted and its uncertainty is much less than that induced by the inherent model dependence. The Bohr–Weisskopf effect for a hydrogen-like ion can be deduced from a measurement by means of the formula

$$\epsilon^{(1s)} = \frac{\Delta E_{\mathrm{Dirac}}^{(1s)} + \Delta E_{\mathrm{QED}}^{(1s)} - \Delta E_{\mathrm{exp}}^{(1s)}}{\Delta E_{\mathrm{Dirac}}^{(1s)}},$$

where

$\Delta E_{\mathrm{Dirac}}^{(1s)}$ is the value of the $1s_{1/2}$ hyperfine-structure splitting including the nuclear-charge distribution,

Table 1.5 Hyperfine-structure splitting of the ground state in hydrogen- and lithium-like bismuth.

	$^{209}Bi^{82+}$	$^{209}Bi^{80+}$
Relativistic one-electron value	5.8393(3)	0.958 49
Finite nuclear size effect	−0.6482(7)	−0.113 8(2)
Bohr–Weisskopf effect	−0.061(27)	−0.013 4(2)
One-electron QED (order α)	−0.0298	−0.005 1(2)
Interelectronic interaction		
of first order in $1/Z$		−0.029 48
of second and higher orders in $1/Z$		0.000 24(12)
estimate for QED correction (eV)		0.000 18(9)
Theory, total (eV)	5.100(27)	0.797 1(2)
Experiment (eV)	5.0840(8) 5.0843(4)	0.820(26)

$\Delta E_{\mathrm{QED}}^{(1s)}$ is the QED correction to the hyperfine-structure splitting for the $1s_{1/2}$ state, and

$\Delta E_{\mathrm{exp}}^{(1s)}$ is the experimental value of the $1s_{1/2}$ hyperfine-structure splitting.

The first and the last of these quantities are well known, and the QED value is also known but can be tested when a similar experiment is carried out on the $2s_{1/2}$ state of a lithium-like ion, where a similar formula has to be applied. The expected ratio for the Bohr–Weisskopf effect is $\epsilon^{(2s)}/\epsilon^{(1s)} = 1.078$ for the case of ^{209}Bi (Shabaev *et al.* 1998b). This quantity has to be tested by experiment. The QED contributions have to be known to a very high precision. The possibility for testing QED in experiments on the hyperfine splitting via the determination of a specific difference of the hyperfine splitting values of hydrogen- and lithium-like bismuth has been recently examined (Shabaev *et al.* 2001). The quoted experimental data are taken from Klaft *et al.* (1994) and Winter *et al.* (1999) for hydrogen-like bismuth and from Beiersdorfer *et al.* (1998) for lithium-like bismuth.

The theory's proposal of evaluating the α^2-corrections has already been taken up and a precision search was started at the GSI (Borneis *et al.* 2000). The search is not yet complete and a high-precision test of the QED contributions to the hyperfine-structure splitting in highly charged ions is not therefore currently available. For ^{209}Bi, theoretical and experimental values are displayed in Table 1.5. For $^{209}Bi^{82+}$, the QED corrections and the relativistic one-electron value including the finite nuclear size effect have been taken from Sunnergren *et al.* (1998). The separate finite nuclear size effect has been obtained by subtracting the corresponding value of a point-like nucleus calculated via Equation (1.25) (neglecting δ, ϵ and χ_{QED}). The numbers for the Bohr–Weisskopf effect presented in Table 1.5 and all values for $^{209}Bi^{80+}$ have been taken from Shabaev *et al.* (2000b).

The lifetime of the upper hyperfine-structure level in $^{209}Bi^{82+}$ was measured to be $\tau_{exp} = 397.5(1.5)$ μs (Winter *et al.* 1999). Employing the experimental value of the HFS transition energy ω together with the transition amplitude for $^{209}Bi^{82+}$, the experimental value of the bound-electron g-factor is found to be 1.7343(33). This result is in remarkably good agreement with the theoretical value of 1.7310 for $^{209}Bi^{82+}$. Also the measured value for $^{207}Pb^{81+}$ (Seelig *et al.* 1998) was found to be in good agreement with the theoretical prediction.

To summarize in brief: QED provides a unique framework for the foundation of a relativistic description of atomic many-electron systems. Being able to account for their effects in few-electron systems to the utmost precision is of great importance for the determination of fundamental constants in experiments with hydrogen or to explore the range of validity in the strong external fields as they occur in highly charged ions. At present, measurements of the one-, two- and three-electron ions, the hyperfine-structure splitting and of the g-factors of bound electrons are considered to be the most promising scenarios for sensitive tests for QED. In all cases radiative corrections have to be taken into account as well. Complete QED calculations for electron correlation effects are necessary in order to describe the measured Lamb shift in lithium-like uranium.

2 Four-Component *Ab Initio* Methods for Atoms, Molecules and Solids

Markus Reiher
Lehrstuhl für Theoretische Chemie,
Friedrich–Alexander-Universität Erlangen–Nürnberg

Juergen Hinze
Theoretische Chemie, Fakultät für Chemie,
Universität Bielefeld

2.1 Introduction

Relativistic effects play an important role in the spectroscopy of atoms and molecules whenever heavy atoms are involved or electronic or nuclear spins become significant, as in ESR and NMR spectroscopy or for the fine and hyperfine structure of electronic states. Also the chemical behaviour of the heavy elements, beyond $Z \sim 50$, is strongly influenced by relativistic effects. As the chemical interactions are affected by the 'slow' valence electrons, relativistic effects were thought to be of minor importance. However, electrons of shells with low l values, especially s electrons, do penetrate the atomic core and experience relativistic retardation effects near highly charged nuclei; these shells therefore contract. On the other hand, electrons in shells with higher l values, $l \geqslant 2$, are screened better, due to the contracted s shells, from the nuclear charge and therefore become destabilized; thus these shells are more expanded than expected.

The mathematical basis of the relativistic quantum mechanical description of many-electron atoms and molecules is much less firm than that of the nonrelativistic counterpart, which is well understood. As we do not know of a covariant quantum mechanical equation of motion for a many-particle system (nuclei plus electrons), we rely on the Dirac equation for the quantum mechanical characterization of a free electron (positron) (Darwin 1928; Dirac 1928, 1929; Dolbeault *et al.* 2000b; Thaller 1992)

$$[c\boldsymbol{\alpha} \cdot \boldsymbol{p} + m_0 c^2 \beta]\Psi(\boldsymbol{r}, t) = \mathrm{i}\hbar \frac{\partial}{\partial t}\Psi(\boldsymbol{r}, t),$$

which is covariant. Here m_0 is the rest mass of the electron and c is the velocity of light in vacuum. We use the notation, which has become the accepted convention, that 3-vectors of objects are set in a bold face, while matrices are in regular type, except

Relativistic Effects in Heavy-Element Chemistry and Physics. Edited by B. A. Hess

for unit matrices, which are set in a bold face. The 4×4 matrices α_k and β are

$$\alpha_k = \begin{pmatrix} 0 & \sigma_k \\ \sigma_k & 0 \end{pmatrix}, \quad k = x, y, z \quad \text{and} \quad \beta = \begin{pmatrix} \mathbf{1} & 0 \\ 0 & -\mathbf{1} \end{pmatrix},$$

with the 2×2 matrices σ_k in the Pauli representation given as

$$\sigma_x = \begin{pmatrix} 0 & 1 \\ 1 & 0 \end{pmatrix}, \qquad \sigma_y = \begin{pmatrix} 0 & -\mathrm{i} \\ \mathrm{i} & 0 \end{pmatrix}, \qquad \sigma_z = \begin{pmatrix} 1 & 0 \\ 0 & -1 \end{pmatrix}.$$

The wave function $\Psi(\boldsymbol{r}, t)$ will have four components, as the operators are 4×4 matrices. Separation of the time leads to the time-independent version of the Dirac equation for a free electron (positron)

$$[c\boldsymbol{\alpha} \cdot \boldsymbol{p} + m_0 c^2 \beta - E\mathbf{1}]\psi(\boldsymbol{r}) = 0, \tag{2.1}$$

which has two types of continuum solutions:

(1) $m_0 c^2 \leqslant E < \infty$ for electrons and

(2) $-m_0 c^2 \geqslant E > -\infty$ for positrons.

The energy, E_I, of the stationary states for electrons bound by an attractive potential, e.g. a positively charged nucleus, must lie in the energy gap $-m_0 c^2 \leqslant E_I \leqslant m_0 c^2$.

However, the interaction potential between two charged particles, nucleus–electron or electron–electron, is not just the Coulomb interaction, since in the relativistic description a retarded, velocity-dependent interaction must be considered. The full and general derivation of these interaction 'potentials' is involved and approximate relativistic corrections to the Coulomb interaction are used in general. The frequency-dependent correction to the electron–electron Coulomb interaction, the Breit operator

$$b_\omega(1, 2) = -\frac{\boldsymbol{\alpha}_1 \cdot \boldsymbol{\alpha}_2 \cos(\omega r_{12})}{r_{12}} + (\boldsymbol{\alpha}_1 \cdot \nabla_1)(\boldsymbol{\alpha}_2 \cdot \nabla_2)\frac{\cos(\omega r_{12}) - 1}{\omega^2 r_{12}}$$

is obtained in the Coulomb gauge as a leading term in a QED perturbation expansion. In general, the frequency-independent form, $\omega \to 0$,

$$b_0(1, 2) = -\frac{\boldsymbol{\alpha}_1 \cdot \boldsymbol{\alpha}_2}{r_{12}} - \tfrac{1}{2}(\boldsymbol{\alpha}_1 \cdot \nabla_1)(\boldsymbol{\alpha}_2 \cdot \nabla_2) r_{12} \tag{2.2}$$

or its equivalent

$$b_0(1, 2) = -\frac{\boldsymbol{\alpha}_1 \cdot \boldsymbol{\alpha}_2}{2r_{12}} - \frac{(\boldsymbol{\alpha}_1 \cdot \boldsymbol{r}_{12})(\boldsymbol{\alpha}_2 \cdot \boldsymbol{r}_{12})}{2r_{12}^3} \tag{2.3}$$

is used.

For the nucleus–electron interaction the Coulomb potential, either for a point nucleus or a finite nucleus, is used directly. The relativistic contributions to the interaction operator are obtained approximately, using the Pauli approximation. They are computed, given a zeroth-order wave function, as a first-order perturbation correction. Much of this can be found for atoms in the textbook by Szasz (1992), or, more generally, for atoms and molecules in the textbook by McWeeny (1992).

With these approximations, the electronic structure can be treated using a relativistic quantum mechanical description, while the nuclei are held fixed. To overcome this clamped-nuclei approximation, an attempt has been made (Parpia *et al.* 1992b) to include relativistic corrected nuclear motion terms and to reach an adiabatic separation of the electronic and nuclear motion at least for atoms.

2.2 General Many-Electron Formalism

For the sake of brevity, we proceed in presenting a pragmatic approach to relativistic electronic structure theory, which is justified by its close analogy to the nonrelativistic theory and the fact that most of the finer relativistic aspects must be neglected for calculations on any atom or molecule with more than a few electrons. For a recent comprehensive account on the foundations of relativistic electronic structure theory we refer to Quiney *et al.* (1998b).

The electron–electron interaction is usually supposed to be well described by the instantaneous Coulomb interaction operator $1/r_{12}$. Also, all interactions with the nuclei whose internal structure is not resolved, like electron–nucleus attraction and nucleus–nucleus repulsion, are supposed to be of this type. Of course, corrections to these approximations become important in certain cases where a high accuracy is sought, especially in computing the term values and transition probabilities of atomic spectroscopy. For example, the Breit correction to the electron–electron Coulomb interaction should not be neglected in fine-structure calculations and in the case of highly charged ions. However, in general, and particularly for standard chemical purposes, these corrections become less important.

Four-component quantum mechanical methods for the calculation of the electronic structure of atoms, molecules and solids are based on the n-electron Hamiltonian

$$H = \sum_{i=1}^{n} [h_{\mathrm{D}}(i) + V_{\mathrm{nuc}}(i)] + \sum_{\substack{i<j,\\ i=1}}^{n} g(i, j),$$

where the one-electron operator $h_{\mathrm{D}}(i)$ used for the description of the kinematics of the electron is, as discussed for Equation (2.1),

$$h_{\mathrm{D}} = c\boldsymbol{\alpha} \cdot \boldsymbol{p} + m_0 c^2 \beta'.$$

The appropriately chosen electron–nucleus potential is denoted as $V_{\mathrm{nuc}}(i)$. Usually, we shift the energy scale with $\beta' = \beta - \mathbf{1}$ to yield energy expectation values which are directly comparable with those obtained from nonrelativistic calculations.

In the case of polynuclear systems like molecules and solids, it is common to use the standard nonrelativistic Born–Oppenheimer approximation for the separation of nuclear and electronic coordinates of the stationary (time-independent) Dirac equation, such that the nonrelativistic Coulomb-type nucleus–nucleus interaction potential is added to the many-electron Hamiltonian H (clamped-nuclei approximation),

$$H_{\text{el}} = H + \sum_{A<B} \frac{Z_A Z_B}{R_{AB}},$$

to yield the electronic Hamiltonian H_{el} used in electronic structure calculations.[1]

Since the exact relativistic many-electron Hamiltonian is not known, the electron–electron interaction operators $g(i, j)$ are taken to be of Coulomb type, i.e. $1/r_{ij}$. As a first relativistic correction to these nonrelativistic electron–electron interaction operators, the Breit correction, Equations (2.2) or (2.3), is used. For historical reasons, the first term in Equation (2.2) is called the Gaunt or magnetic part of the full Breit interaction. Since it is not more complicated than $1/r_{12}$, it is from an algorithmic point of view equivalent to the Coulomb interaction, therefore it has frequently been included in the calculations. The second term, the so-called retardation term, appears to be rather complicated and it has been considered less frequently. In the case of few-electron systems further quantum electrodynamical corrections, like self-energy and vacuum polarization, have also been considered and are reviewed in another part of this book (see Chapter 1).

To set up the total time-independent wave function of the many-electron system the independent-particle model is used in general, resulting in an antisymmetrized Hartree product of four-component orthonormal one-electron functions. Independent of the system (atom, molecule or solid) these four-component single electron functions, the spinors, may be written as

$$\psi_i(\boldsymbol{r}) = \begin{pmatrix} \psi_i^{\text{L}}(\boldsymbol{r}) \\ \psi_i^{\text{S}}(\boldsymbol{r}) \end{pmatrix},$$

with the two-component functions $\psi_i^{\text{L}}(\boldsymbol{r})$ and $\psi_i^{\text{S}}(\boldsymbol{r})$ being its large and small components, respectively. The particular form of the one-electron spinors for atoms, molecules and solids will be detailed in the following sections. An explicit dependence of the many-electron wave function on the interelectronic distance r_{12} as used in the R12 approach of the nonrelativistic theory (for a recent review see Klopper 2000) has not been explicated thus far for four-component many-electron theories.

As the standard ansatz for the many-electron wave function is an antisymmetrized product of one-electron spinors $|\Psi_0\rangle$, it may be written as a Slater determinant or in the language of second quantization (see, for example, Helgaker *et al.* 2000) as

$$|\Psi_0\rangle = \{a_1^\dagger a_2^\dagger \cdots a_{\text{N}}^\dagger\}|0\rangle,$$

where $|0\rangle$ denotes the electron vacuum state on which the electron creation operators $a_i^\dagger$ act. As fermion operators, the creation operators $a_i^\dagger$ and their Hermitian

[1] All operators are given in Hartree atomic units, whereby the numerical values of the charge of an electron e, its rest mass m_0 and $\hbar$ as well as $4\pi\epsilon_0$ are 1 and the speed of light is $c = 137.035\,999\,76(50)$ (Mohr and Taylor 2000).

conjugate annihilation operators a_i, which correspond to the one-electron orbitals (four-component spinors) $\psi_i(\boldsymbol{r})$, are to satisfy the anti-commutation relations,

$$a_i^\dagger a_j^\dagger + a_j^\dagger a_i^\dagger = a_i a_j + a_j a_i = 0 \quad \text{and} \quad a_i^\dagger a_j + a_j a_i^\dagger = \delta_{ij}.$$

With this ansatz the Dirac–Hartree–Fock equations for the determination of the orbitals can be obtained using the variation method. Their form is

$$\begin{pmatrix} \hat{v}(\boldsymbol{r}) - \epsilon_i & c\boldsymbol{\sigma}\cdot\boldsymbol{p} \\ c\boldsymbol{\sigma}\cdot\boldsymbol{p} & \hat{v}(\boldsymbol{r}) - 2m_0c^2 - \epsilon_i \end{pmatrix} \begin{pmatrix} \psi_i^{\mathrm{L}}(\boldsymbol{r}) \\ \psi_i^{\mathrm{S}}(\boldsymbol{r}) \end{pmatrix} = 0, \tag{2.4}$$

where the potential operator $\hat{v}(\boldsymbol{r})$ will depend on the orbitals, and will contain local (electron–nucleus and electron–electron Coulomb interactions) as well as nonlocal (electron exchange and coupling) parts; it may be different for different orbitals. If the Breit operator is included explicitly in the variational determination of the orbitals, additions to the off-diagonal elements also appear. The foundations and implications of this ansatz have been reviewed recently from the viewpoint of quantum electrodynamics (Quiney *et al.* 1998b) (compare also Lindgren 1998; Sucher 1980; Wilson 2001).

For a more accurate treatment, electron correlation has to be taken into account; to this end methods akin to those used in a nonrelativistic description are employed. The total wave function may be considered as a superposition of (all possible) excitations. This can be expressed in terms of a configuration interaction (CI) ansatz

$$|\Psi_{\mathrm{CI}}\rangle = \sum_I |\Psi_I\rangle C_I,$$

which can be used in CI or multiconfiguration self-consistent-field (MCSCF) theory. Alternatively, perturbation theory or a size-consistent coupled-cluster approach (Helgaker *et al.* 2000) may be used with

$$|\Psi_{\mathrm{CC}}\rangle = \exp(T)|\Psi_0\rangle,$$

where the cluster operator T, which generates the excitations, is defined as

$$T = \sum_i T_i,$$

with the i-fold excitation operators T_i given as

$$T_1 = \sum_{ai} t_i^a a_a^\dagger a_i, \qquad T_2 = \frac{1}{4}\sum_{abij} t_{ij}^{ab} a_a^\dagger a_i a_b^\dagger a_j, \qquad \text{etc.}$$

At first sight, it seems that in the relativistic case, it would be only a little more difficult to solve the SCF equations (2.4) based on the Dirac equation (2.1) for the four-component orbitals rather than the one-component SCF equations of the nonrelativistic theory. This could be done numerically, using a finite-difference method, or algebraically with a basis-set expansion to express the orbitals. Extra orbitals generated via MCSCF theory or as 'virtual' orbitals could then be used to include electron correlation effects. Early attempts to do just this with basis functions for atoms proved feasible, provided the basis set was not optimized in the relativistic environment (Kim 1967), while relativistic finite-difference SCF calculations (Desclaux 1975) were quite successful. These difficulties arise because the Dirac Hamiltonian is not bounded from below due to the positronic continuum states. This aspect has been discussed by Talman (1986), who proposed an appropriate mini-max procedure for the extremalization, i.e. the orbital Equations (2.4) should be solved to yield a minimum with respect to the large component and a maximum with respect to the small component (for more recent work see Dolbeault *et al.* 2000a,b,c; Esteban and Sere 1999; Griesemer and Siedentop 1999; Griesemer *et al.* 1999; Kutzelnigg 1997). To prevent the positronic continuum interfering in the solutions of the orbital equations, different projection methods have been suggested.

In the case of finite-basis sets, which are used for the representation of the one-electron spinors, the basis sets for the small component must be restricted such as to maintain 'kinetic balance' (Stanton and Havriliak 1984), which means in terms of the rearranged second equation in the matrix equations (2.4) that

$$\psi_i^{\mathrm{S}}(\boldsymbol{r}) = -[\hat{v}(\boldsymbol{r}) - 2m_0c^2 - \epsilon_i]^{-1} c\boldsymbol{\sigma} \cdot \boldsymbol{p}\psi_i^{\mathrm{L}}(\boldsymbol{r}) \tag{2.5}$$

can be fulfilled by the basis-set expansions at least approximately. The matrix representation of operators in four-component theories and resulting consequences which arise from the basis-set choice for the representation of one-electron spinors have been detailed by Dyall *et al.* (1984). Maintenance of the exact relationship between large and small components (i.e. exact kinetic balance) would require (see the discussion in Kutzelnigg 1997) that

$$\psi_i^{\mathrm{S}}(\boldsymbol{r}) = X\psi_i^{\mathrm{L}}(\boldsymbol{r}),$$

where X fulfils the equation

$$X = \frac{1}{2m_0c^2}[c\boldsymbol{\sigma} \cdot \boldsymbol{p} + [V, X] - cX\boldsymbol{\sigma} \cdot \boldsymbol{p}X].$$

A solution of this equation (Heully *et al.* 1986) for X in closed form is not known and a hierarchy of approximations for X is used in practice instead (see, for example, the 'atomic balance' in Section 2.4).

In the numerical solution of the SCF orbital equations kinetic balance restrictions are not required, as this condition will be satisfied exactly. However, in the numerical solution of MCSCF equations for purely correlating orbitals, difficulties may arise if the 'orbital energy' ϵ_i becomes too negative (Bieroń *et al.* 1994; Indelicato 1995, 1996; Kim *et al.* 1998). Here it is suggested that we use projection operators to eliminate the functions that correspond to the negative continuum.

Apart from these 'one-particle effects', additional complications arise in the case of two electrons, which may dissolve from a bound state to the positronic and electronic continua. A mathematically rigorous approach to avoid this so-called 'continuum dissolution' is to use as basis functions the relativistic Coulomb Sturmians

(Szmytkowski 1997) or equivalently the L spinors (Grant and Quiney 2000a; Quiney *et al.* 1989b), derived for atoms. In light of these contributions, it is not clear at all, whether the 'negative-energy' states can be eliminated in perturbation, coupled-cluster or CI expansions (Grant and Quiney 2000a); the jury is still out.

Independently of the approximations used for the representation of the spinors (numerical or basis expansion), matrix equations are obtained for Equations (2.4) that must be solved iteratively, as the potential $\hat{v}(\boldsymbol{r})$ depends on the solution spinors. The quality of the resulting solutions can be assessed as in the nonrelativistic case by the use of the relativistic virial theorem (Kim 1967; Rutkowski *et al.* 1993), which has been generalized to allow for finite nuclear models (Matsuoka and Koga 2001). The extensive contributions by I. P. Grant to the development of the relativistic theory of many-electron systems has been paid tribute to recently (Karwowski 2001). The higher-order QED corrections, which need to be considered for heavy atoms in addition to the four-component Dirac description, have been reviewed in great detail (Mohr *et al.* 1998) and in Chapter 1 of this book.

A different approach to the solution of the electron correlation problem comes from density functional theory (see Chapter 4). We hasten to add that in a certain approximation of relativistic density functional theory, which is also reviewed in this book, exchange and correlation functionals are taken to replace Dirac–Fock potentials in the SCF equations. Another approach, which we will not discuss here, is the direct perturbation method as developed by Rutkowski, Schwarz and Kutzelnigg (Kutzelnigg 1989, 1990; Rutkowski 1986a,b,c; Rutkowski and Schwarz 1990; Schwarz *et al.* 1991).

2.3 Atomic-Structure Calculations

The electronic structure of atoms has been studied for many decades on the basis of four-component methods (Grant 1994; Kim 1993a,b; Reiher and Hess 2000; Sapirstein 1993, 1998). Nevertheless, significant improvements have been achieved in recent years. Even one-electron atoms still give us new insight into four-component electronic structure theory (Andrae 1997; Autschbach and Schwarz 2000; Chen and Goldman 1993; Chen *et al.* 1994; Pyykkö and Seth 1997). In this section we review methodological improvements as well as new implementations and typical applications.

In general, the many-electron wave function is expressed in terms of antisymmetrized products of one-electron functions and the clamped-nucleus approximation as well as the central-field and equivalence restriction for the orbitals is used. Thus the one-electron spinor takes the form

$$\psi_i(\boldsymbol{r}) = \psi_{n_i\kappa_i m_{j,i}}(\boldsymbol{r}) = \frac{1}{r}\begin{pmatrix} P_{n_i\kappa_i}(r)\chi_{\kappa_i m_{j,i}}(\theta,\phi) \\ \mathrm{i}Q_{n_i\kappa_i}(r)\chi_{-\kappa_i m_{j,i}}(\theta,\phi)\end{pmatrix},$$

where $\chi_{\kappa_i m_{j,i}}(\theta,\phi)$ is the standard two-component spherical spinor, with the orbital angular momentum, l, and spin coupled to j and m_j. The values of the relativistic spherical symmetry quantum number $\kappa = (l - j)(2j + 1)$ are $\mp 1, \mp 2, \ldots$, which correspond to $j = |\kappa| - 1/2$ and $l = j + \mathrm{sgn}(\kappa)/2$. In the central-field approximation for atoms the angular dependence can be integrated analytically and the Dirac-type one-dimensional one-electron equations of the type,

$$\begin{pmatrix} [v(r', r) - \epsilon_{n\kappa}] & c\left[-\dfrac{\mathrm{d}}{\mathrm{d}r} + \dfrac{\kappa}{r}\right] \\ c\left[\dfrac{\mathrm{d}}{\mathrm{d}r} + \dfrac{\kappa}{r}\right] & [v(r', r) - 2c^2 - \epsilon_{n\kappa}] \end{pmatrix} \begin{pmatrix} P_{n\kappa}(r) \\ Q_{n\kappa}(r) \end{pmatrix} = 0, \qquad (2.6)$$

are obtained for the determination of the shell functions. Here the potential, $v(r', r)$, which may depend on $n\kappa$, may be local ($r' = r$) or nonlocal, containing exchange and coupling terms; this depends on the details of the derivation of the one-electron equations. The radial shell functions, $P_{n\kappa}(r)$ and $Q_{n\kappa}(r)$, are expressed either numerically on a mesh of grid points or in terms of basis functions as Slater-type functions, Gauss-type functions or B-splines.

Note that the kinetic balance condition (see preceding section), which prescribes a fixed relationship (to a certain order of $1/c$) between large and small component and which plays a crucial role in finite-basis-set expansions, need not be introduced explicitly if a numerical representation of the spinor components on a mesh is used, since in this case the integro-differential orbital equations of the type of Equation (2.6) are solved directly. Difficulties do arise, however, in the MCSCF determination of those orbitals, which are required to account for the dynamical correlation only and which already have a large number of nodes (Bieroń *et al.* 1994). On the other hand, if B-splines are used as basis functions, the kinetic balance condition cannot be introduced easily and the electron–electron interaction operator is generally protected by projection operators, such as to eliminate the contribution of 'negative-energy' states.

2.3.1 Methods and programs

During the last few decades, general computer programs, based on the Dirac equation, for extensive four-component electronic structure calculations for atoms have been developed by several groups. These programs are being improved and extended continually, incorporating the explicit treatment of the Breit interaction as well as an ever more sophisticated consideration of interelectron correlation and even of QED effects.

Early *numerical* Dirac–Fock SCF programs were implemented by Smith and Johnson (1967) and by Desclaux (1975). Desclaux's program has been extended to a more general MCSCF form and to the possibility of also calculating the small QED contributions (Desclaux 1993; Indelicato and Desclaux 1993). With this code it is now also possible to calculate the Breit interaction variationally as has been demonstrated for caesium ions (Rodrigues *et al.* 2000). The other commonly used general numerical MCSCF atomic structure program GRASP (Dyall *et al.* 1989) is continually improved and has been extended for large-scale CI calculations (Parpia *et al.* 1996) in order to permit a more extensive treatment of electron correlation effects. The GRASP program has also been modified and supplemented such that the total wave function is represented on a determinant basis (Fritzsche and Anton 2000), which has become the foundation for a program extension (Fritzsche *et al.* 2000b) especially suited to the calculation of transition energies and transition moments for atoms. Several additional program parts have been implemented in the course of these developments (Fritzsche 1997, 2000; Fritzsche *et al.* 1998b; Gaigalas and Fritzsche 2001).

In addition, new numerical Dirac–Fock atomic structure programs have been developed independently: one of these can treat ground and excited states (Chernysheva and Yakhontov 1999) and another one is of the general MCSCF type (Reiher 1998), which includes new numerical and algorithmic techniques (Andrae *et al.* 2000a; Reiher and Hinze 1999) for highly accurate calculations. With the latter program the influence of different finite-nucleus models on the computed electronic structure results has been investigated in detail (Andrae 2000; Andrae *et al.* 2000b) and an extensive study on the variationally treated frequency-independent Breit interaction in few-electron systems has been carried out (Reiher and Kind 2001).

A program using Slater-type basis functions (STFs) to expand the spinors with the provision to perform large-scale CI calculations has been used to investigate the problem of what has been called 'continuum dissolution' (Jáuregui *et al.* 1997) by studying the full-CI expansion, including the 'positive' and 'negative' energy spinors, for two-electron ions. Here it is demonstrated, at least for finite-basis sets, that the eigenstates of the full-CI Hamiltonian, which correspond to two-electron states, do not 'dissolve' in the continuum, even though they contain 'positive' and 'negative' one-particle states to all orders (Ley-Koo *et al.* 2000).

Gauss-type basis functions (GTFs), which are the mainstay of molecular electronic structure calculations, are also used for the expansion of the spinors in programs for the relativistic description of atomic states. There appear to be several programs where Gaussian basis functions are used to expand the four-component spinors for the computation of atomic states; however, these programs are not described explicitly. One was developed originally in Ishikawa's group (Ishikawa 1990a,b; Ishikawa and Koc 1997; Ishikawa and Quiney 1993; Ishikawa *et al.* 1992, 1991). Other developments of this type have evolved into molecular programs (see below), which could still be used for atomic wave function calculations. The 'Ishikawa' program is the basis for the coupled-cluster method developed by Eliav and Kaldor (Eliav *et al.* 1994a; Ilyabaev and Kaldor 1992a,b, 1993; Landau *et al.* 1999), which has been used for the calculation of ground-state energies and atomic term values (Eliav *et al.* 1994b,c, 1995, 1996; Eliav (Ilyabaev) *et al.* 1994) and electron affinities (Eliav *et al.* 1997). The 'Ishikawa' program has been extended to account via a single operator MCSCF formulation for the nondynamical correlation (Vilkas *et al.* 1997, 1998a,b). This also yields the virtual functions, which are then used to account for the dynamical correlation using multi-reference Møller–Plesset perturbation theory (Ishikawa *et al.* 2000; Vilkas *et al.* 1998c). The Dirac-SCF part of this program has also been used in a number of papers for the development of what is called 'universal Gaussian basis sets', for the optimization of which a generator coordinate method has been developed (Canal Neto *et al.* 2000; Jorge and da Silva 1996a,b, 1997, 1998; Malli and Ishikawa 1998). These basis sets are developed so that they can be used in relativistic molecular calculations. It seems puzzling, however, that some of the atomic ground-state energies obtained with the 'universal Gaussian basis sets' are as low or even lower than those obtained with the corresponding finite-difference calculation, which is probably due to the fact that the use of the approximate kinetic balance according to Equation (2.5) implies that the resulting SCF equations are correct only to the order $O(c^{-4})$.

Intermediate between the numerical finite-difference method and the use of global expansion functions, STFs and GTFs, is the representation of the spinors by local B-splines. This is used in a relativistic atomic program developed in Johnson and Sapirstein's group, which again is not explicitly documented or cited in the literature, though its evolution can be traced (Chen *et al.* 1993; Johnson *et al.* 1988; Plante *et al.* 1994). On the basis of this program, many-electron atoms, their term values and transition probabilities are computed by including the electron correlation using many-body perturbation theory (MBPT) or coupled-cluster (CC) methods. However, in order to avoid the 'continuum dissolution' in these calculations, the electron interaction operator is projected onto the space spanned by the positive-energy states, to yield what is termed the no-pair approximation. As the one-particle positive energy functions, which span this space, are obtained as solutions of a one-particle Dirac equation with a specified potential, they and thus the space spanned by these functions should depend on the specific potential used. In addition, it is found that certain properties, especially the agreement of transition probabilities computed in their length or velocity form, do require the contribution of negative-energy states (Johnson *et al.* 1995; Safronova *et al.* 1999a). From this, it appears not yet clear how to treat electron correlation in a four-component framework correctly, though some progress in this direction has been made (Kołakowska 1997; Ley-Koo *et al.* 2000).

The relativistic many-body approach has also been developed and advanced by the Göteborg group (Lindgren 1987, 1989; Lindgren and Morrison 1986; Lindroth 1988; Mårtensson-Pendrill *et al.* 1995), whose work concentrated mainly on QED effects which are reviewed in Chapter 1 of this book.

For the relativistic description of electron–atom collision cross-sections, the R-matrix method has been extended to permit the explicit numerical computation of electron–atom scattering on a four-component relativistic target by using the combination of the GRASP program with the DARC[2] (Dirac Atomic R-Matrix Code) package (Norrington and Grant 1987). Here it is possible, without loss of accuracy, to integrate the close-coupling equations in the outer region, using just a regular two-component nonrelativistic description (Schwacke 2000). The theory for the variational R-matrix method, based on the Dirac equation has also been developed (Hamacher and Hinze 1991; Szmytkowski 2001; Szmytkowski and Hinze 1996a,b); however, thus far no general computer programs based on this theory have evolved.

Of these programs only GRASP and to some extent the DARC package have achieved a wide distribution, such that they are being used by a larger number of research groups.

[2] See Norrington's manuals for GRASP version 0.9 and DARC (1994).

The other programs are generally used only by the members of those research groups where these programs have been or are being developed.

Valuable results have been attained with these programs, but much more is possible in the future. For one there are the term values of highly ionized heavy atoms, which are difficult to access experimentally, but also the corresponding transition probabilities important for the explicit simulation of a high-temperature plasma. Another aspect, which has attracted attention, is the hyperfine structure of atomic spectra and with it the determination of nuclear moments in the combination of computation and high-resolution experiments.

2.3.2 Term values

Desclaux's list of Dirac–Fock expectation values for almost all atoms of the periodic table (Desclaux 1973), which has served as a very valuable reference in the past, has been updated and extended (Visscher and Dyall 1997).

An exhaustive study of the four term values that arise out of the configuration $2p^5 3s$ for the ions with $Z = 10$–92 has been performed (Avgoustoglou *et al.* 1995) using a B-spline basis and MBPT with a no-pair Hamiltonian for the electron interaction. In this study the contributions of the various corrections, i.e. correlation, Breit and its correlation and frequency dependence, QED and reduced mass, are investigated in detail in their Z dependence. All these corrections, except the reduced mass correction, are found to be quite significant, particularly for high-Z values of the ions. The final results obtained are in good agreement with the available experimental data and can serve as predictions in those cases where experimental data are not available.

With the CC method, using kinetically balanced Gaussian basis sets, many term values for neutral and low ionized states of lanthanum, actinium and eka-actinium ($Z = 121$) and the corresponding ionization potentials and electron affinities were computed (Eliav *et al.* 1998c). With this method the ionization energies and fine-structure splittings of zinc-like and copper-like ions were also calculated (Vilkas *et al.* 2000) as well as the electronic structure of eka-lead ($Z = 114$) (Landau *et al.* 2001). In those cases, where the experimental data are known, remarkable agreement is obtained, providing credence for the predicted values. The states arising out of the half-filled 5f shell of americium were computed using the numerical MCSCF approach (Johnson *et al.* 1996), while the intrashell excitations of the f^2 shells of Pr^{3+} and U^{4+} could be determined using the CC method (Eliav *et al.* 1995). MBPT for quasidegenerate systems was used to obtain the energy levels of the ions of the oxygen isoelectronic sequence (Vilkas *et al.* 1999).

2.3.3 Transition probabilities and lifetimes

The computation of transition probabilities using MBPT is an arduous task, since an explicit wave function is not obtained in the calculation of the corresponding term values; thus the correlation corrections, implicit in the effective Hamiltonian used, have to be considered again in the computation of the transition moments. The theoretical foundations have been presented in an explicit study for two-electron atoms (Johnson *et al.* 1995), where electric dipole as well as magnetic dipole and quadrupole transition rates were computed for two-electron ions with $Z = 2$–100. It is found that the contribution of the negative-energy states needs to be included in order to attain a gauge independence of the results and agreement between the length and velocity forms of the electric dipole transition moments. A similar study is presented for the term values (Safronova *et al.* 1996, 1997) and the transition moments (Safronova *et al.* 1999b) of the beryllium-like ions for $Z = 6$–100. Likewise for the sodium-like ions for $Z = 11$–16 (Safronova *et al.* 1998) and the alkali atoms Na–Fr (Safronova *et al.* 1999a), not only term values and transition moments are computed to high precision, but also the hyperfine interactions and the dipole polarizabilities are determined. Furthermore, the transition probabilities for beryllium-like ions (Chou 2000), magnesium-like ions (Safronova *et al.* 2000b), nickel-like ions (Safronova *et al.* 2000a) and for the $3d^2$ states of the calcium-like ions (Safronova *et al.* 2001) have been computed. Transition energies and rates of helium-like argon and beryllium-like iron and molybdenum have also been studied with MBPT (Lindroth and Hvarfner 1992; Lindroth and Salomonson 1990). K_α transition energies have been calculated with high accuracy in medium- and high-Z atoms (Indelicato and Lindroth 1992).

As the MBPT and CC methods for electron correlation appear to still be limited to systems with one or two electrons (holes) outside (inside) a closed shell, the more flexible MCSCF method is used to compute several term values of the lithium- through neon-like ions of uranium, thorium and bismuth (Santos *et al.* 1998), for which some observations have been reported recently. With the inclusion of the Breit interaction and QED effects good agreement between computed and experimental data is obtained. With the calculation of the corresponding transition moments many predictions for further observations could be made. In those cases where MBPT data could also be obtained, they are compared with the MCSCF results and the merits of the two methods are assessed. The MCSCF methodology was also employed to compute the spin-forbidden transition probability for the $3s^2(^1S_0)$–$3s3p(^3P_1)$ transition of Al^+ through S^{4+} (Zou and Froese Fischer 2000). In order to resolve earlier discrepancies, the lifetimes of the $3s3p^2$ levels of Au^{66+} and Br^{22+} were reconsidered (Beck and Norquist 2000).

Using the numerical MCSCF description as implemented in the GRASP programs, term values and transition probabilities for many atoms and ions have been computed, i.e. for Nb^+ (Beck and Datta 1995), for the beryllium-like isoelectronic sequence (Jönsson and Froese Fischer 1998; Jönsson *et al.* 1998; Ynnerman and Froese Fischer 1995) and for the magnesium-like isoelectronic sequence (Jönsson and Froese Fischer 1997), for the oxygen isoelectronic sequence (Froese Fischer *et al.* 1998) and for the Cd isoelectronic sequence (Biémont *et al.* 2000) as well as for Na^{6+} and Na^{7+} (Tordoir *et al.* 1999). In a study of the parity forbidden lines of Pb, the magnetic- and electric-quadrupole transition probabilities are computed (Horodecki *et al.* 1999). Using the orbitals determined in the field of an optimized core potential for a large CI expansion, the transition probabilities for several states of Hg were computed (Głowacki *et al.*

2000). The amplitude of the $5p(^2P_j)$–$5d(^2D_{3/2})$ transition of ^{87}Rb was determined (Bayram *et al.* 2000), in order to aid the interpretation of the two-step, two-photon absorption of the $5s(^2S_{1/2})$–$5p(^2P_j)$–$5d(^2D_{3/2})$ process.

Extensive numerical MCSCF studies have been carried out on energy levels, transition probabilities and lifetimes for phosphorus-like and for silicon-like ions (Fritzsche *et al.* 1998a, 1999; Kohstall *et al.* 1998) as well as on electric-dipole emission lines in Ni II (Fritzsche *et al.* 2000a).

2.3.4 Hyperfine structure

The hyperfine structure of the nd^2 states of Sc^+ and of Y^+ has been computed (Bieroń *et al.* 1995), as well as for the $J = 2$ states of the configuration $(4d + 6s)^2$ of $^{139}La^+$ (Datta and Beck 1995). In a similar study the hyperfine structure of the $7^2P_{3/2}$ and $7^2P_{1/2}$ states of Ra^+ has been calculated (Yuan *et al.* 1995). In extensive MCSCF studies using GRASP, the hyperfine structure for a number of states and the nuclear quadrupole moments of ^{45}Sc (Bieroń *et al.* 1997), of ^{90}Y (Bieroń and Grant 1998) and of ^{49}Ti (Bieroń 1999) were computed. The effects of electron correlation, relativity and nuclear structure on the computed hyperfine-structure constants of Be^+ and F^{6+} have been investigated (Bieroń *et al.* 1999).

New high-resolution experimental studies of the hyperfine structure have been used in conjunction with a thorough calculation of these interactions using GRASP to determine the electrical quadrupole moment of the ^{131}Xe nucleus (Paduch and Bieroń 2000). With a large-scale numerical multiconfiguration Dirac–Fock calculation of the diagonal magnetic dipole hyperfine-structure constants A of the lowest $^2S_{1/2}$, $^2P_{1/2}$, and $^2P_{3/2}$ states of lithium, as well as the electric quadrupole constant B of the $^2P_{3/2}$ state, the power of this method has been tested and demonstrated (Bieroń *et al.* 1996). The converged results obtained reproduced the experimental values to high precision. This provided the basis for a similar study of the hyperfine structure of lithium-like ions up to high-Z values (Boucard and Indelicato 2000). In this study information was sought about the Bohr–Weisskopf effect (changes in the hyperfine structure due to the distribution of the magnetic moment in the finite nucleus) versus QED effects, in order to yield information on the nuclear anapole moment as observed in parity nonconservation experiments.

2.3.5 Photoionization and electron–atom scattering

The photoionization cross-sections of 4d shell electrons of Ba have been computed (Band and Trzhaskovskaya 1997), using a modified version of GRASP for the calculation of the atomic and ionic wave functions in an extended average configuration description, and a special procedure was used to integrate the integro-differential equations for the determination of the continuum orbital for the unbound electron. This type of methodology, i.e. computing bound and unbound single electron states by solving the Dirac equation with a suitably chosen effective potential, has also been used to compute differential cross-sections for the elastic scattering of electrons by mercury (Sienkiewicz 1997) as well as the inner-shell hole-state relaxation effects of krypton (Shi *et al.* 1998) and of ytterbium and mercury (Santos *et al.* 1999). A very detailed and exhaustive study of electron–atom scattering, using the R-matrix method as implemented in the combination of the GRASP and DARC program system, has been carried out for two- and five-electron ions, especially for iron (Schwacke 2000). The detailed results obtained are compared with those of more approximate methods, and the influence of the Breit interaction and other QED corrections is investigated.

2.4 Molecular Structure Calculations

It took until the 1980s before relativistic *molecular* calculations were carried out. However, early calculations on the basis of one-centre approximations can be traced back to the early 1970s (Desclaux and Pyykkö 1974). In these calculations wave functions from four-component atomic calculations on anionic centres like Pb^{4-} were used to set up a model for the PbH_4 molecule by surrounding the anion by four protons. Some years later, the first accurate all-electron Dirac–Fock calculations were performed on the linear molecules AgH and AuH using spinor expansion into Slater-type functions (Lee and McLean 1982). This appears to be the first time that the small component basis has been restricted according to 'kinetic balance'. A formalism of a relativistic CI approach for four-component molecular calculations has been devised and applied to Hg and Pb atoms (Esser 1984a,b).

The last decade has seen a vast amount of method and algorithm development to set up computer programs that can be used for efficient four-component calculations of the electronic structure of molecules. These calculations need incredibly large computer resources even for standard noncorrelated methods like Dirac–Hartree–Fock applied to molecules with only one heavy atom.

In the sequel, we describe which program packages have been developed and which algorithmic improvements have been achieved. Afterwards we discuss applications of these programs to few-atom molecules.

2.4.1 Molecular one-electron functions

The expansion of four-component one-electron functions into a set of global basis functions can be done in several ways independent of the particular choice of the type of the basis functions. For instance, four independent expansions may be used for the four components. However, we might also relate the expansion coefficients of the four components to each other. In contrast to these expansions, the molecular spinors can also be expressed in terms of 2-spinor expansions. This latter ansatz appears to be quite common and will be described in greater detail now. Obviously, analogous thoughts also apply for the first two possibilities (for a detailed discussion compare Dyall *et al.* 1991a).

The most frequently used ansatz for the representation of molecular one-electron spinors is a basis expansion into Gauss-type spinors (where we have adopted the notation used in Quiney *et al.* (1998b))

$$\psi_i(\boldsymbol{r}) = \begin{pmatrix} \sum_\mu c^{\mathrm{L}}_{\mu i}\, M^{\mathrm{L}}_\mu(\boldsymbol{r}_A) \\ \sum_\mu c^{\mathrm{S}}_{\mu i}\, M^{\mathrm{S}}_\mu(\boldsymbol{r}_A) \end{pmatrix},$$

with the spherical two-component basis functions

$$M^{\mathrm{L}}_\mu(\boldsymbol{r}_A) = \frac{1}{r_A} f^{\mathrm{L}}_\mu(r_A)\chi_{\kappa_\mu m_\mu}(\vartheta_{A_\mu}, \varphi_{A_\mu}), \tag{2.7}$$

$$M^{\mathrm{S}}_\mu(\boldsymbol{r}_A) = \frac{\mathrm{i}}{r_A} f^{\mathrm{S}}_\mu(r_A)\chi_{-\kappa_\mu m_\mu}(\vartheta_{A_\mu}, \varphi_{A_\mu}), \tag{2.8}$$

containing Gauss-type radial basis functions

$$f^{\mathrm{L}}_\mu(r_A) = N^{\mathrm{L}}_\mu r_A^{l_\mu+1} \exp(-\lambda_\mu r_A^2),$$
$$f^{\mathrm{S}}_\mu(r_A) = N^{\mathrm{S}}_\mu [(\kappa_\mu + l_\mu + 1) - 2\lambda_\mu r_A^2] r_A^{l_\mu} \exp(-\lambda_\mu r_A^2),$$

and standard two-component spin-angular functions $\chi_{\kappa_\mu m_\mu}(\vartheta_{A_\mu}, \varphi_{A_\mu})$ (Grant and Quiney 1988). The small component's radial function has been fixed according to the kinetic balance condition (Stanton and Havriliak 1984), which has its origin in the coupled nature of Dirac's first-order differential equations and is introduced to keep the method variationally stable. The index A denotes the coordinates of the nucleus's centre $\boldsymbol{R}_A$ of atom A, to which the basis function M_μ is attached, i.e. $\boldsymbol{r}_A = \boldsymbol{r} - \boldsymbol{R}_A$.

As an alternative, Cartesian Gaussians,

$$f^{\mathrm{L}}_\mu(\boldsymbol{r}_A) = N^{\mathrm{L}}_\mu x^{\alpha_\mu} y^{\beta_\mu} z^{\gamma_\mu} \exp(-\lambda^{\mathrm{L}}_\mu r_A^2), \tag{2.9}$$

$$f^{\mathrm{S}}_\mu(\boldsymbol{r}_A) = N^{\mathrm{S}}_\mu x^{\alpha_\mu} y^{\beta_\mu} z^{\gamma_\mu} \exp(-\lambda^{\mathrm{S}}_\mu r_A^2), \tag{2.10}$$

may be used. The exponents α_μ, β_μ and γ_μ are connected to the angular quantum number. The approximate kinetic balance condition,

$$\psi_i^{\mathrm{S}} = -\frac{c}{\hat{v} - \epsilon_i - 2mc^2} \boldsymbol{\sigma} \cdot \boldsymbol{p} \psi_i^{\mathrm{L}} \approx \frac{1}{2mc} \boldsymbol{\sigma} \cdot \boldsymbol{p} \psi_i^{\mathrm{L}},$$

is to be fulfilled here as well, which means that first derivatives of Cartesian primitive Gaussians,

$$\frac{\partial}{\partial x} x_\mu^\alpha \exp(-\lambda_\mu r_A^2) = [\alpha_\mu x^{\alpha_\mu - 1} - 2\lambda_\mu x^{\alpha_\mu + 1}] \exp(-\lambda_\mu r_A^2),$$

have to be evaluated to find the exponents λ^{S}_μ for the small component from the exponents of the large components λ^{L}_μ. In the case of contraction of Gaussian basis functions, the kinetic balance condition becomes even less rigorous. For these cases an atomic balance procedure has been developed (Visscher *et al.* 1991a,b).

The relativistic one- and two-electron integrals are evaluated using standard techniques well known from nonrelativistic quantum chemistry. For a detailed discussion on how to evaluate efficiently two-electron integrals over Gaussian basis functions see (Helgaker *et al.* 2000). The choice and generation of basis sets has been addressed by many authors (Da Silva *et al.* 1993; Dyall and Enevoldsen 1999; Dyall and Fægri 1996; Hu *et al.* 1999; Ishikawa *et al.* 1997, 1992; Jorge and da Silva 1998; Malli *et al.* 1992, 1993, 1994; Okada *et al.* 1990; Quiney *et al.* 1989a). Dyall and Fægri (1996) stress that the selection of the number of basis functions used for the representation of a shell $\{n_i\kappa_i\}$ should not be made on the grounds of the nonrelativistic shell classification $\{n_il_i\}$ but on the natural basis of j quantum numbers resulting in basis sets of similar size for, e.g., $s_{1/2}$ and $p_{1/2}$ shells, while the $p_{3/2}$ basis may be chosen to be smaller. As a consequence, if, for instance, the $p_{1/2}$ and $p_{3/2}$ shells are treated on the $\{n_il_i\}$ footing, the number of contracted basis functions may be doubled (at least in principle).

Apart from the expansion into Gauss-type functions the use of Slater-type functions has been discussed (Grant and Quiney 1988), although the analytic evaluation of integrals becomes as hopeless as in the nonrelativistic theory. Therefore, these STFs are only a good choice for atoms, linear molecules, or for four-component density functional calculations, where integrals over the total electron density are evaluated numerically.

2.4.2 Program development

In principle, all four-component molecular electronic structure codes work like their nonrelativistic relatives. This is, of course, due to the formal similarity of the theories where one-electron Schrödinger operators are replaced by four-component Dirac operators enforcing a four-component spinor basis. Obviously, the spin symmetry must be treated in a different way, i.e. it is replaced by the time-reversal symmetry being the basis of Kramers' theorem. Point group symmetry is replaced by the theory of double groups, since spatial and spin coordinates cannot be treated separately.

Until now several program packages for *ab initio* four-component molecular electronic structure calculations have been developed. Because their number is comparatively small and the algorithmic development of these programs is still ongoing, it is worthwhile to describe briefly their particular features.

MOLFDIR has been developed since the mid 1980s and represents a pioneering landmark since it is the first general program for four-component molecular calculations. On the basis of the Dirac–Fock kernel of the package (Aerts 1986; Aerts and Nieuwpoort 1986; Visser *et al.* 1991c), correlation methods have been introduced. Within the configuration interaction (CI) framework a complete open shell CI (COSCI) and a relativistic variant of the restricted active space CI method have been implemented (Nieuwpoort *et al.* 1994; Visscher *et al.* 1994, 1995a). As a size-consistent correlation method, a coupled-cluster code has been developed and incorporated for CCSD calculation with perturbatively estimated triples contributions (Visscher *et al.* 1995b,

1996a). The CCSD program also provides in turn the total energies from Møller–Plesset perturbation theory to second order (MP2). Cartesian Gaussians as given in Equations (2.9) and (2.10) are used as basis functions. The program has been parallelized for the calculation of two-electron integrals for the Dirac–Fock SCF method and for the SCF procedure itself (Pernpointner *et al.* 2000). Recently, the coupled-cluster program has also been parallelized for CCSD calculations (Pernpointner and Visscher 2001b). The magnetic part of the frequency-independent Breit interaction can be evaluated either self-consistently or as a first-order perturbation. The program supports all double groups which can be constructed from the point group O_h and its subgroups. It appears now that all future developments are not going to be incorporated in the MOLFDIR package but in DIRAC, which will be described in the following.

DIRAC contains probably the most efficient and elegant implementation of Dirac–Fock theory as a direct SCF method (Saue *et al.* 1997) in terms of quaternion algebra (Saue and Jensen 1999; Visscher and Saue 2000). The program was first tested for the CsAu molecule (Saue *et al.* 1997). For the treatment of electron correlation, MP2 has been implemented in its direct Kramers-restricted RMP2 version (Laerdahl *et al.* 1997). Additionally, the formalism of a relativistic MCSCF procedure based on a Newton–Raphson procedure for the determination of orbital and state rotation parameters has also been developed within this framework (Jensen *et al.* 1996). An implementation according to this formalism has now been completed (Thyssen 2001). This MCSCF module accounts for near degeneracies which occur more frequently than in nonrelativistic theory because of the spin–orbit splitting of shells. The MCSCF program has been supplemented by a direct CI module for the treatment of large-determinant expansions (Fleig *et al.* 2001; Thyssen 2001).

The coupled-cluster and CI modules from the MOLFDIR program suite have been included in the DIRAC program. The Breit interaction has not yet been implemented in the DIRAC package. DIRAC supports point group symmetries up to D_{2h} symmetry.

Apart from these programs, the following computer codes with less general functionality but particular individual features have been developed.

DREAMS is a program that has evolved as a Dirac–Fock code (Dyall 1994c; Dyall *et al.* 1991a) and has been extended to the RMP2 approach for the estimation of correlation energies for closed and open-shell systems (Dyall 1994a).

BERTHA can perform Dirac–Fock and RMP2 calculations (Grant and Quiney 2000b; Quiney *et al.* 1998b,c). The two-electron integrals are evaluated using the well-established McMurchie–Davidson algorithm. Spherical Gaussians, as given in Equations (2.7) and (2.8), are used as basis functions. The evaluation of matrix elements for electromagnetic properties within the quantum electrodynamical framework has been described (Quiney *et al.* 1997). Test calculations have been performed for the water molecule (Quiney *et al.* 1998c).

FSRCC is a multi-reference Fock-space relativistic coupled-cluster program by Eliav and Kaldor for correlated calculations on the ground and excited states of molecules.

The Dirac–Fock wave function generated by MOLFDIR is used as the reference function. The principles of the method are described in Eliav *et al.* (1994b, 1995, 1996), Eliav (Ilyabaev) *et al.* (1994) and Ilyabaev and Kaldor (1992a,b). The hydrides SnH_4 (Eliav and Kaldor 1996) and CdH (Eliav *et al.* 1998a) have been studied with this approach. The latest development is the so-called intermediate Hamiltonian Fock-space coupled-cluster method (Landau *et al.* 1999), which yields highly accurate results due to an increased size of the model space. The relativistic Fock-space coupled-cluster by the Tel Aviv group is now being incorporated into the MOLFDIR and DIRAC packages, respectively, to become generally available (E. Eliav 2000, personal communication).

Additionally, molecular four-component codes have also been developed in several laboratories but they have not yet reached the functionality and general applicability of those discussed above. For example, Mohanty, Clementi and Parpia developed Dirac–Fock programs (Mohanty 1992; Mohanty and Clementi 1990, 1991; Parpia and Mohanty 1995). Mohanty (1992) treated the frequency-dependent and frequency-independent Breit interaction on the same footing to be included in the self-consistent field procedure. Pisani and Clementi implemented the unrestricted (open-shell) Dirac–Fock method (Pisani and Clementi 1993, 1994a, 1995b) and Matsuoka *et al.* also set up a Dirac–Fock program (Fujimura and Matsuoka 1992; Matsuoka 1992, 1993; Watanabe and Matsuoka 1998). An efficient algorithm has been developed for the calculation of two-electron repulsion integrals (Yanai *et al.* 2001) and molecular symmetry has also been discussed for Dirac–Fock theory in Cao *et al.* (1998).

The programs described so far use basis-set expansions for the one-electron spinors. The fully numerical approach, which is still a challenging task for *general* molecules in nonrelativistic theory (Andrae 2001), has also been tested for Dirac–Fock calculations on diatomics (Düsterhöft *et al.* 1994, 1998; Kullie *et al.* 1999; Sundholm 1987, 1994; Sundholm *et al.* 1987; v. Kopylow and Kolb 1998; v. Kopylow *et al.* 1998; Yang *et al.* 1992). The finite-element method (FEM) was tested for Dirac–Fock and Kohn–Sham calculations by Kolb and co-workers in the 1990s. However, this approach has not yet been developed into a general method for systems with more than two atoms; only test systems, namely few-electron *linear* molecules at some fixed internuclear distance, have been studied with the FEM. Nonetheless, these numerical techniques are able to calculate the Dirac–Fock limit and thus yield reference data for comparisons with more approximate basis-set approaches. The limits of the numerical techniques are at hand:

(i) calculations of general molecules with more than two atoms have not yet been achieved,

(ii) molecular structure optimization techniques will be very computer time consuming because gradients and approximates to the Hessian matrix for the total electronic energy must be calculated numerically,

(iii) virtual orbitals, which are needed for correlated methods, are not obtained automatically as in basis-set approaches, and finally

(iv) convergence is usually worse for fully numerical methods (compare the quadratically convergent Newton–Raphson approach in basis-set MCSCF algorithms).

In addition to the *ab initio* approach to relativistic electronic structure of molecules, four-component Kohn–Sham programs, which approximate the electron–electron interaction by approximate exchange-correlation functionals from density functional theory, have also been developed (Liu *et al.* 1997; Sepp *et al.* 1986). However, we concentrate on the *ab initio* methods and refer the reader to Chapter 4, which treats relativistic density functional theory (RDFT).

2.4.3 Avoiding $(SS \mid SS)$ integrals

The main bottleneck of four-component calculations has its origin in the kinetic balance condition that generates a very large basis set for the small component, which contributes for moderate nuclear charge numbers Z only little to expectation values.[3] To reduce the tremendous computational effort of the evaluation of two-electron integrals for small components, i.e. for $(SS \mid SS)$-type integrals, approximations have been developed. Of course, the number of integrals can be reduced by using increased thresholds for these integrals to be evaluated or by neglecting these integrals completely. However, this would change the relativistic Roothaan equations in a somehow uncontrollable manner, although it might work in certain cases where light atoms are bound to heavy ones (Dyall 1992; Dyall *et al.* 1991b; Pisani and Clementi 1994a; Styszyński *et al.* 1997). A much more consistent way of avoiding these integrals has been suggested by Visscher (1997), who found that a simple point-charge model (the so-called simple Coulombic correction (SCC)) can be used very successfully to correct for the complete neglect of $(SS \mid SS)$ integrals. The reason for this is that the molecular small component density can be approximated by a superposition of atomic small component densities. The working procedure of the SCC method is surprisingly simple: calculate the potential energy curve without $(SS \mid SS)$ integrals and correct *a posteriori* the energy at every point by adding half of the point-charge interaction of the small component's atomic charges, which can be obtained from numerical atomic Dirac–Fock calculations, at the given internuclear distance plus the difference in total energy of the atoms calculated with and without $(SS \mid SS)$ integrals. Results for I_2 and At_2 in particular demonstrate that for DF and CCSD(T) calculations, bond distances, harmonic frequencies, total energies and dissociation energies are completely recovered by using the SCC method to correct for the neglect of $(SS \mid SS)$ integrals. The main advantage of the SCC approach is that it works independently of the particular electronic structure method employed.

---

[3] Note that this changes for high-Z atoms in molecules, where the so-called 'small' component becomes large and thus its contribution to expectation values and integrals is increased.

2.4.4 The nonrelativistic limit within the basis set approach

The nonrelativistic limit appears to be clearly defined as the limiting case for the speed of light approaching infinity. However, the several approximations introduced in practical calculations may lead to a wrong limit if the speed of light is just increased to a sufficiently high value of, say, 10^5. If contracted basis sets are used and the contraction coefficients are determined from Dirac–Fock atomic calculations, they only represent the optimum contraction coefficients in the relativistic case, but not in the nonrelativistic limit. In the corresponding nonrelativistic calculation the atomic orbitals might not be well represented by the relativistic contraction coefficients. It is therefore necessary to redetermine the contraction coefficients for the large value of the speed of light using the same primitive set of basis functions. We might also think about the case where the exponential coefficients of the basis set are dependent on the speed of light. Only in complete basis sets, which are out of reach in practical calculations, would these problems not occur.

The correct nonrelativistic limit as far as the basis set is concerned is obtained for uncontracted basis sets, which obey the strict kinetic balance condition and where the same exponents are used for spinors to the same nonrelativistic angular momentum quantum number; for examples, see Parpia and Mohanty (1995) and also Parpia *et al.* (1992a) and Laaksonen *et al.* (1988). The situation becomes more complicated for correlated methods, since usually many relativistic configuration state functions (CSFs) have to be used to represent the nonrelativistic CSF analogue. This has been discussed for LS and jj coupled atomic CSFs (Kim *et al.* 1998).

2.4.5 Electronic structure calculations

In general, only small molecules, usually diatomics, have been studied with four-component methods. Often, correlation effects have not yet been taken into account. Those larger molecules, which have also been studied to some extent, exhibit high symmetry like O_h or T_d consisting of only two symmetry-inequivalent atoms. Therefore, hydrides, oxides and halides are by far the most extensively studied molecules.

Table 2.1 gives an overview over those molecules which have been investigated so far. The table is not complete in the sense that some molecules which have already been mentioned and which primarily served as 'guinea pigs' for testing a new technique or implementation—particularly when noncorrelated, i.e. when Dirac–Fock calculations have been performed—are missing. (We did not include recent calculations on molecules with superheavy atoms like element (111) (Liu and van Wüllen 1999; Seth and Schwerdtfeger 2000), but refer to the review by Kratz and Pershina in this volume.)

One purpose of these calculations is to understand the effect of a four-component treatment for different types of molecules to evaluate the reliability of more approximate treatments like two-component or one-component methods. In other words, those cases must be identified where only four-component calculations yield sufficiently accurate results. In all other cases, more approximate methods, which do not suffer from large-basis expansions due to the fulfilment of the kinetic balance condition, may be used. Obviously, this decision depends on the accuracy wanted. For highly accurate calculations four-component methods are always the right choice if the basis set can be chosen to be large enough. This assessment of the reliability of approximate methods, which enable high-throughput calculations on large molecules, is important, for example, for the computational approach to the nuclear waste problem (compare Ismail *et al.* 1999). Very recently, effective core potentials (ECPs) were generated by adjusting the parameters to four-component MCSCF calculations (Metz *et al.* 2000b). The assessment of the accuracy of ECPs for heavy atoms can be based on comparison with DF results (Schwerdtfeger *et al.* 2000).

For molecules the evaluation of the Breit correction to the Coulomb-type electron–electron interaction operator becomes computationally highly demanding and cannot be routinely evaluated, not even on the Dirac–Fock level. To test the significance of the Breit interaction, the Gaunt term is evaluated as a first-order perturbation. It turned out that it can be neglected in most cases as can be seen from the DF + B_{mag} calculations cited in Table 2.1.

Molecules with more than one heavy atom are seldom studied. For example, heavy diatomics of the sixth row of the periodic table like Pt_2, Au_2, Tl_2, Pb_2 and Bi_2 (Varga *et al.* 2000b) and the mercury dimer (Bastug *et al.* 1995) have only been investigated with a four-component method within the framework of DFT. Since present-day DFT calculations are not able to reproduce adequately the van der Waals interaction, it is preferable to treat the mercury dimer on the grounds of post-Dirac–Fock methods like relativistic coupled-cluster using a sufficiently large basis set. This becomes computationally extremely demanding, so that at the present time the most thorough study on the mercury dimer has been performed using scalar relativistic methods with spin–orbit corrections (Dolg and Flad 1996a; Flad and Dolg 1996a). Attempts have been made to improve on the results of Munro *et al.* (2001) and Schwerdtfeger *et al.* (2001), but four-component all-electron correlated calculations on the mercury dimer have not yet been done. As a test calculation on a large molecule, germanocene has been calculated within Dirac–Fock theory using a small double zeta quality basis set (Quiney *et al.* 1998b). Point group symmetry could not be exploited and structure optimization was also not possible. For sure, results from Dirac–Fock calculations on such molecules can only be understood as pure test calculations, since they lack a suitable treatment of correlation and are, from this point of view, as uncertain as nonrelativistic Hartree–Fock calculations, particularly with small basis sets.

Since relativity affects spinor energies, four-component methods perform best for the calculation of properties for which this has direct consequences, like ionization energies. In contrast with this direct effect on properties, it is not obvious and straightforward to draw conclusions for physical quantities which are given relative to a reference system, like D_e or r_e (see Rutkowski and Schwarz 1990; Rutkowski *et al.* 1992; Schwarz 1987, 1991; Schwarz *et al.* 1996c). For example, if the relativistic effects are almost the same in an A–B molecule and in the corresponding atoms A and B and the bonding is hardly affected by relativity, the four-component calculation would only shift the absolute energy scale while the relative energetic quantities remain

Table 2.1 List of molecules studied with four-component methods. The fourth column lists quantities, which have been investigated; primary data P = {total electronic energies $\langle E \rangle$, orbital energies ϵ_i, population analyses PA}, ionization energies IE, electron affinities EA, atomization energies A, spectroscopic data S = {equilibrium distance r_e, dissociation energy D_e, frequencies/wave numbers ω_e, bond angles ϕ}, electric properties E = {dipole moment μ, quadrupole moment Θ, dipole polarizability α}, infrared intensities I, excited states ES, electric field gradients EFG, energetics of reaction R.

class	molecule	type	calc. of	ref.
(metal)	YbH	RDFT	S, ES, μ	[1]
hydrides	PtH	DF/CI	S, ES	[2]
	Pt{$H_{1,2}$}	DF	S, PA, IE, μ	[3]
	PdH	DF/CI	ES, PA, r_e	[4]
	{Cu,Ag,Au}H	DF/MP2	S, PA	[5]
	{Si,Ge,Sn,Pb}{$H_{2,4}$}	DF	$\langle E \rangle$, S, μ, I, R	[6]
	{O,S,Se,Te,Po}H	DF	r_e, A, IE	[7]
	{O,S,Se,Te,Po}H_2	DF	r_e, A, ϵ_i	[8]
	{C,Si,Ge,Sn,Pb}H_4	DF + B_{mag}	r_e, D_e	[9]
	{Si,Ge,Sn,Pb}H_4	DF	r_e, ω_e	[10]
	{I,At,Uus}H	DF	r_e, ω_e, μ	[11]
	TlH	DF + B_{mag}/CC	S	[12]
	{La,Lu,Ac,Lr}{$H_{1,3}$}	DF/MP2	r_e, ϕ, ω_e	[13]
hydrogen	H{F,Cl,Br,I}	DF/CC	$\langle E \rangle$, EFG	[14]
halides	HCl	DF/CI	S, ES	[15]
	HBr	DF/CI	S, ES	[16]
	H{F,Cl,Br,I,At}	DF/CC	S	[17]
di- and	I_2	DF + B_{mag}/CC/CI	S, IE, ES	[18]
inter-	F_2, Cl_2, Br_2, I_2, At_2	DF/CC	S	[19]
halogens	XY∈{F,Cl,Br,I}	DF/CC	S, E	[20]

[1] Liu *et al.* (1998), [2] Visscher *et al.* (1993), [3] Dyall (1993a), [4] Sjøvoll *et al.* (1997), [5] Collins *et al.* (1995), [6] Dyall (1992), [7] Pisani and Clementi (1995a), [8] Pisani and Clementi (1994b), [9] Visser *et al.* (1992a), [10] Dyall *et al.* (1991b), [11] Saue *et al.* (1996), [12] Fægri and Visscher (2001), [13] Laerdahl *et al.* (1998), [14] Visscher *et al.* (1998), [15] Ellingsen *et al.* (2000), [16] Matila *et al.* (2000), [17] Visscher *et al.* (1996b), [18] de Jong *et al.* (1997), [19] Visscher and Dyall (1996), [20] de Jong *et al.* (1998).

approximately the same. Therefore, it is most interesting to discuss those molecules where relativity affects these 'secondary' quantities significantly. From the list given in Table 2.1 we will discuss in more detail those results which yield large relativistic

Table 2.1 *Cont.*

class	molecule	type	calc. of	ref.
metal	{Ge,Sn,Pb}O	DF	S, I, μ, ϵ_i, PA	[21]
oxides	YbO	RDFT	S, ES, μ, PA	[22]
	ThO	DF	$\langle E \rangle$, S	[23]
	UO_2^{2+}, PaO_2^+, ThO_2	DF	r_e, ω_e, ϵ_i, PA	[24]
	UO_2^{2+}	DF/CC	ϵ_i, PA, EFG	[25]
	UO_6	DF	S, PA	[26]
	EuO_6^{9-}	DF/CI	ES	[27]
(metal)	YbF	RDFT	S, ES, μ	[22]
halides	GdF	DF	P, IE, EA, S	[28]
	{Cu,Ag,Au}F	RMP2	S	[29]
		DF	EFG	[30]
	{La,Lu,Ac,Lr}F	DF/MP2	r_e, ω_e	[31]
	GdF_2	DF	P, IE, EA, S, A	[32]
	Hg{$F_{2,4}$,$Cl_{2,4}$}	RDFT	S	[33]
	ThF_4	DF + B_{mag}	P	[34]
	UF_6	DF + B_{mag}	P	[35]
		DF + B_{mag} − CI	P, r_e, EA	[36]
	CuCl	DF − CI	S, ES	[37]
	{Ti,Zr,Hf,Rf}Cl_4	RDFT	S	[38]
	$RfCl_4$	DF + B_{mag}	P	[39]
	{Nb,Ta,Ha}Cl_5	RDFT	ϵ_i, PA, IE, EA	[40]
	Xe{$F_{1,2,4,6}$}	DF + B_{mag}	$\langle E \rangle$, r_e, D_e	[41]
	{Tl,(113)},{At,(117)}	DF	S, PA, ϵ_i, μ	[42]
	{Al,Ga}{F,Cl,Br}	DF/CCSD(T)	EFG	[43]

[21] Dyall (1993b), [22] Liu *et al.* (1998), [23] Watanabe and Matsuoka (1997), [24] Dyall (1999), [25] de Jong *et al.* (1999), [26] Pyykkö *et al.* (2000), [27] Visser *et al.* (1992b), [28] Tatewaki and Matsuoka (1997), [29] Laerdahl *et al.* (1997), [30] Pernpointner *et al.* (1998), [31] Laerdahl *et al.* (1998), [32] Tatewaki and Matsuoka (1998), [33] Liu *et al.* (1999), [34] Malli and Styszyński (1994), [35] Malli and Styszynski (1996), [36] de Jong and Nieuwpoort (1998), [37] Sousa *et al.* (1997), [38] Varga *et al.* (2000a), [39] Malli and Styszynski (1998), [40] Pershina and Fricke (1993), [41] Styszyński *et al.* (1997), [42] Fægri and Saue (2001), [43] Pernpointner and Visscher (2001a).

effects that can be determined only on the basis of four-component methods instead of scalar or two-component ones. Obviously, this situation is fulfilled for molecules containing actinide or transactinide elements.

A prominent example is UF_6, for which seemingly only four-component methods give the full relativistic effect (Malli and Styszynski 1996). However, this conclusion may be questioned since the Dirac–Fock method by definition does not contain correlation effects. But correlation usually plays a decisive role and cannot be neglected. To safely draw a conclusion on whether four-component methods are the only means to arrive at reliable energetic and structural data for UF_6, a calculation on a correlated level must be performed and compared with its thoroughly calculated nonrelativistic limit by setting the speed of light to about 10^5 and taking into account that several approximations depend implicitly on the speed of light, as discussed above.

The UF_6 molecule has also been studied extensively using a more elaborate method, namely configuration interaction, to assign the experimental photoelectron spectrum (de Jong and Nieuwpoort 1998). The qualitative analysis of chemical bonding exhibits that the U–F bond is more ionic in the relativistic framework (de Jong and Nieuwpoort 1998). The 6s orbital of uranium remains atom-like in the molecule due to relativistic contraction and does not contribute to chemical bonding, while it contributed in nonrelativistic Hartree–Fock theory.

While relativity stabilizes UF_6, stabilization need not always occur. Recently, it has been found that UO_6 is no local minimum within Dirac–Fock theory, while it is stable in quasirelativistic single- and multi-reference calculations (Pyykkö *et al.* 2000). Only four-component multi-reference calculations will give the final answer to the stability of this molecule, in which uranium is in the extraordinary formal oxidation state +XII.

Apart from studies on single molecules or homologous molecules for the analysis of vertical trends in the periodic table of the elements, some interesting chemical and physical effects—such as lanthanide contraction, phosphorescence phenomena and parity violation—that are perfect areas to be tackled by four-component methods have been investigated. Some of the latest results are discussed in the following subsections.

2.4.6 Lanthanide and actinide contraction

The long-known lanthanide contraction has been recently investigated with four-component methods to determine the percentage of the relativistic effect (Laerdahl *et al.* 1998). The filling of inner f shells in lanthanide and actinide atoms, respectively, is accompanied by a steady decrease of the size of the atom. As a result, the 'size' of atoms in the same group in the second and third transition-metal series of the periodic table is similar. Of course, the lanthanide contraction is also observable in the nonrelativistic framework, but it is amplified through 'relativity'. Although the four-component approach, for which Dirac–Fock and, to account somehow for correlation effects, relativistic MP2 calculations have been utilized, is not supposed to give surprisingly new results and insight into the problem already studied using relativistic effective core potential methods (Küchle *et al.* 1997; Seth *et al.* 1995), the four-component study is more rigorous and appealing as far as the treatment of relativity is concerned. It has been found (Laerdahl *et al.* 1998) that between 10% and 30% of the lanthanide contraction and between 40% and 50% of the actinide contraction is caused by relativity in monohydrides, trihydrides and monofluorides of La, Lu and Ac, Lr, respectively.

2.4.7 Phosphorescence

A major advantage of four-component methods, in which not only the ground state but also excited states are accessible (CI, MCSCF or Fock-space CC methods), is that electronic transitions, which are spin forbidden in nonrelativistic theory, can be studied due to the implicit inclusion of spin–orbit coupling. Four-component methods are thus able to describe phosphorescence phenomena adequately. However, only a little work has been done for this type of electronic transitions and almost all studies utilize approximate descriptions of spin–orbit coupling (see, for instance, Christiansen *et al.* 2000).

2.4.8 Parity violation

Probably one of the most appealing features of four-component methods is the possibility of testing fundamental physical symmetries through accurate electronic structure calculations of molecules. Parity nonconservation (PNC) effects are produced by electroweak interactions and lead to asymmetries on a macroscopic level. The parity-violation processes in nature, discovered by Lee and Yang (1956), attribute a very small energy difference to two enantiomeric molecules. This energy difference has been considered (and controversially discussed) as a possible reason for the existence of homochirality, i.e. the natural existence of L-amino acids and D-monosaccharides in our biosphere (Berger *et al.* 2000; Bonner 2000). The behaviour of a system under inversion (reflection through the origin of the coordinate system) is denoted by its parity. The wave functions of two enantiomeric molecules are not eigenfunctions of the parity operator any more because the parity operator converts one into the other. The energy difference, which can be attributed to both isomers through a parity-odd interaction operator, has been given and discussed, for instance, in Quiney *et al.* (1998b).

The electric dipole transition from shell ns to $(n+1)$s, for example, is parity forbidden; only the magnetic-dipole and electric-quadrupole transitions are allowed. However, taking account a 'model' interaction Hamiltonian for the exchange of an intermediate Z^0 vector boson between the nucleus and an electron introduces (odd) matrix operators that exchange upper and lower components, which have different parity. Through PNC interaction the spinors become parity mixed such that a nonzero transition amplitude results (Hartley and Sandars 1991). This PNC effect has first been discussed for the electronic structure of atoms (see, for example, Hartley and Sandars (1991), Chriplovic (1991) and Khriplovich and Lamoreaux (1997) for reviews, Wood *et al.* (1997) and Bennett and Wieman (1999) for recent experimental and Johnson *et al.* (1993b) for theoretical work). Only very recently has the PNC effect been extensively studied in electronic structure calculations on molecules using nonrelativistic wave functions (see, for instance, Bakasov and Quack 1999; Bakasov *et al.* 1998; Berger and Quack 2000a,b).

To assess the parity-violation effect in chiral molecules, the four-component methods approach their limit as far as the treatment of electron correlation and the size of the basis sets are concerned. Therefore, pioneering work on molecules using single-determinantal Dirac–Fock calculations for TlF (Laerdahl *et al.* 1997; Quiney *et al.* 1998a), YbF (Quiney *et al.* 1998d), H_2X_2 (with X = O, S, Se, Te, Po) (Laerdahl and Schwerdtfeger 1999), and chiral halogenides of methane and its higher homologues (Laerdahl *et al.* 2000a) is only preliminary. Four-component coupled-cluster results have only recently been presented for H_2O_2 and H_2S_2, demonstrating that electronic correlation contributions are small in these cases but depend critically on the molecular structure (Thyssen *et al.* 2000). A preference for one enantiomeric form is not evident (Laerdahl *et al.* 2000b). In a very recent combined experimental-theoretical study, the first evidence for an energy difference between chiral molecules in a crystal has possibly been found: namely, for iron complexes containing the $Fe(phen)_3^{2+}$ metal fragment (Lahamer *et al.* 2000).

2.4.9 Calculation of properties from response theory

Apart from primary structural and energetic data, which can be extracted directly from four-component calculations, molecular properties, which connect measured and calculated quantities, are sought and obtained from response theory. In a pilot study, Visscher *et al.* (1997) used the four-component random-phase approximation for the calculation of frequency-dependent dipole polarizabilities for water, tin tetrahydride and the mercury atom. They demonstrated that for the mercury atom the frequency-dependent polarizability (in contrast with the *static* polarizability) cannot be well described by methods which treat relativistic effects as a perturbation. Thus, the variationally stable one-component Douglas–Kroll–Hess method (Hess 1986) works better than perturbation theory, but differences to the four-component approach appear close to spin-forbidden transitions, where spin–orbit coupling, which the four-component approach implicitly takes care of, becomes important. Obviously, the random-phase approximation suffers from the lack of higher-order electron correlation.

Of particular importance in chemistry is the response of a molecular system to an external magnetic field as applied in routinely performed NMR experiments for the identification of compounds, the analysis of reaction mechanisms, and reaction control. Theoretical tools must provide spin–spin coupling constants and shielding tensors in order to calculate quantities, which can be related to experimental data. Needless to say, coupling constants and chemical shifts calculated from shielding tensors can only be obtained from accurate four-component methods for heavy nuclei. The theory of relativistic calculations of magnetic properties has recently been analysed in great detail (Aucar *et al.* 1999).

On the molecular level, magnetic fields arise also from nuclei with nonzero spin resulting in nuclear spin–spin and electronic-spin–nuclear-spin interactions.

An interaction operator for the interaction of an electron at position $\boldsymbol{r}$ with an external magnetic field $\boldsymbol{B}_{\text{ext}}$ and with a nucleus at position $\boldsymbol{R}_A$ ($\boldsymbol{r}_A = \boldsymbol{r} - \boldsymbol{R}_A$) is of the form (Quiney *et al.* 1998b)

$$H_I = c\boldsymbol{\alpha} \cdot \boldsymbol{A} = c\boldsymbol{\alpha} \cdot \left[\tfrac{1}{2}\boldsymbol{B}_{\text{ext}} \times \boldsymbol{r} + \frac{1}{c^2}\frac{\boldsymbol{\mu} \times \boldsymbol{r}_A}{r_A^3} \right],$$

with the four-potential $\boldsymbol{A}$ corresponding to the magnetic field $\boldsymbol{B}_{\text{tot}} = \nabla \times \boldsymbol{A}$. The first term in this interaction operator accounts for the applied laboratory magnetic field, while the second reflects the interaction with the magnetic moment of nucleus A with spin not equal to zero. In first-order perturbation theory, matrix elements of the unperturbed electronic wave function with this interaction operator may be evaluated.

The components of the shielding tensor are defined by (Quiney *et al.* 1998b)

$$\sigma_{qq'} = \left. \frac{\partial^2 E(\boldsymbol{B}, \boldsymbol{\mu})}{\partial B_q \partial \mu_{q'}} \right|_{\boldsymbol{B}=\boldsymbol{0},\ \boldsymbol{\mu}=\boldsymbol{0}}.$$

The isotropic (spherically averaged) part of this tensor is used for the calculation of chemical shifts, i.e. shifts relative to some reference molecule, which can be compared with shifts obtained from NMR measurements in solution (compare, for instance, the results obtained for H_2O and NH_3 (Quiney *et al.* 1998b)). An extensive study of NMR shielding and indirect nuclear spin–spin coupling tensors for hydrogen halides can be found in Visscher *et al.* (1999). Four-component calculations of indirect spin–spin coupling constants within the random-phase approximation have also been performed on MH_4 with M = C, Si, Ge, Sn, Pb and $Pb(CH_3)_3H$ (Enevoldsen *et al.* 2000). It was found that for GeH_4 and PbH_4 the relativistic increases in the coupling constant for the one-bond coupling are 12% and 156%, respectively.

2.5 Electronic Structure of Solids

Relativistic electronic structure calculations on solid-state systems have usually been performed using scalar relativistic approximations to the full four-component Dirac–Fock and post-Dirac–Fock theory (see, for example, Andersen 1975; Boettger 1998a; Geipel and Hess 1997; Shick *et al.* 1997; Wood and Boring 1978). Four-component approaches that would shed light on the errors of the above-mentioned more approximate methods are rare. Although the extension of the standard Dirac–Fock model to solid-state calculations appears to be straightforward (see, for example, Ladik 1997), calculations have only been carried out on somewhat artificial test systems like a one-dimensional periodic chain of selenium atoms (Hu *et al.* 1998, 2000; Schmidt and Springborg 2000).

Methods for solid-state calculations have been devised on the basis of the Dirac equation (bei der Kellen and Freeman 1996; Shick *et al.* 1999; Wang *et al.* 1992). Very recent progress has been achieved in the framework of four-component density functional theory for solids (Theileis and Bross 2000) (compare also the review on the calculation of magnetic properties in this volume). For the basis expansion in Theileis and Bross (2000), the modified augmented plane wave (MAPW) method is used: the one-electron spinor is represented by four-component plane waves well known from the relativistic treatment of free electrons. These plane waves are augmented by four-component spinors for bound states—of the type discussed in the preceding two sections—from which, finally, an expansion into spherical waves is subtracted. The MAPW method has been applied to the electronic structure of FCC gold and platinum reproducing the experimental lattice constants very well, but overestimating the compressibility by a few per cent (Theileis and Bross 2000).

2.6 Concluding Remarks and Perspective

Four-component theories for the calculation of electronic structures as described in detail in this review have become mature particularly in the last decade as a highly accurate tool for any kind of system be it an atom, molecule or solid. Theoretical as well as methodological understanding gave detailed insight into the foundations of relativistic electronic structure theory. Some fundamental questions are still open as we have explicated, especially in the first two sections of this review, and they will certainly be tackled and answered in the years to come. Methodological advances will also continue to be made.

For atomic structure calculations, the four-component MCSCF approach was the method of choice for a long time. The implementation of methods for treating very large CSF spaces, particularly Davidson-type diagonalization techniques, produced a tool to compete with highly precise experimental measurements. The coupled-cluster method—or MBPT approach as it is usually called in the physics community—turned out to be a valuable, alternative, size-consistent method.

Dirac–Fock calculations were the standard four-component method for electronic structure calculations on molecules during the last decade. However, they are still very demanding or completely infeasible if applied to large unsymmetric molecules with several heavy atoms. In addition, taking properly care of electron correlation increases the computational effort tremendously. Future work will certainly continue the development of relativistic correlation methods, which will be far less expensive.

Finally, four-component methods will reach a high degree of applicability such that the relativistic approaches will become the *standard* tool for electronic structure calculations in the next decades. The four-component theories provide the general framework, in which more approximate methods—such as elimination methods for the small components, reduction methods to one-component wave functions, and also the nonrelativistic approaches—elegantly fit. This function of the four-component theories as the theoretical basis will certainly be reflected in algorithms and computer codes to be developed.

3 Relativistic Quantum Chemistry with Pseudopotentials and Transformed Hamiltonians

Bernd Artur Hess
Chair of Theoretical Chemistry,
Friedrich–Alexander-Universität Erlangen–Nürnberg

Michael Dolg
Institute of Physical and Theoretical Chemistry,
Rheinische Friedrich–Wilhelms-Universität Bonn

3.1 Introduction

Relativistic effects have a marked influence on the electronic structure of heavy atoms and molecules. After very early pioneering work (Kołos and Wolniewicz 1964; Ladik 1959), they were brought to the attention of the community by Pyykkö (1978), Pyykkö and Desclaux (1979) and Pitzer (1979). The development of the field since has been documented in many review papers, and we only mention a few more recent ones contributed to by authors of the 'Schwerpunkt' (Dolg and Stoll 1995; Hess and Marian 2000; Hess *et al.* 1995; Marian 2000). The bibliography of the field is very well documented in the books by Pyykkö (1986, 1993, 2000), an up-to-date version of the bibliography being available online (Pyykkö 2001).

Relativistic effects in atoms and molecules are commonly separated in kinematical effects, which do not cause a splitting of energy levels due to the spin degrees of freedom, and the effects of spin–orbit coupling. This separation is not unambiguously defined (Visscher and van Lenthe 1999), but is nevertheless extremely convenient when discussing effects on phenomenology.

Kinematical relativistic effects are caused by the fact that in the vicinity of the nucleus the electrons acquire high velocities, at a substantial fraction of the velocity of light. The direct influence of the relativistic kinematics (the so-called *direct* relativistic effect) is thus largest in the vicinity of the nucleus. However, as far as their impact on chemistry is concerned, relativistic effects are most important in the valence shells, which despite the small velocities of outer electrons are still strongly affected by relativistic kinematics (Schwarz *et al.* 1989). In particular, valence s and p orbitals possess inner tails: they are *core-penetrating orbitals*, which means that there is a nonvanishing probability of finding their electrons close to the nucleus and thus

Relativistic Effects in Heavy-Element Chemistry and Physics. Edited by B. A. Hess

exposed to the direct relativistic effect. The shells with higher angular momentum, d and f, are not core-penetrating orbitals due to their centrifugal barrier. They are subjected to an *indirect* relativistic effect, which is due to the relativistic relaxation of the other shells (in the first place the contraction of s and p orbitals), which will alter the shielding of the d and f electrons. In particular, the contraction of the s and p semicore (i.e. the s and p shells with the same quantum numbers and about the same spatial extent as the shell in question, albeit with very different energy) will lead to a more effective shielding of the d and f orbitals and thus to an energetic destabilization. Thus, a good rule of thumb is that s and p shells are energetically stabilized (with a concomitant higher ionization potential and electron affinity for ionization or attachment of an electron in these shells) and that (valence) d and f shells experience relativistic destabilization.

This destabilization may lead, in turn, to an indirect stabilization of the next higher s and p shells with spatial extent similar to the d shell in question. This situation occurs in the case of the late transition metals and leads to the 'gold maximum' of relativistic effects and the unusually large relativistic effects in the elements of groups 10–12. If the d shell is only weakly occupied, as is the case in the early transition metals, the direct effect on the s and p shells is partly balanced by the indirect effect on those shells, and the relativistic effects are generally much smaller.

The spin-dependent relativistic effects are connected with the spin degrees of freedom, which also enter as dynamical quantities in the relativistic theory and thus couple with the orbital motion, leading to spin–orbit coupling phenomena. Since they are of symmetry-breaking nature, they may also be of importance in the lightest elements and their compounds (Hess *et al.* 1995; Langhoff and Kern 1977; Minaev and Ågren 1996). Their magnitude scales with the fourth power of the nuclear charge, and becomes comparable with the effects of the Coulomb interaction in the sixth row of the periodic system.

The atomic relativistic effects on the orbital energies and thus on excitation energies, ionization potentials and electron affinities have a direct influence on chemically relevant data, namely structure, electronic spectra and force constants of molecules. For molecules containing heavy elements, it is thus mandatory that relativistic effects be included in the computational methods to determine the electronic structure. The canonical theory to accomplish this is the Dirac theory of the relativistic electron (Dirac 1928), which makes use of a wave function with four components, two for the two spin degrees of freedom of the electron and two more for the spin degrees of freedom for the charge-conjugated partner of the electron, namely the positron. Methods for the calculation of the electronic structure using this kind of four-component wave function are covered in Chapter 2.

While theoretically most appealing, the four-component methods are very expensive in terms of computational resources. This is, of course, because the charge-conjugated degrees of freedom are treated as dynamical variables and thus require their own basis set in the calculation. For technical reasons, the basis set for the small component tends to be even larger than the large-component basis. Since the charge-conjugated degrees of freedom are not excited at energies typical for the valence shell of neutral or mildly ionized atoms and molecules, it is desirable to integrate them out at the very beginning. This leads to a transformed Hamiltonian, operating on a two-component wave function for the electronic degree of freedom. Moreover, it is also possible for the Dirac equation to separate off spin-dependent terms rigorously (Dyall 1994b; Visscher and van Lenthe 1999), so that in many cases we can use a spin-averaged one-component wave function, calculated from a Hamiltonian transformed from the spin-free (scalar relativistic) part of the Dirac equation. The transformed Hamiltonians are obtained by means of a unitary transformation that annihilates the coupling between the electron-like and the positron-like degrees of freedom. Their wave functions still formally have four components. Since, however, there is no coupling any more between the states of positive energy (the electrons) and the states of negative energy (the positrons), we now have the possibility of focusing on the former and working only with two-component wave functions. While spin–orbit coupling is described in the 'Dirac-like' (four-component) representation by a purely algebraic structure (the Clifford algebra of the Dirac matrices), there is a space part of the spin–orbit coupling operator in the decoupled representation.

In the next two sections we shall present the theory of transformed Hamiltonians and applications obtained in the framework of the 'Schwerpunkt' of the German Science Foundation on Relativistic Effects on Heavy-Element Chemistry and Physics.

However, we can even go a step further in the development of efficient calculational tools for the electronic structure of molecules. Noting that the inner-shell electrons are also relatively inert to the interactions of the electrons in the valence shell, we may strive to include the relativistic effects in an effective core potential (ECP) and treat only the valence electrons explicitly. Section 3.4 of this chapter will deal with various methods to describe relativistic effects in molecules by means of ECPs.

3.2 Transformed Hamiltonians: Theory

The wave function in the Dirac equation is a multi-component quantity, two components describing the spin degrees of freedom (spin up and spin down) of the electron, and two more describing the spin degrees of freedom for a charge-conjugated particle, loosely speaking, a positron. The Dirac Hamiltonian operates on these four components by means of 4×4 matrices, which do not depend on the dynamical variables. It is most convenient to group the four components two-by-two, and the four Dirac matrices,

$$\boldsymbol{\alpha} = \begin{pmatrix} 0 & \boldsymbol{\sigma} \\ \boldsymbol{\sigma} & 0 \end{pmatrix}, \qquad \beta = \begin{pmatrix} 1 & 0 \\ 0 & -1 \end{pmatrix},$$

are used to formulate the operators describing the observables. Every entry in the matrices above is to be interpreted as a 2×2 matrix, in particular the $\boldsymbol{\sigma}$ matrices are the familiar Pauli spin matrices

$$\sigma_x = \begin{pmatrix} 0 & 1 \\ 1 & 0 \end{pmatrix}, \qquad \sigma_y = \begin{pmatrix} 0 & -\mathrm{i} \\ \mathrm{i} & 0 \end{pmatrix}, \qquad \sigma_z = \begin{pmatrix} 1 & 0 \\ 0 & -1 \end{pmatrix}.$$

The one-particle Dirac equation for a particle with spin 1/2 in the external potential of the nucleus V_{ext} can then be written as

$$D\psi_{\text{D}} = E\psi_{\text{D}}, \qquad D = c\boldsymbol{\alpha p} + \beta mc^2 + V_{\text{ext}} \tag{3.1}$$

or in *split notation* as

$$\left.\begin{aligned} c\boldsymbol{\sigma p}\psi_{\text{S}} + mc^2\psi_{\text{L}} + V_{\text{ext}}\psi_{\text{L}} &= E\psi_{\text{L}}, \\ c\boldsymbol{\sigma p}\psi_{\text{L}} - mc^2\psi_{\text{S}} + V_{\text{ext}}\psi_{\text{S}} &= E\psi_{\text{S}}, \end{aligned}\right\} \tag{3.2}$$

In this notation the presence of two *upper* and two *lower* components of the four-component Dirac spinor ψ_{D} is emphasized. For solutions with positive energy and weak potentials, the latter is suppressed by a factor $1/c^2$ with respect to the former, and therefore commonly dubbed the *small component* ψ_{S}, as opposed to the *large component* ψ_{L}. While a Hamiltonian for a many-electron system like an atom or a molecule requires an electron interaction term (in the simplest form we add the Coulomb interaction and obtain the Dirac–Coulomb–Breit Hamiltonian; see Chapter 2), we focus here on the one-electron operator and discuss how it may be transformed to two components in order to integrate out the degrees of freedom of the charge-conjugated particle, which we do not want to consider explicitly.

A representation of the one-electron Dirac equation which is decoupled in the electronic and the charge-conjugated degrees of freedom is achieved by a unitary transformation of (3.1)

$$H^{\text{decoupled}} = U^\dagger D U = \begin{pmatrix} h_+ & 0 \\ 0 & h_- \end{pmatrix} \tag{3.3}$$

with

$$UU^\dagger = 1, \qquad U = \begin{pmatrix} (1+X^\dagger X)^{-1/2} & (1+X^\dagger X)^{-1/2}X^\dagger \\ X(1+XX^\dagger)^{-1/2} & (1+XX^\dagger)^{-1/2} \end{pmatrix},$$

and D denoting a Dirac-type Hamiltonian.

The operator X maintains the exact relationship between the large and the small component

$$\phi_{\text{S}} = X\phi_{\text{L}}$$

for any trial function for the small component ϕ_{S} and for the large component ϕ_{L}. The operator X is not known in general. For exact eigenfunctions of the one-electron Dirac equation, dubbed $(\psi_{\text{L}}, \psi_{\text{S}})$, it could in principle be determined by expressing the small component in terms of the large component in (3.2) using the expression for ψ_{S} from the lower equation

$$\psi_{\text{S}} = \frac{1}{2mc^2}\left(1 + \frac{E-V}{2mc^2}\right)^{-1} c\boldsymbol{\sigma p}\psi_{\text{L}}. \tag{3.4}$$

An expansion in $(E-V)/2mc^2$ is the basis of the method of elimination of the small component, which in its classical version is only of limited use because the expansion is valid only for $E - V(r) < 2mc^2$, a condition which does not hold close to the nucleus.

In the general case, X must fulfil the nonlinear equation (Chang *et al.* 1986; Heully *et al.* 1986; Kutzelnigg 1997)

$$X = \frac{1}{2mc^2}(c\boldsymbol{\sigma p} - [X, V] - X(c\boldsymbol{\sigma p})X).$$

Obviously, the solution of this equation for X is as complex as the solution of the Dirac equation itself, and approximations have to be employed.

Since the transformed large component, now describing electron states only, should be normalized to one, the equation contains renormalization terms $(1 + X^\dagger X)^{-1/2}$ to take the change from the Dirac normalization prescription for any four-component wave function Φ

$$\begin{aligned}\langle\Phi|\Phi\rangle &= \langle\phi_{\mathrm{L}}|\phi_{\mathrm{L}}\rangle + \langle\phi_{\mathrm{S}}|\phi_{\mathrm{S}}\rangle \\ &= \langle\phi_{\mathrm{L}}|\phi_{\mathrm{L}}\rangle + \langle X\phi_{\mathrm{L}}|X\phi_{\mathrm{L}}\rangle\end{aligned}$$

into account. Closed-form solutions for X are known only for a restricted class of potentials (Nikitin 1998). A very important special case is, however, the free particle, defined by $V \equiv 0$. In this case, we find a closed-form solution

$$X^{V=0} = \left(mc^2 + \sqrt{m^2c^4 + p^2c^2}\right)^{-1} c\boldsymbol{\sigma p}.$$

This defines the *exact* Foldy–Wouthuysen transformation for the free particle. Note that the square root is not expanded here.

Early attempts to reduce the Dirac and Dirac–Coulomb–Breit Hamiltonian to the electronic degrees of freedom are the Foldy–Wouthuysen transformation (Foldy and Wouthuysen 1950) and the elimination of the small component. Both methods lead to first order in c^{-2} to the Breit–Pauli Hamiltonian. Even in first order a singular operator obtains, featuring a $\delta(r)$ function in the so-called Darwin term and a p^4 term with negative sign as relativistic correction to the kinetic energy, both of which preclude their use in a variational calculation. Beyond first order the expansion leads to increasingly singular and even undefined expressions (Morrison and Moss 1980). We shall not further describe these older approaches and also do not discuss formulations leading to energy-dependent or nonHermitian operators. Instead, we focus on the variationally stable transformed Hamiltonians that have appeared more recently in the literature (Chang *et al.* 1986; Douglas and Kroll 1974; Hess 1986).

A very well-studied technique to arrive at regular expansions was developed in the mid 1980s (Chang *et al.* 1986; Heully *et al.* 1986). The essential point is to rewrite the prefactor of $c\boldsymbol{\sigma p}$ in (3.4) as

$$\frac{2mc^2}{2mc^2 - V}\left(1 + \frac{E}{2mc^2 - V}\right)^{-1} \tag{3.5}$$

and expand the term in parentheses. This expansion is the basis of the so-called Chang–Pélissier–Durand (CPD) Hamiltonian and the *regular approximations*, which were developed by the Amsterdam group (van Lenthe *et al.* 1995, 1996) to a workable method for electronic-structure calculations.

A truncation of the expansion (3.5) defines the zero- and first-order regular approximation (ZORA, FORA) (van Lenthe *et al.* 1993). A particular noteworthy feature of ZORA is that even in the zeroth order there is an efficient relativistic correction for the region close to the nucleus, where the main relativistic effects come from. Excellent agreement of orbital energies and other valence-shell properties with the results from the Dirac equation is obtained in this zero-order approximation, in particular in the scaled ZORA variant (van Lenthe *et al.* 1994), which takes the renormalization to the transformed large component approximately into account, using

$$\frac{1}{\sqrt{1+X^{\dagger}X}} \approx \frac{1}{\sqrt{1+\langle\phi_{\mathrm{L}}X^{\dagger}|X\phi_{\mathrm{L}}\rangle}}.$$

The analysis (van Leeuwen *et al.* 1994) shows that in regions of high potential the zero-order Hamiltonian reproduces relativistic energies up to an error of order $-E^2/c^2$. On the other hand, in regions where the potential is small, but the kinetic energy of the particle high, the ZORA Hamiltonian does not provide any relativistic correction.

The main disadvantage of the method is its dependence on the zero point of the electrostatic potential, i.e. gauge dependence. This occurs because the potential enters nonlinearly (in the denominator of the operator for the energy), so that a constant shift of the potential does not lead to a constant shift in the energy. This deficiency can, however, be approximately remedied by suitable means (van Lenthe *et al.* 1994; van Wüllen 1998).

The second major method leading to two-component regular Hamiltonians is based on the Douglas–Kroll transformation (Douglas and Kroll 1974; Hess 1986; Jansen and Hess 1989). The classical derivation makes use of two successive unitary transformations

$$H_{+} = U_1^{\dagger}U_0^{\dagger}\, D\, U_0U_1$$

of the Dirac operator, the first being a free-particle FW transformation characterized by the parametrization $U_0 = \exp(\mathrm{i}S)$ for a Hermitian operator S, which is chosen to annihilate the coupling term between the upper and the lower component in the free-particle Dirac equation. The second transformation is parametrized

$$U_1 = \sqrt{1+W_1^2} + W_1$$

for a suitably chosen skew-Hermitian operator W_1. While this prescription seems to be rather ad hoc, a transparent explanation has recently been given by Kutzelnigg (1999) from the theory of effective Hamiltonians.

We consider a Hamiltonian operating on two subspaces characterized by projectors P and $Q = 1 - P$. This partitioning defines projected Hamiltonians

$$H_{++} = PHP, \qquad H_{+-} = QHP,$$
$$H_{-+} = PHQ, \qquad H_{--} = QHQ,$$

which we write in a shorthand notation similar to Equation (3.3) as

$$H = \begin{pmatrix} H_{++} & H_{+-} \\ H_{-+} & H_{--} \end{pmatrix}.$$

We now look for an operator $L = W^\dagger H W$ with $W^\dagger W = 1$ such that L is block diagonal,

$$L = \begin{pmatrix} L_{++} & 0 \\ 0 & L_{--} \end{pmatrix}.$$

We are interested in the *effective Hamiltonian* L_{++} only, which acts on the *model space* described by the projector P. This operator can be obtained if the equation

$$L_\mathrm{N} = L_{+-} + L_{-+} = (W^\dagger H W)_\mathrm{N} = 0$$

can be solved, where the subscript 'N' denotes the *nondiagonal* part of the corresponding operators (Kutzelnigg 1982). If H is given by means of a perturbative ansatz $H = H_0 + \lambda V$ (with corresponding partitioning of H_0 and V in terms of the projectors P and Q), L is obtained with perturbative contributions

$$L_0 = H_0, \qquad L_1 = V_{++}, \qquad L_2 = W_1^\dagger H_0 W_1 + W_1^2 H_0 + H_0 W_1^2, \qquad \text{etc.},$$

and W_1 is obtained by solving

$$[H_0, W_1] + V_\mathrm{N} = 0.$$

If we employ this formalism for the decoupling of D according to Equation (3.3), we have different choices for H_0 at our disposal, which in turn define different transformations:

$H_0 = \beta mc^2$	Foldy–Wouthuysen transformation,
$H_0 = -\frac{1}{2}\Delta + V_\mathrm{ext}$	direct perturbation theory (Kutzelnigg 1989),
$H_0 = \beta mc^2 + c\boldsymbol{\alpha p}$	Douglas–Kroll transformation.

It is clear from H_0 that the Douglas–Kroll transformation makes use of a model space of relativistic free-particle spinors, and that it is defined by a perturbative expansion with the external potential as perturbation. Indeed, using the formulas given above, we get the familiar expressions for the second-order Douglas–Kroll-transformed Dirac operator, which is often dubbed Douglas–Kroll–Hess (DKH) operator

$$H_+ = L_0 + L_1 + L_2,$$

$$L_0 = H_0 = E_p = \sqrt{m^2c^4 + p^2c^2},$$

$$L_1 = V_{++}(\boldsymbol{p}, \boldsymbol{p}') = A_p R_p \hat{V}(\boldsymbol{p}, \boldsymbol{p}') R_{p'} A_{p'},$$

$$V_N(\boldsymbol{p}, \boldsymbol{p}') = A_p R_p \hat{V}(\boldsymbol{p}, \boldsymbol{p}') A_{p'} - A_p \hat{V}(\boldsymbol{p}, \boldsymbol{p}') R_{p'} A_{p'},$$

$$W_1(\boldsymbol{p}, \boldsymbol{p}') = \frac{V_N(\boldsymbol{p}, \boldsymbol{p}')}{E_{p'} + E_p},$$

where $\hat{V}(\boldsymbol{p}, \boldsymbol{p}')$ denotes the Fourier transform of the external potential $V_{\mathrm{ext}}(\boldsymbol{x})$ and we have given the kernels of the respective nonlocal operators. A_p and R_p are given by

$$A_p = \sqrt{\frac{E_p + mc^2}{2E_p}}, \qquad R_p = \frac{c\boldsymbol{\sigma}\,\boldsymbol{p}}{E_p + mc^2} \equiv \boldsymbol{\sigma}\,\boldsymbol{P}_p.$$

If a multiparticle system is considered and the electron interaction is introduced, we may use the Dirac–Coulomb–Breit (DCB) Hamiltonian which is given by a sum of one-particle Dirac operators coupled by the Coulomb interaction $1/r_{ij}$ and the Breit interaction B_{ij}. Applying the Douglas–Kroll transformation to the DCB Hamiltonian, we arrive at the following operator (Hess 1997; Samzow and Hess 1991; Samzow *et al.* 1992), where an obvious shorthand notation for the indices p_i has been used:

$$H_+ = \sum_i E_i + \sum_i V_{\mathrm{eff}}(i) + \frac{1}{2}\sum_{i \neq j} V_{\mathrm{eff}}(i, j),$$

with

$$\begin{aligned}
V_{\mathrm{eff}}(i) &= A_i R_i [\hat{V}_i + (\boldsymbol{\sigma}_i \boldsymbol{P}_i)\hat{V}_i(\boldsymbol{\sigma}_i \boldsymbol{P}_i)] A_i R_i,\\
V_{\mathrm{eff}}(i, j) &= A_i A_j \Bigg[\frac{1}{r_{ij}} + (\boldsymbol{\sigma}_i \boldsymbol{P}_i)\frac{1}{r_{ij}}(\boldsymbol{\sigma}_i \boldsymbol{P}_i) + (\boldsymbol{\sigma}_j \boldsymbol{P}_j)\frac{1}{r_{ij}}(\boldsymbol{\sigma}_j \boldsymbol{P}_j)\\
&\qquad + (\boldsymbol{\sigma}_i \boldsymbol{P}_i)(\boldsymbol{\sigma}_j \boldsymbol{P}_j)\frac{1}{r_{ij}}(\boldsymbol{\sigma}_i \boldsymbol{P}_i)(\boldsymbol{\sigma}_j \boldsymbol{P}_j) + B_{ij}(\boldsymbol{\sigma}_i \boldsymbol{P}_i)(\boldsymbol{\sigma}_j \boldsymbol{P}_j)\\
&\qquad + (\boldsymbol{\sigma}_i \boldsymbol{P}_i) B_{ij} (\boldsymbol{\sigma}_j \boldsymbol{P}_j) + (\boldsymbol{\sigma}_j \boldsymbol{P}_j) B_{ij} (\boldsymbol{\sigma}_i \boldsymbol{P}_i)\\
&\qquad + (\boldsymbol{\sigma}_i \boldsymbol{P}_i)(\boldsymbol{\sigma}_j \boldsymbol{P}_j) B_{ij} \Bigg] A_i A_j,\\
B_{ij} &= -\frac{1}{2}\frac{1}{r_{ij}}\left[\boldsymbol{\sigma}_i \cdot \boldsymbol{\sigma}_j + \left(\boldsymbol{\sigma}_i \cdot \frac{\boldsymbol{r}_{ij}}{r_{ij}}\right)\left(\boldsymbol{\sigma}_j \cdot \frac{\boldsymbol{r}_{ij}}{r_{ij}}\right)\right].
\end{aligned}$$

Making repeated use of the Dirac relation

$$(\boldsymbol{\sigma}\boldsymbol{u})(\boldsymbol{\sigma}\boldsymbol{v}) = \boldsymbol{u}\boldsymbol{v} + \mathrm{i}\boldsymbol{\sigma}(\boldsymbol{u} \times \boldsymbol{v}),$$

which is valid for operators $\boldsymbol{u}$ and $\boldsymbol{v}$ not containing $\boldsymbol{\sigma}$ matrices, terms linear in either one of the $\boldsymbol{\sigma}$ matrices are extracted:

$$\begin{aligned}
H^{\mathrm{SO}} &= \sum_i \sum_\alpha Z_\alpha f_1(\boldsymbol{p}_i) \boldsymbol{s}_i \left(\frac{\boldsymbol{r}_{i\alpha}}{r_{i\alpha}^3} \times \boldsymbol{p}_i\right) f_1(\boldsymbol{p}_i)\\
&\quad - \sum_{i \neq j} f_2(\boldsymbol{p}_i, \boldsymbol{p}_j)\,\boldsymbol{\sigma}_i \frac{\boldsymbol{r}_{ij} \times \boldsymbol{p}_i}{r_{ij}^3} f_2(\boldsymbol{p}_i, \boldsymbol{p}_j)\\
&\quad - 2\sum_{i \neq j} f_2(\boldsymbol{p}_i, \boldsymbol{p}_j)\,\boldsymbol{\sigma}_i \frac{\boldsymbol{r}_{ij} \times \boldsymbol{p}_j}{r_{ij}^3} f_2(\boldsymbol{p}_j, \boldsymbol{p}_i)\\
&\equiv \sum_i H_1^{\mathrm{SO}}(i) + \sum_{i \neq j} H_2^{\mathrm{SO}}.
\end{aligned} \tag{3.6}$$

In this expression, $H_1^{SO}(i)$ denotes the one-electron operator for electron i, and $H_2^{SO}(i, j)$ the two-electron operator for electron pair (i, j). The quantities f_1 and f_2 are constants equal to $(2mc^2)^{-1}$ in the case of the Breit–Pauli operator, which defines the spin–orbit operator obtained in second order from the FW transformation. By contrast, they are momentum-dependent operator functions in the case of the spin–orbit part of the Douglas–Kroll-transformed Hamiltonian, which effectively regularize the $1/r^3$ singularity of the Breit–Pauli Hamiltonian:

$$f_1(\boldsymbol{p}_i) = \frac{A_i}{E_i + mc^2}, \qquad f_2(\boldsymbol{p}_i, \boldsymbol{p}_j) = \frac{A_i A_j}{E_i + mc^2}.$$

Since the factors $(E_i + mc^2)^{-1}$ grow asymptotically (for $|p_i| \to \infty$, i.e. $r_i \to 0$) like $1/|\boldsymbol{p}_i|$, all contributions of momentum operators in the numerator (leading to the $1/r^3$ divergence in the case of the Breit–Pauli operator) are cancelled asymptotically, and only a Coulomb singularity remains. Recently, Brummelhuis *et al.* (2002) have formally proved that the operator is variationally stable.

The Breit–Pauli operator may be recovered by expanding in powers of c^{-2}

$$\frac{A_i}{E_i + mc^2} = \frac{1}{2mc^2} - \frac{3p_i^2}{16m^3c^4} + \cdots$$

and keeping only the lowest-order term.

Since the operators f_1 and f_2 occur only at the level of the calculation of the spatial spin–orbit integrals over atomic orbitals, Breit–Pauli spin–orbit coupling operators and DKH spin–orbit coupling operators can be discussed on the same footing as far as their matrix elements between multi-electron wave functions are concerned. These terms constitute, by definition, the spin–orbit interaction part of the operator H_+ (Hess *et al.* 1995). The spin-independent terms characteristic of relativistic kinematics define the *scalar relativistic* part of the operator, and terms with more than one $\boldsymbol{\sigma}$ matrix (not considered here) contribute to spin–spin coupling phenomena.

In the case of singlet ground states well separated from the rest of the spectrum, it is often convenient to use the spin-averaged approximation and treat the spin–orbit coupling operator in a second step, be it perturbatively or variationally in a spin–orbit configuration interaction procedure with two-component spinors. In most applications (see, however, Park and Almlöf 1994; Samzow *et al.* 1992) the Douglas–Kroll transformation of the external potential V is limited to its one-electron part while the two-electron terms are left in their Coulomb form. This leads to the most frequently used spin-averaged one-component DKH operator

$$H_{\mathrm{DKH}} = \sum_i E_{p_i} + \sum_i V_{\mathrm{eff}}(i) + \sum_{i<j} \frac{1}{r_{ij}}.$$

A numerical analysis of the energy values (Hess 1986; Molzberger and Schwarz 1996) and also perturbation theory (Kutzelnigg 1997) shows that the eigenvalues of the DKH Hamiltonian for a single particle agree with the results of the Dirac equation to order c^{-4}. Note that this is the same order in which deviations in the matrix representation of the Dirac equation itself are expected (Kutzelnigg 1997; Stanton and Havriliak 1984).

Implementations of the spin-free DKH Hamiltonian exist by now for many standard quantum chemistry packages like MOLECULE-SWEDEN, COLUMBUS, TURBOMOLE, MOLCAS and NWCHEM. The method has also been implemented in several programs for the calculation of periodic structures, in particular crystals (Boettger 1998b; Fehrenbach and Schmidt 1997; Geipel and Hess 1997).

The DKH Hamiltonian has been implemented and also applied in the context of the density-functional theory of molecules. Important contributions have been made in the last decade by Notker Rösch's group in Munich, who implemented the formalism (Häberlen and Rösch 1992; Rösch *et al.* 1996) in their LCGTO-FF scheme (linear combination of Gaussian-type orbitals with fitting functions). Rösch's group was also among the first to implement relativistic exchange-correlation functionals (compare Chapter 4) and tested their performance in practical applications on AuH, AuCl, Au_2, Ag_2 and Cu_2 (Mayer *et al.* 1996). Since the influence of relativistic corrections to the exchange-correlation functional on structural parameters was found to be small, the authors came to the conclusion that the common practice of neglecting them is justified, at least for compounds without superheavy elements.

For the first time, analytic gradients have been devised and implemented for the DKH approach by Nasluzov and Rösch (1996), which is a prerequisite for efficient geometry optimizations of molecules with more than just a few atoms.

The modules for relativistic electronic structure theory were all integrated in the program PARAGAUSS, which is a parallelized implementation of the LCGTO-FF-DFT method (Belling *et al.* 1999a,b).

The Douglas–Kroll transformation can be carried out to higher orders, if desired (Barysz *et al.* 1997). In this way, arbitrary accuracy with respect to the eigenvalues of D can be achieved.

3.2.1 Two-component all-electron methods for spin–orbit coupling

The evaluation of the spin–orbit one- and two-electron operators is very demanding in terms of computing power. Recently, a method has been developed which makes it possible that effective one-electron integrals are used as spin–orbit operators. This approach depends on the observation that the two-electron part provides an effective screening of the one-electron contributions, and the approach is thus termed a spin–orbit mean-field (SOMF) approach (Hess *et al.* 1996). To date, it appears to be one of the most rigorous approaches towards the definition of an effective one-electron spin–orbit operator. In effect, it constitutes an extension of the frozen-core approximation and reproduces matrix elements of the full one- and two-particle operator excellently, even in light molecules, where the two-electron contributions to the total matrix element amount to about 50% of the total spin–orbit splitting (Danovich *et al.* 1998; Hess *et al.* 1996; Marian and Wahlgren 1996; Tatchen and Marian 1999).

Since the many-electron wave function can be expanded in a linear combination of Slater determinants, its matrix element with a spin–orbit coupling operator of the form of Equation (3.6) can be expressed as a sum of matrix elements of the operator between Slater determinants. For a matrix element between Slater determinants which differ in exactly one spin orbital (i.e. which are singly excited from $i \to a$ with respect to each other), the matrix element is

$$H_{ia}^{\mathrm{SO}} = \langle i|H_1^{\mathrm{SO}}|a\rangle + \frac{1}{2}\sum_k n_k\{\langle ik|H_2^{\mathrm{SO}}|ak\rangle - \langle ik|H_2^{\mathrm{SO}}|ka\rangle - \langle ki|H_2^{\mathrm{SO}}|ak\rangle\}, \tag{3.7}$$

n_k denoting the occupation number of the kth orbital.

In an independent-particle model, Equation (3.7) defines a Fock operator describing valence electrons moving in a field generated by the electrons in orbitals k. Using this relationship, we define an approximate operator

$$\begin{aligned} H_{ia}^{\text{mean field}} &= \langle i|H_1^{\mathrm{SO}}|a\rangle \\ &\quad + \frac{1}{2}\sum_{k,\ \text{fixed}\ \{n_k\}} n_k\{\langle ik|H_2^{\mathrm{SO}}|ak\rangle - \langle ik|H_2^{\mathrm{SO}}|ka\rangle - \langle ki|H_2^{\mathrm{SO}}|ak\rangle\} \end{aligned}$$

with fixed occupation numbers, which can be taken as an effective average over the two-electron contribution of the valence shell in Equation (3.7).

In general, a mean-field approximation is defined by any set of occupation numbers $\{n_k\}$ by means of a corresponding Fock operator matrix element, and the dependence of the results on the specific set of occupation numbers turned out to be very weak in practical calculations. This approximation has also been developed independently by Berning *et al.* (2000).

The fact that the construction of the molecular mean field necessitates the evaluation of two-electron spin–orbit integrals in the complete AO basis represents a serious bottleneck in large applications. Based on earlier observations by Richards *et al.* (1981), we realized the possibility that all multi-centre two-electron integrals are neglected as an additional approximation. This results in considerable speed-up, since it provides the possibility of devising special programs which evaluate one centre only (Schimmelpfennig 1996) if one-electron two-centre terms are also neglected. Indeed, thorough investigations into a variety of molecules show that multi-centre one- and two-electron contributions partly compensate in a systematic manner (Danovich *et al.* 1998; Rakowitz 1999; Tatchen and Marian 1999). Even more efficiency may be gained if the spin-independent core–valence interactions are also replaced by atom-centred effective core potentials (ECPs), and we shall come back to this approach in Section 3.4.

Current versions of the SomfEcp program (Rakowitz 1999) can process Breit–Pauli and spin–orbit integrals based on the Douglas–Kroll transformation from the general-utility integral programs within the BnSoc package (Hess *et al.* 2000) and one-centre integrals from Amfi (Schimmelpfennig 1996). The use of the spin–orbit

Hamiltonian resulting from the Douglas–Kroll transformation is particularly indicated for heavy elements and in variational calculations, because it is bounded from below (Samzow *et al.* 1992). The raw integrals are by now combined with AO and MO information from a variety of standard program packages (MOLECULE-SWEDEN, COLUMBUS, TURBOMOLE) and SOMF integrals are provided for BNSOC, COLUMBUS, MOLCAS and LUCIAREL.

Electron correlation effects on spin–orbit interactions

In light molecules, spin–orbit coupling predominantly affects spectral properties such as fine-structure and transition probabilities. In heavy-element compounds spin–orbit interaction is also of concern for bond distances and binding energies. Independent of the spin–orbit interaction scheme, it is indispensable to employ methods which take electron correlation and relativistic effects into account. Recent review articles give an overview of progress in this field (Hess and Marian 2000; Marian 2000).

In compounds containing heavy main-group elements electron correlation depends on the particular spin–orbit component. The j–j coupled $6p_{1/2}$ and $6p_{3/2}$ of, for example, thallium exhibit very different radial amplitudes. As a consequence, electron correlation in the p shell, which has been computed at the spin-free level, is not transferable to the spin–orbit coupled case. This feature is named *spin polarization*. It is best recovered in spin–orbit CI procedures where electron correlation and spin–orbit interaction can be treated on the same footing, at least in principle (Hess and Marian 2000; Rakowitz and Marian 1996).

In practice, configurations are selected according to some criterion such as excitation class or energy. Unfortunately, electron correlation contributions are slowly convergent. They originate mainly from double and higher excitations, while spin–orbit coupling is dominated by singly excited configurations. The SPDIAG spin–orbit CI program (Hutter 1994) within the BNSOC package (Hess *et al.* 2000) is based on the MRD-CI approach. The latter makes use of a correlation energy criterion for configuration selection and estimates the contribution of the discarded configurations to the spin-free correlated energy by means of Epstein–Nesbet perturbation theory (Buenker and Peyerimhoff 1974). If we assume that a quasidegenerate perturbation theory expansion in the basis of the most important LS-contracted CI vectors represents a decent approximation to the spin–orbit CI solution, the MRD-CI extrapolation scheme can easily be extended to the spin–orbit coupled case (Rakowitz 1999; Rakowitz and Marian 1997). Also, other approaches towards a balanced treatment of spin–orbit interaction and electron correlation are based on a manipulation of the spin-free energies and wave functions (Balasubramanian 1988; DiLabio and Christiansen 1998; Llusar *et al.* 1996; Teichteil *et al.* 1983; Vallet *et al.* 2000). In the so-called spin-free state shifted (SFSS) spin–orbit CI method, diagonal energies are shifted by means of a projector on the set of LS-contracted CI states (Llusar *et al.* 1996). Rakowitz *et al.* (1998) employed this method to the spectrum of the Ir^+ ion. They demonstrated that the heavily spin–orbit-perturbed spectrum of this ion can be obtained in good agreement with experiment at the single excitation level if higher-level correlated electrostatic energies are used to determine the energy shifts.

Density functional approaches

The quantum-chemical determination of electronically excited states of chromophores with 100–200 valence electrons is not feasible with standard *ab initio* correlation methods. Recently, a combined density functional and single-excitation CI (DFT/SCI) approach was proposed by Grimme (1996). The SCI is based on Kohn–Sham orbitals, employs scaled Coulomb integrals for diagonal and off-diagonal elements of the CI matrix and an empirical shift function for diagonal elements, utilizing five global empirical parameters in all. The method, which is constrained to singly excited states and geometries close to the equilibrium, was successfully applied to a series of spectroscopical problems (Bulliard *et al.* 1998; Grimme *et al.* 1998; Pulm *et al.* 1997). To treat multiply excited states and bond dissociation, the program was developed further towards a general-utility one-component CI (Grimme and Waletzke 1999). Double counting of dynamic correlation is avoided by exponential scaling of off-diagonal matrix elements. In order to ease the computational effort, resolution of the identity (RI)-approximated two-electron integrals are employed as available from the TURBOMOLE package (Weigend and Häser 1997; Weigend *et al.* 1998).

In the DFT/SCI and DFT/MRCI programs, matrix elements of spin-free one- and two-electron Hamiltonians are evaluated according to the ansatz by Wetmore and Segal. These authors introduced patterns which reduce the number of possible spin-couplings to an extent that all coupling patterns can be calculated in advance and stored (Segal *et al.* 1978; Wetmore and Segal 1975). Kleinschmidt and Marian (2000) showed that this procedure can easily be extended to effective one-electron spin–orbit Hamiltonians. By means of the Wigner–Eckart theorem the effort can be reduced further. For states of equal multiplicity ($\Delta S = 0$), only matrix elements of the $\hat{s}_0$ operator have to be considered explicitly. Likewise, it is sufficient to set up arrays for $\hat{s}_+$ if $\Delta S = \pm 1$. Routines for the computation of these arrays have already been implemented in a spin–orbit extended version of the DFT/MRCI program.

An implementation of spin–orbit coupling in the framework of the LCGTO-FF method was developed by Rösch's group. Their first approach makes use of a transformation of the kinetic energy and the external potential only (Mayer 1999), featuring an extension of the programs to complex two-component spinors in a double-group formalism. Very recently (Mateev *et al.* 2002; Mayer *et al.* 2002), the classical Coulomb interaction was also included in the transformation, which made accurate calculation of the spin–orbit splitting and binding energies of actinoid complexes feasible.

3.3 Transformed Hamiltonians: Applications

3.3.1 Small molecules

In order to show that transformed Hamiltonians are useful even for very heavy systems, numerous case studies have been carried out in the last decade. Due to space limitations, only a small number can be reviewed explicitly.

Table 3.1 Equilibrium distance of the Au_2 molecule obtained in a CCSD(T) calculation using the DKH Hamiltonian with a basis set including up to i functions; 34 electrons are correlated, and the results have been counterpoise corrected. In the lower part of the table we give the deviations from the best result which are obtained if a less demanding treatment is undertaken (Hess and Kaldor 2000).

	R_e (pm)	ω_e (cm^{-1})	D_0 (eV)
CCSD(T)	248.8	186.9	2.19
experiment	247.2	191	2.29
	ΔR_e	$\Delta\omega_e$	ΔD_0
neglect of semicore correlation	1.0	−2.7	−0.05
neglect of triples	0.6	−0.3	−0.21
omitting counterpoise correction	−1.8	7.4	
omitting g, h, i FCTS	1.4	−4.5	−0.10
omitting h, i FCTS	0.4	−0.3	−0.01
omitting i FCTS	0.1	−0.2	0.00

A comparison of different methods was undertaken for the hydride of element 111 (Seth *et al.* 1996). The conclusion of this study was that Dirac–Fock calculations, all-electron DKH calculations and relativistic pseudopotential calculations give very similar results, showing that relativistic effects are also well described in the more approximate methods. A large relativistic bond length contraction of about 50 pm was found, which makes the bond length of (111)H even slightly shorter than that of AuH, which is 152.4 pm, with a relativistic effect of the order of 20 pm (see Kaldor and Hess 1994).

A recent study on the benchmark molecule Au_2 (Hess and Kaldor 2000) shows that in particular the proper treatment of correlation is crucial for also getting reliable structural data for heavy molecules. Even a basis set including up to i functions was found to lead to a deviation of 1.6 pm from experiment for the equilibrium distance. Semicore (5p) correlation, triples contribution and counterpoise correction prove to be important for a reliable determination of the electronic structure of Au_2 (see Table 3.1). The paper features a sizeable bibliography of earlier benchmark calculations on this molecule carried out with a large variety of methods.

A pilot calculation on CdH using one-, two- and four-component Fock space relativistic coupled-cluster methods has been published by Eliav *et al.* (1998b). The calculated values obtained were in very good agreement with experiment. While the four-component method gives the best results, one- and two-component calculations include almost all the relativistic effects.

The LCGTO-DF method was used to calculate electronic and spectroscopic properties of the monoxides and monocarbonyls of Ni, Pd and Pt (Chung *et al.* 1995). Substantial relativistic effects have been found for the metal–ligand distance in the Pt compounds. At the relativistic level the Pt–O distance was calculated to be 172 pm, which is 20 pm shorter than in the nonrelativistic calculations, comparing well with the experimental result of 173 pm. The bond in PtO is found to be 9 pm shorter than in the corresponding Pd compounds, for which the relativistic effects are smaller (6–7 pm), albeit nonnegligible. The shortening of the bond is in line with a considerable strengthening of the bond (e.g. by 2.57 eV in PtO). Similar results were obtained for PtCO, with a relativistic shortening of roughly the same size. For the carbonyls of Ni, Pd and Pt, a study has been carried out (Chung *et al.* 1996a) that carefully analyses their bonding mechanisms. The bond is found to be dominated by the π back-donation mechanism, which for Pd and Pt is considerably reinforced by relativistic effects.

3.3.2 Metal clusters and metal complexes

The scalar relativistic LCGTO-DF method mentioned above was used to study a large variety of metal clusters in order to investigate the development of bulk properties when the clusters get larger and larger. Gold clusters with up to 147 atoms have been investigated using this self-consistent all-electron method (Häberlen *et al.* 1997). For the mean bond length and atomization energy, scaling properties were determined which were found to be similar to those of the lighter transition metals Ni and Pd (Krüger *et al.* 1997; Xiao *et al.* 1999).

Relativistic effects on metal–ligand interactions were studied in the case of Mo–N bonds on $R_3Mo(III)$ complexes with N_2 (Neyman *et al.* 1997). An unusually large (for a second-row transition metal) relativistic effect was found, leading to strong Mo–N bonds with a marked lowering of the reaction barrier for N_2 cleavage.

A more complex situation was found for Na_6Pb. The unusual stability of this cluster was explained by comparison with the analogous Mg compound (Albert *et al.* 1995) and was found to originate in a larger charge transfer to the more electronegative lead and a larger polarizability of the Pb atom. In a combined experimental and theoretical study it was shown for the clusters Na_xAu and Cs_xAu that certain properties of the bulk are qualitatively present at the level of small clusters (Heiz *et al.* 1995). While the Na compounds show metallic behaviour, and the electronic structure can be described by means of the jellium model, in the Cs–Au clusters an ionic bond is most prominent.

Several gold cluster compounds were investigated theoretically in collaboration with experimental projects in the ‘Schwerpunkt’ (compare Chapter 7). Again, the scalar relativistic LCGTO-DF method was used. In the case of trigold oxonium the dimerization behaviour dependent on the ligands was studied (Chung *et al.* 1996b). Preparation and calculation of the electronic structure of a novel anionic gold–indium cluster was reported in Gabbaï *et al.* (1997). Geometry and electronic structure of gold phosphine thiolate complexes were reported by Krüger *et al.* (2000).

In the investigations of clusters with iron carbonyl fragments (Albert *et al.* 1996; Sinzig *et al.* 1998; Stener *et al.* 1999) the magnetic properties were at the focus of interest. In the case of $[M_4\{Fe(CO)_4\}_4]^{4-}$, M = Cu, Ag, Au, a sizeable reduction of magnetism of the bare metal clusters M_4Fe_4 upon carbonylation was found, in analogy with the Ni carbonyl clusters. The planar quadratic structure of the metal core was explained by the oxidation state I of the noble metal. $[Pt_3Fe_3(CO)_{15}]^-$ and $[Ag_{13}\{Fe(CO)_4\}_8]^{4-}$ were identified as interesting candidates for molecular magnetism. In a combined experimental and theoretical study it was shown that the magnetic moment of these compounds is not due to unpaired spins of d electrons, but rather due to unpaired electrons which are delocalized over the whole metal cluster.

3.3.3 Properties depending on spin–orbit coupling

Nuclear magnetic resonance and *g* tensors

The so-called *heavy-atom chemical shift* of light nuclei in nuclear magnetic resonance (NMR) had been identified as a spin–orbit effect early on by Nomura *et al.* (1969). The theory had been formulated by Pyykkö (1983) and Pyper (1983), and was previously treated in the framework of semi-empirical MO studies (Pyykkö *et al.* 1987). The basis for the interpretation of these spin–orbit effects in analogy to the Fermi contact mechanism of spin–spin coupling has been discussed by Kaupp *et al.* (1998b).

For these spin–orbit and spin–spin effects, Nakatsuji *et al.* (1995) have formulated a UHF-based theory using the Breit–Pauli Hamiltonian and recently extended it to include scalar relativistic effects using the Douglas–Kroll transformation (Ballard *et al.* 1996). A large number of applications have since been reported by the Kyoto group. It already turns out that for the chlorine compounds, the inclusion of spin-coupling terms is necessary to bring the calculations into agreement with experiments, the largest contributions being due to the Fermi contact term.

The calculation of spin–orbit corrections to NMR shielding constants has also been recently implemented in the framework of density functional theory (Malkin *et al.* 1996) and developed further in the groups of Kaupp and Malkin. Within the DEMON code, the third-order perturbation method for DFT-IGLO (density functional theory – individual gauge for localized orbitals) calculations of spin–orbit corrections to NMR chemical shifts was extended to include the full one- and two-electron spin–orbit operators (Malkina *et al.* 1998) as well as the atomic mean-field approximation (Hess *et al.* 1996) by including the AMFI program (Schimmelpfennig 1996). As shown in Table 3.2 for hydrogen halides and methyl halides, the corresponding approximation allows us to calculate the spin–orbit contributions with excellent accuracy at almost negligible extra cost over one-electron operators. Basis-set and gauge-origin effects are found to exceed the error introduced by the mean-field approximation by more than an order of magnitude. As expected from analogy with spin–orbit splittings, the two-electron contributions are substantial (30–35%) in the lighter compounds, and considerably smaller for the heavier halogens (6–7% for iodine).

Numerous recent applications have been reported which were carried out either with the all-electron mean-field spin–orbit approach or using pseudopotentials; see the recent reviews by Kaupp *et al.* (1998a) and Bühl *et al.* (1999).

The analogy with the Fermi contact interaction predicts particularly large spin–orbit effects when a large-s character of the bonding of the NMR atom to the heavy atom is present. This has been shown—e.g. for the PI_4^+ cation (Kaupp *et al.* 1999),

Table 3.2 Spin–orbit corrections to NMR shieldings (in ppm)[a].

		σ_{nonrel}	1e[b]	1e + 2e[c]	mean-field[d]	total	exp.
^1H	HF	29.09	0.17	0.12	0.12	29.21	28.89
^1H	HCl	30.98	0.91	0.74	0.73	31.71	31.06
^1H	HBr	31.27	4.86	4.38	4.38	35.65	34.96
^1H	HI	31.74	13.30	12.38	12.37	44.11	43.86
^{13}C	CH_3F	119.98	0.66	0.47	0.50	120.48	116.8
^{13}C	CH_3Cl	162.78	1.95	1.62	1.62	164.40	162.5
^{13}C	CH_3Br	171.09	10.73	9.74	9.66	180.75	178.5
^{13}C	CH_3I	188.76	21.92	20.58	20.56	209.32	212.1

[a]Malkina *et al.* (1998).
[b]One-electron Breit–Pauli spin–orbit operator.
[c]One- and two-electron Breit–Pauli spin–orbit operators.
[d]Mean-field approximation.

for which the calculations predicted an unprecedented high-field ^{31}P NMR shift of below −500 ppm—with an SO contribution of more than −700 ppm. The spin–orbit-induced high-field shift has subsequently been confirmed by solid-state NMR on a number of salts of this cation. It turned out that diiodine bridges reduce the phosphorus s character of the P–I bonds and thus the large SO shifts for *noninnocent* counteranions, and thus the spin–orbit shifts are probes of the bonding conditions. Earlier claims for the existence of PI_5 could be refuted.

The mean-field method for calculating screened spin–orbit integrals has also been employed for the calculation of g tensors, the electron paramagnetic resonance (EPR) analogue of NMR chemical shifts (Malkina *et al.* 2000). In close analogy with the NMR chemical shift implementation, second-order perturbation theory is employed in the framework of the DEMON code. The dominant paramagnetic contribution to the g tensor is due to the cross term between spin–orbit coupling and the orbital Zeeman term. Therefore, the proper treatment of spin–orbit coupling is mandatory even for the treatment of g tensors in light-element compounds. In addition to this cross term, the one-electron contribution to the gauge-correction term (analogous to the second-order term for spin–orbit corrections to NMR chemical shifts) and the relativistic mass correction to the spin Zeeman term have been included.

While this work is based on density-functional theory, recent *ab initio* work on g tensors comprises contributions from Bruna *et al.* (1997) and Engström *et al.* (1998).

Electronic spectra of thioketones

In order to test the validity of the inherent approximations in the spin–orbit mean-field and the DFT/MRCI approaches, electronic spectra and transition rates for spin-allowed as well as spin-forbidden radiative processes were determined for two thioketones, namely dithiosuccinimide and pyranthione (Tatchen 1999; Tatchen *et al.* 2001). In either case absorption and emission spectra as well as depletion rates for the first excited triplet state (T_1) have been measured experimentally (Meskers *et al.* 1995; Sinha *et al.* 1993; Szymanski *et al.* 1987; Taherian and Maki 1983). Spin-independent properties such as electronic excitation energies and dipole (transition) moments are computed by means of the DFT/MRCI program (Grimme and Waletzke 1999). Spin–orbit coupling matrix elements were calculated with the BnSoc package (Hess *et al.* 2000) for LS-coupled MRD-CI states and phosphorescence lifetimes are determined at the level of quasidegenerate perturbation theory.

The DFT/MRCI approach reproduces excitation energies and other spin-independent properties of experimentally known electronic states of pyranthione and dithiosuccinimide excellently. As far as phosphorescence lifetimes of dithiosuccinimide are concerned, calculations have not yet been completed. For the T_1 state of pyranthione we find that phosphorescence and nonradiative decay via intersystem-crossing to the S_0 state are concurrent processes occurring at approximately equal rates in the range of 10^4 s^{-1}, in good accord with experimental data. The $T_1 \rightarrow S_0$ radiative transition borrows its intensity from two sources:

(1) direct spin–orbit coupling of the S_0 and T_1 levels combined with the large dipole moment difference between these states, and

(2) the strong $S_2 \rightarrow S_0$ fluorescence.

The computed spin–orbit splitting in the T_1 state of $D = -18$ cm^{-1} is mainly due to interaction with the close by T_2 state. A rapid depletion of the S_1 state via intersystem crossing to the T_1 state can be mediated by the T_2 state if spin relaxation within the triplet levels is fast.

3.4 Valence-Only Effective Hamiltonians

A further reduction of the computational effort in investigations of electronic structure can be achieved by the restriction of the actual quantum chemical calculations to the valence electron system and the implicit inclusion of the influence of the chemically inert atomic cores by means of suitable parametrized effective (core) potentials (ECPs) and, if necessary, effective core polarization potentials (CPPs). Initiated by the pioneering work of Hellmann and Gombas around 1935, the ECP approach developed into two successful branches, i.e. the model potential (MP) and the pseudopotential (PP) techniques. Whereas the former method attempts to maintain the correct radial nodal structure of the atomic valence orbitals, the latter is formally based on the so-called pseudo-orbital transformation and uses valence orbitals with a simplified radial nodal structure, i.e. pseudovalence orbitals. Besides the computational savings due to the elimination of the core electrons, the main interest in standard ECP techniques results from the fact that they offer an efficient and accurate, albeit approximate, way of including implicitly, i.e. via parametrization of the ECPs, the major relativistic effects in formally nonrelativistic valence-only calculations. A number of reviews on ECPs has been published and the reader is referred to them for details (Balasubramanian 1998; Bardsley 1974; Chelikowsky and Cohen 1992; Christiansen *et al.* 1988; Cundari *et al.* 1996; Dixon and Robertson 1978; Dolg 2000; Ermler *et al.* 1985; Frenking *et al.* 1996; Gropen 1988; Hibbert 1982; Huzinaga 1991, 1995; Kahn 1984; Krauss and Stevens 1984; Laughlin and Victor 1988; Pickett 1989; Pitzer 1984; Pyykkö and Stoll 2000; Seijo and Barandiarán 1999; Weeks *et al.* 1969).

In ECP theory an effective Hamiltonian approximation for the all-electron no-pair Hamiltonian $\mathcal{H}_{\mathrm{np}}$ is derived which (formally) only acts on the electronic states formed by n_{v} valence electrons in the field of N frozen closed-shell atomic-like cores:

$$H_v = \sum_{i}^{n_{\mathrm{v}}} h_{\mathrm{v}}(i) + \sum_{i<j}^{n_{\mathrm{v}}} g_{\mathrm{v}}(i,j) + V_{\mathrm{cc}} + V_{\mathrm{CCP}}. \tag{3.8}$$

The subscripts 'c' and 'v' denote core and valence, respectively. h_{v} and g_{v} are effective one- and two-electron operators, V_{cc} represents the repulsion between all cores and nuclei of the system, and V_{CCP} denotes the CPPs. The total number of electrons in the neutral system n and the number of valence electrons n_{v} are related by the charges of the nuclei Z_λ and the corresponding core charges Q_λ:

$$n_{\mathrm{v}} = n - \sum_{\lambda}^{N} (Z_\lambda - Q_\lambda).$$

Both scalar-quasirelativistic (one-component) and quasirelativistic (two-component) ECPs use a formally nonrelativistic model Hamiltonian

$$h_{\mathrm{v}}(i) = -\tfrac{1}{2}\Delta_i + V_{\mathrm{cv}}(i) \quad \text{and} \quad g_{\mathrm{v}}(i,j) = \frac{1}{r_{ij}}.$$

Relativistic contributions merely result only from the parametrization of the ECP V_{cv}, which describes the interaction of a valence electron with all nuclei and cores of the system. A simple superposition of atomic ECPs is usually applied to model the molecular ECP, with the Coulomb attraction between point charges as the leading term

$$V_{\mathrm{cv}}(i) = \sum_{\lambda}^{N} \left(-\frac{Q_\lambda}{r_{\lambda i}} + \Delta V_{\mathrm{cv}}^{\lambda}(\boldsymbol{r}_{\lambda i}) \right) + \cdots .$$

Similarly, the point-charge approximation is the leading term in the interaction between nuclei and cores

$$V_{\mathrm{cc}} = \sum_{\lambda<\mu}^{N} \left(\frac{Q_\lambda Q_\mu}{r_{\lambda\mu}} + \Delta V_{\mathrm{cc}}^{\lambda\mu}(r_{\lambda\mu}) \right) + \cdots . \tag{3.9}$$

Experience shows that a suitable parametrization of $\Delta V_{\mathrm{cv}}^{\lambda}$ and $\Delta V_{\mathrm{cc}}^{\lambda\mu}$ is usually able to compensate for all underlying approximations with sufficient accuracy.

3.4.1 Model potentials

The most straightforward approach to constructing an ECP is to use the Fock operator $\mathcal{F}_\mathrm{v}$ of a valence orbital φ_a^V and to model the effective one-particle Hartree–Fock potential by a simpler operator V_cv using the following identity:

$$-\sum_{\lambda}^{N}\frac{Z_\lambda}{r_{\lambda i}}+\sum_{c}(2J_c(i)-K_c(i))=\sum_{\lambda}^{N}\left(-\frac{Q_\lambda}{r_{\lambda i}}+\Delta V_\mathrm{cv}^{\lambda}(\boldsymbol{r}_{\lambda i})\right). \tag{3.10}$$

On the left-hand side the first sum runs over all nuclei λ with charge Z_λ and the second over all core orbitals c. J_c and K_c denote the usual Coulomb and exchange operators. Under the assumption of nonoverlapping cores the second sum on the left-hand side can also be regarded as a superposition of one-centre terms. Thus the above equality can be applied for each atom λ individually and, in the sense of the frozen-core approximation, before the molecular calculation. Another approximation follows from the goal that relativistic effects should be treated implicitly. Therefore, not only V_cv is approximated but also an additive relativistic correction term V_rel. In order to obtain the relevant atomic potentials $V_\mathrm{rel}^{\lambda}+V_\mathrm{cv}^{\lambda}$ a two- or one-component quasirelativistic atomic all-electron calculation is performed. The most widely used variant of the method are the *ab initio* model potentials (AIMP) of Seijo, Barandiarán and co-workers (Barandiarán and Seijo 1992, 1994; Casarubios and Seijo 1998, 1999; Díaz-Megías and Seijo 1999; Seijo 1995; Seijo *et al.* 2001), where the quasirelativistic Hamiltonian proposed by Wood and Boring (1978), dubbed WB subsequently, for density functional calculations is used in the framework of Hartree–Fock theory according to the scheme outlined by Cowan and Griffin (1976), which in turn will be denoted by CG. The WB and CG approaches correspond essentially to the use of an energy-dependent one-particle Hamiltonian, which results from the elimination of the small components from the Dirac equation, within the Hartree–Fock scheme, disregarding any resulting nonorthogonality between orbitals of equal lj.

The AIMP method in its present form starts from a quasirelativistic all-electron Hartree–Fock calculation for the atom under consideration in a suitable electronic state and approximates the operators on the left-hand side of Equation (3.10) for an atomic core λ as described in the following.

The long-range local Coulombic (C) part is spherical and is represented by a linear combination of Gaussians with prefactors $1/r$, i.e. a local radial MP

$$-\frac{Z_\lambda-Q_\lambda}{r_{\lambda i}}+2\sum_{c\in\lambda}J_c^{\lambda}(i)=\frac{1}{r_{\lambda i}}\sum_{k}C_k^{\lambda}\mathrm{e}^{-\alpha_k^{\lambda}r_{\lambda i}^2}=\Delta V_\mathrm{C}^{\lambda}(i).$$

The exponents α_k^{λ} and coefficients C_k^{λ} are adjusted to the all-electron potential in a least-squares sense under the constraint that $\sum_k C_k^{\lambda}=Z_\lambda-Q_\lambda$ in order to enforce the correct asymptotic behaviour of the MP. Since the evaluation of integrals over such a local potential is not costly, any desired accuracy can be easily achieved by using a sufficiently long expansion. The nonlocal exchange (X) part is substituted by its spectral representation in the space defined by a set of functions χ_p^λ centred on core λ

$$-\sum_{c\in\lambda} K_c^\lambda(i) = \sum_{p,q} |\chi_p^\lambda(i)\rangle A_{pq}^\lambda \langle \chi_q^\lambda(i)| = \Delta V_{\mathrm{X}}^\lambda(i). \tag{3.11}$$

This operator yields the same one-centre integrals as the true core exchange operator as long as the basis functions can be represented by the set of the χ_p^λ. Two- and three-centre integrals are approximated. Since, in contrast to the Coulomb part, the exchange part is short ranged, a moderate number of functions χ_p^λ is needed and the one-centre approximation is expected to be very good, at least for not-too-large cores. In practical applications the basis used in the spectral representation is chosen to be identical to the primitive functions of the valence basis set used for the atom under consideration and the A_{pq}^λ are calculated during the input processing of each AIMP calculation.

With the Coulomb and exchange parts of the MP discussed so far, the core-like solutions of the valence Fock equation would still fall below the energy of the desired valence-like solutions. In order to prevent the valence-orbitals collapsing into the core during a variational treatment and to retain an Aufbau principle for the valence electron system, the core-orbitals are moved to higher energies by means of a shift operator

$$P^\lambda(i) = \sum_{c\in\lambda} (D_c^\lambda) |\varphi_c^\lambda(i)\rangle\langle\varphi_c^\lambda(i)|.$$

Here the φ_c^λ denote the core orbitals localized on core λ. For practical calculations they are represented by a sufficiently large (all-electron) basis set. In principle, only $D_c^\lambda \to \infty$ would effect a strict orthogonality between core and valence orbitals; however, the more or less arbitrary choice $D_c^\lambda = -2\epsilon_c^\lambda$ is usually made for numerical reasons. With this choice there is no strict orthogonality between core and valence orbitals, but the resulting errors are expected to be small.

The scalar-relativistic and relativistic extensions of the AIMP approach are called CG-AIMP (Cowan–Griffin) and WB-AIMP (Wood–Boring), respectively. In the CG-AIMP approach the mass–velocity and Darwin operators are cast together with the exchange terms into their spectral representation Equation (3.11). The valence orbital energies $\epsilon_{n\kappa}$ are kept fixed during the extraction process and are also used for any semi-core orbitals of the same κ, which are included in the AIMP valence space. A similar strategy is followed for dealing with the first derivative of the valence orbital in the Darwin term. It should be noted, however, that due to the use of relativistic core orbitals and core orbital energies, relativistic contributions are also present in the Coulomb and shift terms of the AIMP. The WB-AIMP method adds to this a representation of the SO operator in the form

$$\Delta V_{\mathrm{cv,so}}^\lambda(i) = \sum_l \left(\sum_k \frac{B_{lk}^\lambda}{r_{\lambda i}^2} \mathrm{e}^{-\beta_{lk}^\lambda r_{\lambda i}^2} \right) \mathcal{P}_l^\lambda(i) \boldsymbol{l}_{\lambda i} \boldsymbol{s}_i \mathcal{P}_l^\lambda(i),$$

Table 3.3 Bond lengths R_e (Å), vibrational constants ω_e (cm^{-1}) and binding energies D_e (eV) of group 5 monoxides from *ab initio* model potential (AIMP) MCPF calculations with an explicit treatment of relativistic effects in the valence shell using the Douglas–Kroll–Hess (DKH) Hamiltonian. Comparison is made with corresponding all-electron DKH MCPF results using basis sets of the same quality (Rakowitz *et al.* 1999a), *ab initio* energy-consistent pseudopotential (EC-PP) ACPF results (Dolg *et al.* 1993b) and experimental data (Exp.).

Molecule	State	Method	R_e	ω_e	D_e
NbO	$^4\Sigma^-$	AIMP	1.680	1022	7.21
		EC-PP	1.675	1033	6.91
		AE DKH	1.676	1022	7.23
		Exp.		989	7.93 ± 0.26
TaO	$^4\Sigma^-$	AIMP	1.706	981	7.11
		EC-PP	1.701	1004	6.91
		AE DKH	1.710	990	7.03
	$^2\Delta$	AIMP	1.689	1026	7.53
		EC-PP	1.691	1023	7.67
		Exp.	1.686	1030	8.24 ± 0.13

where $\boldsymbol{l}_{\lambda i} = \boldsymbol{r}_{\lambda i} \times \boldsymbol{p}_i$ and $\boldsymbol{s}_i$ denote the operators of orbital angular momentum and spin, respectively, and $\mathcal{P}_l^\lambda$ is the projection operator onto the subspace of angular quantum number l with respect to core λ. The coefficients B_{lk}^λ and exponents β_{lk}^λ are determined by means of a least-squares fit to the radial components of the Wood–Boring SO term. AIMP parameters and corresponding basis sets are available on the Internet.[1]

The MP approach shifts the core orbitals in the virtual orbital space and yields valence orbitals with the same nodal structure as the all-electron orbitals. It is therefore possible to combine the MP approach with an explicit treatment of relativistic effects in the valence shell, for example, in the framework of the DKH no-pair Hamiltonian (Rakowitz *et al.* 1999a,b). Corresponding *ab initio* model potential parameters are available on the Internet.[2] It remains to be seen if the additional effort of an explicit treatment of (essentially the direct relativistic effects) in the valence shell leads to a higher accuracy when compared with the implicit treatment using model potentials derived from DKH atomic calculations. Although a strict separation between direct and indirect relativistic contributions is not possible, it is obvious that the indirect effects originating from the core are still provided by the MP. A recent comparison between AIMP calculations with an explicitly relativistic valence model Hamiltonian and corresponding AE DKH calculations on group 5 transition-metal monoxides demonstrates the accuracy of the AIMP approach (Table 3.3). Older PP results using basis sets of slightly lower quality are also included for comparison.

[1] http://www.qui.uam.es/Data/AIMPLibs.html

[2] http://www.thch.uni-bonn.de/tc/TCB.download.html

3.4.2 Pseudopotentials

The analytical forms of the modern PPs used today have little in common with the formulas we obtain by a strict derivation of the theory (Dolg 2000). Formally, the pseudo-orbital transformation leads to nodeless pseudovalence orbitals for the lowest atomic valence orbitals of a given angular quantum number l (one-component) or lj (two-component). The simplest and historically the first choice is the local ansatz for $\Delta V_{\text{cv}}^{\lambda}$ in Equation (3.4). However, this ansatz turned out to be too inaccurate and therefore was soon replaced by a so-called semilocal form, which in two-component form may be written as

$$\Delta V_{\text{cv}}^{\lambda}(\boldsymbol{r}_{\lambda i}) = \sum_{l=0}^{L-1} \sum_{j=|l-1/2|}^{l+1/2} (V_{lj}^{\lambda}(r_{\lambda i}) - V_{L}^{\lambda}(r_{\lambda i}))P_{lj}^{\lambda}(i) + V_{L}^{\lambda}(r_{\lambda i}).$$

P_{lj}^{λ} denotes a projection operator on spinor spherical harmonics centred at the core λ

$$P_{lj}^{\lambda}(i) = P_{l,l\pm 1/2}^{\lambda}(i) = P_{\kappa}^{\lambda}(i) = \sum_{m_j=-j}^{j} |\lambda l j m_j(i)\rangle\langle\lambda l j m_j(i)|.$$

For scalar-quasirelativistic calculations, i.e. when spin–orbit coupling is neglected, a one-component form may be obtained by averaging over the spin

$$\Delta V_{\text{cv,av}}^{\lambda}(\boldsymbol{r}_{\lambda i}) = \sum_{l=0}^{L-1} (V_{l}^{\lambda}(r_{\lambda i}) - V_{L}^{\lambda}(r_{\lambda i}))P_{l}^{\lambda}(i) + V_{L}^{\lambda}(r_{\lambda i}).$$

The projection operator P_{l}^{λ} refers now to the spherical harmonics centred at the core λ

$$P_{l}^{\lambda}(i) = \sum_{m_l=-l}^{l} |\lambda l m_l(i)\rangle\langle\lambda l m_l(i)|.$$

An SO operator may be defined as

$$\Delta V_{\text{cv,so}}^{\lambda}(\boldsymbol{r}_{\lambda i}) = \sum_{l=1}^{L-1} \frac{\Delta V_{l}^{\lambda}(r_{\lambda i})}{2l+1}[l P_{l,l+1/2}^{\lambda}(i) - (l+1)P_{l,l-1/2}^{\lambda}(i)],$$

which contains essentially the difference between the two-component PPs

$$\Delta V_{l}^{\lambda}(r_{\lambda i}) = V_{l,l+1/2}^{\lambda}(r_{\lambda i}) - V_{l,l-1/2}^{\lambda}(r_{\lambda i}).$$

For some calculations (see below) it is advantageous to separate space and spin

$$\Delta V_{\text{cv,so}}^{\lambda}(\boldsymbol{r}_{\lambda i}) = \sum_{l=1}^{L-1} \frac{2\Delta V_{l}^{\lambda}(r_{\lambda i})}{2l+1} P_{l}^{\lambda}(i)\boldsymbol{l}_{\lambda i}\boldsymbol{s}_{i} P_{l}^{\lambda}(i).$$

The potentials V_{lj}^{λ} and V_l^{λ} ($l = 0$ to $l = L$) and ΔV_l^{λ} ($l = 1$ to $l = L - 1$) are either represented as a linear combination of Gaussians multiplied by powers of the electron–core distance or alternatively cast into a nonlocal representation in a (nearly) complete auxiliary basis set.

Relativistic PPs to be used in four-component Dirac–Hartree–Fock and subsequent correlated calculations can also be successfully generated and used (Dolg 1996a); however, the advantage of obtaining accurate results at a low computational cost is certainly lost within this scheme. Nevertheless, such potentials might be quite useful for modelling a chemically inactive environment in otherwise fully relativistic all-electron calculations based on the Dirac–Coulomb–(Breit) Hamiltonian.

3.4.3 Shape-consistent pseudopotentials

The origin of shape-consistent PPs (SC-PPs) (Christiansen *et al.* 1979; Durand and Barthelat 1975) lies in the insight that only the admixture of core orbitals to valence orbitals in order to remove the radial nodes leads to pseudovalence orbitals that are too contracted and finally as a consequence to poor molecular results, for example, to bond distances that are too short. About 25 years ago it was recognized that it is indispensable to have the same shape of the pseudovalence orbital and the original valence orbital in the spatial valence region, where chemical bonding occurs. Formally, this also requires an admixture of virtual orbitals. Since these are usually not obtained in finite-difference atomic calculations, another approach was developed. The starting point is an atomic all-electron calculation at the nonrelativistic, scalar-relativistic or quasirelativistic Hartree–Fock or the Dirac–Hartree–Fock level. In the latter case the small components are discarded and the large components of the energetically lowest valence shell of each quantum number lj are considered as valence orbitals after renormalization. To generate the pseudovalence orbitals $\varphi_{\mathrm{p},lj}$ the original valence orbitals $\varphi_{\mathrm{v},lj}$ are kept unchanged outside a certain matching radius r_{c} separating the spatial core and valence regions (shape-consistency; exactly achieved only for the reference state), whereas inside the matching radius the nodal structure is discarded and replaced by a smooth and, in the interval $[0, r_{\mathrm{c}}]$, nodeless polynomial expansion $f_{lj}(r)$:

$$\varphi_{\mathrm{v},lj}(r) \rightarrow \varphi_{\mathrm{p},lj}(r) = \begin{cases} \varphi_{\mathrm{v},lj}(r) & \text{for } r \geqslant r_{\mathrm{c}}, \\ f_{lj}(r) & \text{for } r < r_{\mathrm{c}}. \end{cases}$$

The free parameters in f_{lj} are determined by normalization and continuity conditions, for example, matching of f_{lj} and $\varphi_{\mathrm{v},lj}$ as well as their derivatives at r_{c}. The choice of r_{c} as well as the choice of f_{lj} is in certain limits arbitrary and a matter of experience.

Having a nodeless and smooth pseudovalence orbital $\varphi_{\mathrm{p},lj}$ and the corresponding orbital energy $\epsilon_{\mathrm{v},lj}$ at hand, the corresponding radial Fock equation

$$\left(-\frac{1}{2}\frac{\mathrm{d}^2}{\mathrm{d}r^2} + \frac{l(l+1)}{2r^2} + V_{lj}^{\mathrm{PP}}(r) + W_{\mathrm{p},lj}[\{\varphi_{p',l'j'}\}]\right)\varphi_{\mathrm{p},lj}(r) = \epsilon_{\mathrm{v},lj}\varphi_{\mathrm{p},lj}(r)$$

can be solved pointwise for the unknown PP V_{lj}^{PP} for each combination lj of interest. The term $W_{\mathrm{p},lj}$ stands for an effective valence Coulomb and exchange potential for $\varphi_{\mathrm{p},lj}$. Relativistic effects enter the potentials implicitly via the value of the orbital energy $\epsilon_{\mathrm{v},lj}$ and the shape of the pseudovalence orbital outside the matching radius. The resulting potentials V_{lj}^{PP} are tabulated on a grid and are usually fitted to a linear combination of Gaussian functions. SC-PPs including SO operators based on Dirac–Hartree–Fock calculations using the Dirac–Coulomb Hamiltonian have been generated by Christiansen, Ermler and co-workers (Blaudeau and Curtiss 1997; Ermler *et al.* 1991; Hurley *et al.* 1986; LaJohn *et al.* 1987; Nash *et al.* 1997; Pacios and Christiansen 1985; Ross *et al.* 1994, 1990; Wallace *et al.* 1991; Wildman *et al.* 1997). The potentials and corresponding valence basis sets are available on the Internet.[3] A similar set for main group and transition elements based on scalar-relativistic Cowan–Griffin all-electron calculations was published by Hay and co-workers (Hay 1983; Hay and Martin 1998; Hay and Wadt 1985a,b; Wadt and Hay 1985). Another almost complete set has been published by Stevens and co-workers (Cundari and Stevens 1993; Stevens *et al.* 1984, 1992).

3.4.4 Energy-consistent pseudopotentials

Energy-consistent *ab initio* PPs (EC-PPs) developed from energy-adjusted semi-empirical PPs, i.e. pseudopotentials which were fitted to reproduce the experimental low-energy atomic spectrum. Since it is usually not possible to account with sufficient accuracy for valence correlation effects, such semi-empirical energy-adjustment was only applicable for one-valence-electron systems. Results for alkaline and alkaline-earth systems obtained with one- and two-valence-electron PPs augmented by CPPs were excellent, especially for atoms and relatively weakly bound molecules, for example, dimers or clusters. However, due to the limited validity of the frozen-core approximation when going from a highly charged one-valence-electron ion to a neutral atom or nearly neutral ion, the one-valence-electron adjustment was bound to fail for other elements, especially for transition metals where a small core has to be chosen. Nevertheless, the idea to fit exclusively to quantum mechanical observables like total valence energies instead of relying on quantities like orbitals and orbital energies, which are only meaningful in an approximate one-particle picture, is very appealing. Therefore, the approach was extended to a many-valence-electron adjustment within a purely *ab initio* framework (Dolg *et al.* 1987). Essentially, any (relativistic) Hamiltonian, coupling scheme and valence correlation treatment may be chosen to generate the all-electron reference data, provided that the same quality of the valence-only wave function is used during the PP adjustment. Moreover, the formalism can be used to generate one-, two- and also four-component PPs at any desired level of relativity (nonrelativistic Schrödinger, or relativistic Wood–Boring, Douglas–Kroll–Hess, Dirac–Coulomb or Dirac–Coulomb–Breit Hamiltonian; implicit or explicit treatment of relativity in the valence shell).

[3] http://www.clarkson.edu/~pac/reps.html

Table 3.4 Bond lengths R_e (Å), vibrational constants ω_e (cm^{-1}) and binding energies D_e (eV) of halogen dimers from *ab initio* energy-consistent (EC-PP) CCSD(T) calculations including a core-polarization potential and corrections for spin–orbit effects (Dolg 1996b). Comparison is made to experimental values (Exp.). Total valence correlation energies E_c (mH) from EC-PP and nonrelativistic all-electron (AE) calculations as well as differential correlation contributions ΔE_c (mH) to the binding energies D_e (Dolg 1996c). In all cases extended all-electron basis sets including up to g-type functions were applied.

	F_2	Cl_2	Br_2	I_2	At_2
R_e (EC-PP)	1.409	1.982	2.281	2.668	2.979
R_e (Exp.)	1.412	1.988	2.281	2.666	
D_e (EC-PP)	1.66	2.44	1.95	1.57	0.80
D_e (Exp.)	1.66	2.51	1.99	1.56	
ω_e (EC-PP)	927	561	324	215	117
ω_e (Exp.)	917	560	325	215	
E_c (EC-PP)	615.0	505.7	424.0	386.1	
E_c (AE)	606.9	478.7	406.9	361.3	
ΔE_c (EC-PP)	107.0	45.9	42.6	35.9	
ΔE_c (AE)	105.7	46.5	41.7	34.1	

The current version of the EC-PP approach uses reference data derived from finite-difference state-averaged all-electron multiconfiguration Dirac–Hartree–Fock calculations based on the Dirac–Coulomb or Dirac–Coulomb–Breit Hamiltonian (Dolg *et al.* 1993b, 2001; Metz *et al.* 2000a,b; Stoll *et al.* 2001). These calculations are performed for a multitude of electronic configurations, states or levels I of the neutral atom and the low-charged ions. The total valence energies E_I^{AE} derived from these calculations define the PP parameters for a given ansatz in a least-squares sense. A corresponding set of finite-difference valence-only calculations (especially the same coupling scheme and correlation treatment has to be applied) is performed to generate the total valence energies E_I^{PP}, and the parameters are varied in such a way that the sum of weighted squared errors in the total valence energies becomes a minimum, i.e.

$$\sum_I w_I (E_I^{PP} - E_I^{AE})^2 := \min.$$

Parameters of *ab initio* EC-PPs and corresponding valence basis sets are available for almost all elements of the periodic table, including some superheavy elements (Andrae *et al.* 1990; Bergner *et al.* 1993; Dolg *et al.* 1987, 1989a,b, 1993a,b; Häussermann *et al.* 1993; Küchle *et al.* 1991; Metz *et al.* 2000a,b; Seth *et al.* 1997; Stoll *et al.* 2001). They are also available on the Internet.[4] Besides tests performed in the original publications, a number of additional molecular calibration studies has been carried

[4] http://www.theochem.uni-stuttgart.de

out in the past for EC-PPs, i.e. XH_4 (X = C, Si, Ge, Sn, Pb) (Steinbrenner *et al.* 1994), X_2H_6 (X = Si, Ge, Sn, Pb) (Nicklass and Stoll 1995), InCl and $InCl_3$ (Schwerdtfeger *et al.* 1995a), MF (M = K, Rb, Cs) (Leininger *et al.* 1996b), InX (X = H, F, Cl) (Leininger *et al.* 1996a), HX and X_2 (X = F, Cl, Br, I, At) (Dolg 1996b), MX (M = La, Lu, Ac, Lr; X = H, O, F) (Küchle *et al.* 1997), AuH (Schwerdtfeger *et al.* 2000). A selection of the results for the homonuclear halogen dimers is listed in Table 3.4.

Special attention was given to the accuracy of valence correlation energies obtained with pseudovalence orbitals, since due to the simplified nodal structure these quantities tend to be too large (Dolg 1996c,d). Although in single extreme cases the valence correlation energies are significantly overestimated, for example, by 20% in F^{5+}, the errors in total valence energies of neutral and low-charged systems are usually less than 10%. The correlation contributions to energy differences as binding energies, ionization potentials or electron affinities exhibit errors of less than 0.1 eV, even when large-core EC-PPs or SC-PPs are applied. The accuracy of total and differential valence correlation energies obtained from PP methods appears to be comparable with those of MP approaches. Note that in contrast to all-electron (frozen-core) calculations in both of these methods the core orbitals are present in the orbital space used in the correlation treatment.

3.4.5 Core–core/nucleus repulsion correction

In the case of large cores a correction to the point-charge repulsion model in Equation (3.9) is needed. A Born–Mayer-type ansatz

$$\Delta V_{\mathrm{cc}}^{\lambda\mu}(r_{\lambda\mu}) = B_{\lambda\mu}\exp(-b_{\lambda\mu}r_{\lambda\mu})$$

proved to be quite successful for the parametrization of core–core and core–nucleus repulsion corrections (CCRC, CNRC). For the CNRC the parameters $B_{\lambda\mu}$ and $b_{\lambda\mu}$ can be obtained by fitting directly to the electrostatic potential of the atomic core electron system λ and multiplying the resulting coefficient with the charge of nucleus μ, whereas for the CCRC the deviation from the point-charge model has to be determined by molecular Hartree–Fock or Dirac–Hartree–Fock calculations for each pair of frozen cores.

3.4.6 Core polarization potentials

Although the frozen-core approximation underlies all ECP schemes discussed so far, both static (polarization of the core at the Hartree–Fock level) and dynamic (core–valence correlation) polarization of the core may accurately and efficiently be accounted for by a core polarization potential (CPP). The CPP approach was originally used by Meyer and co-workers (Müller *et al.* 1984) for all-electron calculations and adapted by the Stuttgart group (Fuentealba *et al.* 1982) for PP calculations. The

CPP term accounting for the dipole polarizability α_λ of the core λ is given as

$$V_{\mathrm{CCP}} = -\frac{1}{2}\sum_{\lambda} \alpha_\lambda f_\lambda^2 + \sum_{\lambda,i} V^\lambda(i).$$

Here f_λ is the electric field at core λ generated by all other cores and nuclei as well as all valence electrons. Since the validity of the underlying multipole expansion breaks down for small distances from the core λ, the field has to be multiplied by a cut-off function:

$$f_\lambda = -\sum_i \frac{\boldsymbol{r}_{i\lambda}}{r_{i\lambda}^3}(1 - \exp(-\delta_{\mathrm{e}}^\lambda r_{i\lambda}^2))^{n_{\mathrm{e}}} + \sum_{\mu \neq \lambda} Q_\mu \frac{\boldsymbol{r}_{\mu\lambda}}{r_{\mu\lambda}^3}(1 - \exp(-\delta_c^\lambda r_{\mu\lambda}^2))^{n_c}. \quad (3.12)$$

In cases where ns and np valence orbitals are present together with $(n-1)$d and $(n-2)$f valence orbitals, for example, for Cs, it proved to be more accurate to augment the core-polarization potential by a short-range local potential (Dolg 1996a)

$$V^\lambda(i) = C^\lambda \exp(-\gamma^\lambda r_{i\lambda}^2).$$

The use of an l-dependent cut-off function in Equation (3.12) might even lead to superior results (Foucrault *et al.* 1992).

The use of CPPs to account for core–valence correlation effects of inner shells in combination with accurate relativistic small- or medium-core ECPs (Yu and Dolg 1997) may be a useful direction for future developments, especially in view of the large computational effort for an explicit treatment of core–valence correlation in case of d and/or f shells and the significant basis-set superposition errors occurring at the correlated level (Dolg *et al.* 2001).

3.4.7 Choice of the core

A critical point when adjusting and applying an ECP is the proper choice of the core. It is clear that the computational savings become larger for increasing size of the core; however, due to the limitations of the underlying frozen-core approximation the accuracy decreases at the same time. Whenever computationally feasible, small-core ECPs should be applied, for example, for transition metals the $(n-1)$spd and ns shells should be treated as valence shells (11–20 valence electrons for groups 3–12). The failures of large-core ECPs (3–12 valence electrons for groups 3–12) independent of the actual ECP approach are illustrated by a recent calibration study of AuH (Table 3.5).

Small-core ECPs are also found to be more reliable for alkaline and alkaline-earth elements (9 and 10 valence electrons for groups 1 and 2, respectively (Leininger *et al.* 1996b)) as well as for the early post-d-group elements (21 and 22 valence electrons for groups 13 and 14, respectively (Leininger *et al.* 1996a; Metz *et al.* 2000a,b)). Since the underlying d-shell becomes more core-like the later post-d-group elements may also be treated with large-core ECPs (4–8 valence electrons for groups 14–18);

Table 3.5 Bond length R_e (Å), vibrational constant ω_e (cm^{-1}) and binding energy D_e (eV) of gold hydride AuH at the Hartree–Fock level obtained with different pseudopotentials and all-electron (AE) approaches (Seth and Schwerdtfeger 2000). The parentheses following the acronym PP denote the reference data and the number of valence electrons for the Au PP.

	R_e		ω_e		D_e	
method	HF	MP2	HF	MP2	HF	MP2
AE, DC	1.570	1.484	2095	2521	1.79	3.21
AE, DKH	1.576	1.486	2068	2517	1.74	3.03
EC-PP(WB,19)[a]	1.575	1.480	2071	2523	1.70	3.16
SC-PP(DHF,19)[b]	1.579	1.484	2064	2518	1.69	3.14
SC-PP(DHF,19)[c]	1.570	1.477	2088	2542	1.72	3.19
SC-PP(CG,19)[d]	1.561	1.472	2130	2587	1.77	3.31
SC-PP(DHF,11)[e]	1.592	1.519	2024	2303	1.66	2.52
SC-PP(CG,11)[f]	1.626	1.547	1897	2156	1.43	2.45
AE nonrel.	1.830	1.694	1475	1716	1.10	1.96

[a]Andrae *et al.* (1990), [b]Stevens *et al.* (1992), [c]Ross *et al.* (1990), [d]Hay and Wadt (1985b), [e]Ross *et al.* (1990), [f]Hay and Wadt (1985a).

however a CPP has to be added to account for core–valence correlation in accurate calculations (Dolg 1996b; Nicklass *et al.* 1995; Steinbrenner *et al.* 1994).

3.5 Effective Core Potentials: Applications

The ECPs discussed so far, accounting for the most important relativistic effects including SO interaction, should be applied together with high-level wave-function-based correlation treatments in order to arrive at results close to experimental data. In contrast to scalar-relativistic calculations, in which after the initial integral evaluation step no differences occur compared with a nonrelativistic treatment, we have essentially three different strategies for correlated calculations including SO interaction: the SO contributions may be taken into account before, during or after the treatment of electron correlation effects. In the limit of complete one- and many-electron basis sets the results of all approaches have to be the same; however, the convergence with respect to this limiting result is quite system dependent. In practice, the first method is the most rigorous for heavy elements, but it is usually also the most expensive. Although the last approach is often computationally the cheapest, it has the smallest flexibility of the wave function with respect to SO contributions and is therefore mostly limited to not too heavy atoms. Finally, in view of discussions of SO contributions to molecular constants obtained from various valence-only and all-electron computational schemes, we have to make sure that the scalar relativistic level is defined in exactly the same way (see, for example, Dolg *et al.* 2001).

Table 3.6 Bond length R_e (Å), vibrational constant ω_e (cm^{-1}) and binding energy D_e (eV) of Eka-Au hydride (111)H without (with) counterpoise correction of the basis-set superposition error. All-electron (AE) values based on the Dirac–Coulomb–Hamiltonian (Seth and Schwerdtfeger 2000) are compared with valence-only results obtained with energy-consistent (EC) (Dolg *et al.* 2001) and shape-consistent (SC) (Han and Hirao 2000) pseudopotentials (PP). The numbers 19 and 34 in parentheses denote the number of valence electrons for the Eka-Au PP.

method	R_e	ω_e	D_e
AE, DHF	1.521	2743	1.56
EC-PP(19), SO, HF[a]	1.518(1.518)	2779(2779)	1.69(1.69)
SC-PP(19), SO, HF[a]	1.516	2786	1.59
AE, DHF+CCSD(T)	1.523	2674	2.83
EC-PP(19), SO, ACPF[a]	1.525(1.532)	2648(2616)	2.79(2.73)
EC-PP(19), ACPF + SO[b]	(1.531)	(2622)	(2.76)
EC-PP(19), CCSD(T) + SO[b]	(1.529)	(2642)	(2.83)
SC-PP(19), SO, CCSD(T)[a]	1.506	2721	3.16
SC-PP(34), SO, CCSD(T)[a]	1.512	2668	2.87

[a] Kramers-restricted two-component HF and subsequent correlation treatment.
[b] Scalar-relativistic one-component HF and correlation treatment, spin–orbit corrections added.

Small molecules

The most accurate and computationally demanding approach is to use relativistic ECPs, including the SO terms, already at the independent particle level (Esser 1984b; Esser *et al.* 1981; Hafner and Schwarz 1979). Due to the spin degrees of freedom, the spinor basis is twice as large as in the nonrelativistic case and usually complex. This complicates the subsequent integral transformation and correlation treatment. Kramers symmetry and double point group symmetry can be exploited both at the one- and many-electron function level to reduce the computational effort. The approach offers the largest flexibility of the wave function and is conceptually closest to all-electron treatments based on the Dirac–Coulomb–(Breit) Hamiltonian. It is certainly the best approach for systems containing atoms in which SO interaction leads to significant differences in the radial shape and energy of the $j = l - \frac{1}{2}$ and $j = l + \frac{1}{2}$ spinors, for example, in order to study the splitting of the $^2P_{1/2}$ and $^2P_{3/2}$ components of the Tl ground state (Visscher and Saue 2000). Recently correlated all-electron and shape-consistent as well as energy-consistent pseudopotential calculations of this type were performed for the spectroscopic constants of the monohydride of the superheavy element 111 (Table 3.6).

In many cases, less costly treatments of the SO interaction may be successful as well. Scalar-relativistic ECPs are employed at the independent particle level to generate a set of (real) orbitals which are used for the integral transformation. The SO term of the ECP is included in the calculations at the correlated level, i.e. electron correlation and spin–orbit effects are treated on an equal footing. Double group symmetry may again be applied to reduce the computational effort (Chang and Pitzer

1989; Tilson *et al.* 2000; Yabushita *et al.* 1999). This approach performs very well for systems with atoms in which SO interaction, despite a large energetic splitting between the $j = l - \frac{1}{2}$ and $j = l + \frac{1}{2}$ spinors, leads only to relatively small differences in their radial shape, for example, the f shells of lanthanide or actinide atoms. Usually, the space spanned by the many-electron function basis in the SO case is much smaller than what is feasible in calculations without SO coupling, i.e. the electron correlation treatment is less accurate. To account for these deficiencies, SO corrections derived from limited calculations with and without SO ECP may be added to scalar-relativistic energies obtained from high-quality correlation treatments, or spin-free state-specific shifts derived from such calculations may be added to the Hamiltonian matrix with SO coupling before diagonalization (Llusar *et al.* 1996).

Finally, the most efficient treatment, which works well in all cases where SO effects are small, is the inclusion of the SO part of the ECPs in the calculations at the latest possible stage, i.e. after the correlation treatment performed with scalar-relativistic ECPs in the nonrelativistic coupling scheme (Alekseyev *et al.* 1994b; Buenker *et al.* 1998; Teichteil and Spiegelmann 1983; Vallet *et al.* 2000). A low-dimensional ($O(10^2)$) complex Hamiltonian matrix including SO interaction is set up and diagonalized in the basis of the many-electron states obtained without SO interaction. Applying this approach, during the last decade Buenker and co-workers studied ground and excited states of numerous, mainly diatomic, main-group molecules and molecular ions, for example, BiF (Alekseyev *et al.* 1993), BiH (Alekseyev *et al.* 1994b), BiO (Alekseyev *et al.* 1994a) and BiS (Lingott *et al.* 1999). With a few exceptions shape-consistent PPs were used in connection with the MRD-CI correlation treatment, including perturbative selection of variationally treated configurations, energy-extrapolation to a full variational treatment and a multi-reference Davidson size-extensivity correction, to generate the wave functions and energies in the nonrelativistic coupling scheme.

The ECPs described so far are often applied in DFT calculations. Although there is no guarantee that ECPs designed for wave function-based treatments perform well in DFT calculations, experience shows that this seems to be the case, at least for small-core potentials. A recent study of Han and Hirao investigated the transferability of both shape- and energy-consistent nonrelativistic and scalar-relativistic small-core PPs for Au to DFT calculations of AuH, AuCl and Au_2 using LDA and GGA functionals (Han and Hirao 2000). For example, at the DFT level the maximum absolute deviations for R_e, ω_e and D_e from nonrelativistic (scalar-relativistic DKH) all-electron data obtained with corresponding 19-valence-electron energy-consistent PPs were 0.001 (0.007) Å, 1 (2) cm^{-1} and 0.02 (0.05) eV for Au_2. These maximum deviations are only slightly larger than those obtained at the HF and CCSD level, i.e. 0.001 (0.003) Å, 0 (1) cm^{-1} and 0.01 (0.02) eV.

Metal clusters

The extension of the applicability of quantum Monte Carlo (QMC) calculations to systems with heavy elements depends critically on the availability and accuracy of large-core PPs, possibly augmented by CPPs. Besides the usual problems of QMC,

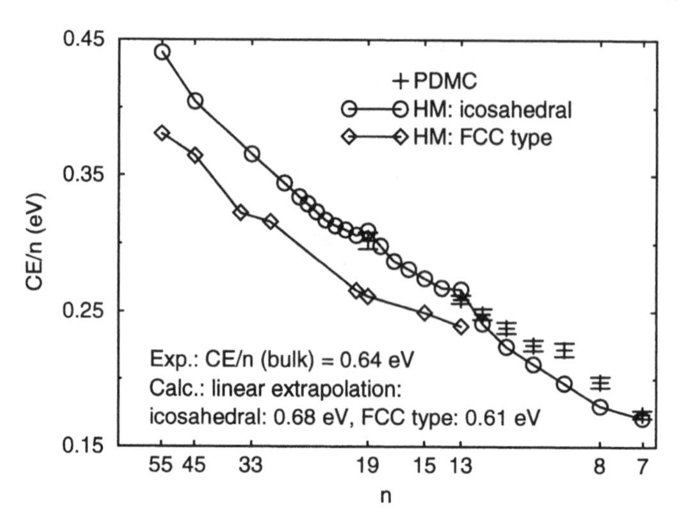

Figure 3.1 Size dependence of cohesive energies per atom (CE/n) of mercury clusters Hg_n from calculations using a large-core EC-PP and CPP for Hg. Valence correlation is accounted for either within the hybrid model approach (HM) by a pair-potential adjusted for Hg_2 or by pure-diffusion quantum Monte Carlo (PDMC) calculations (Wang *et al.* 2000).

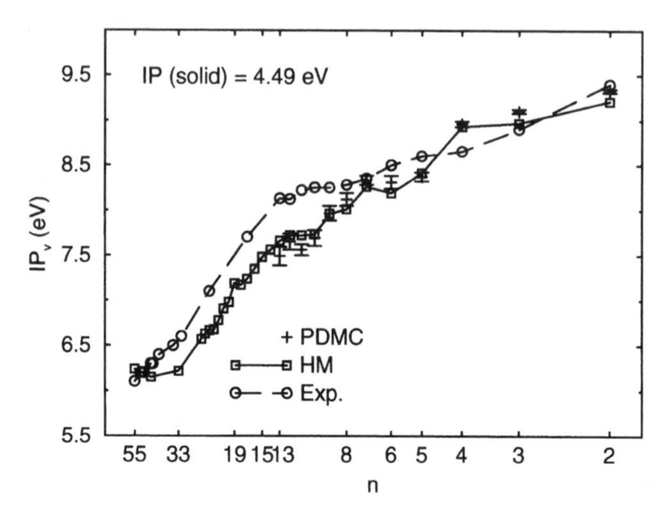

Figure 3.2 As Figure 3.1, but for ionization potentials.

for example, fixed-node errors in standard pure diffusion QMC, a local representation of the generally semilocal PP has to be generated using some suitable many-electron trial wave function. Quite accurate results have been obtained by Mitas (1994) for excitation energies of the Fe atom in the LS-coupling scheme, whereas Flad *et al.* (1997) demonstrated for Pb that with the help of a coordinated sampling scheme in variational QMC also accurate spin–orbit splittings can be obtained. Flad *et al.* also

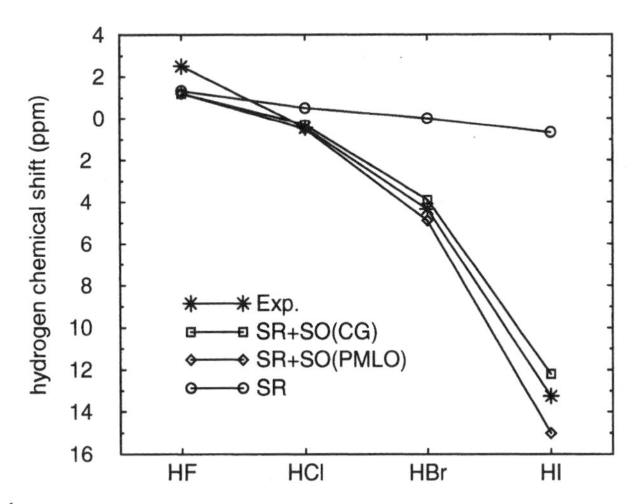

Figure 3.3 ^{1}H-NMR chemical shifts in HX (X = F, Cl, Br, I) from scalar-relativistic (SR) and relativistic (SR+SO) EC-PP DFT calculations for common gauge (CG) and individual gauge for Pipek-Mezey (PM) localized orbitals (LO) in comparison to experimental gas-phase values.

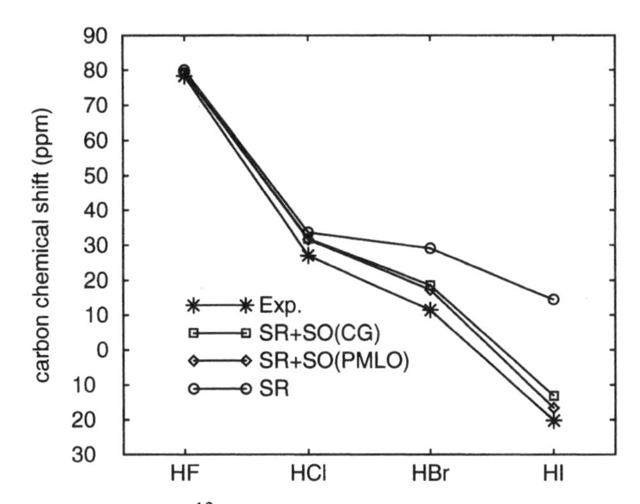

Figure 3.4 As Figure 3.3, but ^{13}C-NMR chemical shifts in CH_3X (X = F, Cl, Br, I). The calculated values are compared with experimental gas phase (X = F) and neat liquid (X = Cl, Br, I) values.

showed that QMC in combination with large-core EC-PPs and CPPs for Hg can be used to study the weak bonding (mainly van der Waals) interaction in Hg_2 (Dolg and Flad 1996b; Flad and Dolg 1996b) as well as the size dependence of properties and the bonding type in small- to medium-size Hg clusters (Wang *et al.* 2000). A simple combination of PP+CPP+CCRC Hartree–Fock calculations with a pairwise additive valence correlation correction derived from highly correlated Hg_2 calculations (hybrid model, HM) yields results in favourable agreement to PP+CPP+CCRC pure diffusion QMC calculations and experimental data, for example, for the size dependence of cohesive energies (Figure 3.1) or the ionization potentials (Figure 3.2).

Properties depending on spin–orbit coupling

In addition to the determination of molecular geometries, vibrational frequencies, binding energies, dipole moments and polarizabilities, excitation energies, ionization potentials and electron affinities, ECPs can also be useful for the derivation of NMR and EPR properties of heavy-element compounds. Besides the relativistic changes of geometrical parameters, the relativistic effects on the electron distribution at a given geometry are also important. A disadvantage of the PP approach is that due to the pseudo-orbital transformation, a straightforward calculation of NMR parameters of the heavy-element centres themselves is not possible. Often, however, it is the NMR of light nuclei in systems containing heavy elements that is of interest. It was demonstrated by Kaupp, Malkina, Malkin and co-workers that energy-consistent PPs in connection with gradient-corrected DFT can be applied successfully to evaluate ^{1}H, ^{13}C, ^{17}O and ^{31}P NMR chemical shifts of the ligands of heavy-metal complexes (Vaara *et al.* 2001). Self-consistent KS calculations are performed with scalar-relativistic PPs, followed by a third-order perturbation calculation for SO corrections, evaluated with SO PPs within the DFT-IGLO (individual gauges for localized orbitals) scheme. Recent results of calibration calculations for small systems, for example, ^{1}H-chemical shifts for halogen halides (Figure 3.3) and ^{13}C-chemical shifts for methyl halides (Figure 3.4), as well as applications to ^{13}C-chemical shifts of transition-metal carbonyls and methyl mercury compounds demonstrate that the experimental trends are very well reproduced, although the numerical accuracy of the results is not yet quantitative. The simultaneous inclusion of scalar relativistic and SO effects at moderate cost should also allow future applications to even larger systems of chemical interest, for example, complexes with several heavy atoms. Similarly, scalar-relativistic PPs and SO PPs have been applied to the evaluation of electronic g-tensors, the EPR analogue of the NMR chemical shift. The PP approach appears to be particularly suitable for the evaluation of g-tensors, since these are largely a valence property. Presently, the methodology is also extended to spin–spin and hyperfine coupling constants.

4 Relativistic Density Functional Theory

E. Engel, R. M. Dreizler
Institut für Theoretische Physik, J. W. Goethe Universität Frankfurt
S. Varga, B. Fricke
Fachbereich Physik, Universität Kassel

4.1 Introduction

Over the last decade density functional theory (DFT) has left its traditional realm, condensed matter theory, and has attracted widespread interest in quantum chemistry, material science and biophysics (see, for example, Barnett and Landman 1993; Becke 1992; Chetty *et al.* 1995; Eichinger *et al.* 1999; Johnson *et al.* 1993a; Krajci *et al.* 1997; Morgan *et al.* 1999; Moroni *et al.* 1997). The question of a relativistic generalization (RDFT) thus emerges quite naturally. While the basic concepts of RDFT were introduced quite some time ago (MacDonald and Vosko 1979; Rajagopal 1978; Rajagopal and Callaway 1973), their practical implementation has taken a lot longer. Both the advancement of the RDFT formalism and its implementation have been the subject of our contribution to the programme *Relativistic Effects in Heavy-Element Chemistry and Physics* (REHE) of the Deutsche Forschungsgemeinschaft.

In this chapter we summarize the various projects pursued in this context. We place some emphasis on an overview of the various formulations of RDFT in the literature. Starting from quantum electrodynamics (QED), we are directly led to the covariant form of RDFT, in which the ground-state four-current $j^{\nu} = (cn, \boldsymbol{j})$ plays the role of the basic density variable. This RDFT variant is ideally suited to a discussion of the basic existence theorem, questions of gauge invariance and the field-theoretical form of the effective single-particle equations (Engel and Dreizler 1996; Engel *et al.* 1995b, 1998a; Facco Bonetti *et al.* 1998). In practical calculations for magnetic systems, on the other hand, an RDFT version which depends on the magnetization density $\boldsymbol{m}$, rather than on $\boldsymbol{j}$, is utilized (MacDonald and Vosko 1979; Ramana and Rajagopal 1981a). Applications of this relativistic 'spin-density' functional approach are given in Chapter 5. As one of our projects within the REHE programme, a stable algorithm for the investigation of open-shell atoms has been developed on this basis (Engel *et al.* 2001a).

Most frequently, however, a purely density-dependent version of RDFT is used. In this context we have examined the role of relativistic corrections to the exchange-correlation (xc) energy functional. In view of the limited accuracy of the relativistic local density approximation (RLDA) (Das *et al.* 1980; Engel *et al.* 1995a; Ramana *et*

Relativistic Effects in Heavy-Element Chemistry and Physics. Edited by B. A. Hess

al. 1982), the generalized gradient approximation (GGA) has been extended into the relativistic domain (RGGA) (Engel *et al.* 1996, 1998b). Applications of the RGGA to gold compounds as well as to the bulk showed, however, that, in spite of the obvious improvements obtained for atoms, the gradient terms overcorrect the errors of the RLDA for equilibrium distances and binding energies in molecules or solids with heavy constituents (Schmid *et al.* 1998; Varga *et al.* 1999). This deficiency of the RGGA indicates the need for truly nonlocal xc-functionals.

The prototype of such a functional is the exact exchange of RDFT, which not only includes the relativistic kinematics of the electrons but also the (retarded) Breit interaction among them. As an explicit functional of the auxiliary single-particle spinors of RDFT, the exact exchange is an implicit density functional, for which the multiplicative Kohn–Sham (KS) potential must be evaluated indirectly via the relativistic version of the optimized potential method (OPM) (Engel *et al.* 1995a, 1998a; Kreibich *et al.* 1998). Within the REHE programme we have obtained a variety of exchange-only (x-only) ROPM results and put forward an accurate semi-analytical approximation to the full ROPM. A corresponding orbital-dependent correlation functional has been derived from perturbation theory on the basis of the KS Hamiltonian (Engel *et al.* 1998a) and studied for atoms and molecules (Engel *et al.* 2000a; Facco Bonetti *et al.* 2001).

In order to facilitate the application of the ROPM to more complex systems we have constructed relativistic pseudopotentials on the basis of the exact exchange (Engel *et al.* 2001c; Höck and Engel 1998), using a relativistic extension of the Troullier–Martins approach to norm-conserving pseudopotentials (Engel *et al.* 2001b). On this basis the antiferromagnetic (AFM) ground states of transition-metal oxides have been studied (Schmid 2000).

As a fully relativistic density functional approach to the electronic and geometric structures of molecules containing heavy elements, the relativistic discrete variational method (RDVM) (Rosén and Ellis 1975) has been successively improved to a new level of quality within the REHE programme. Individual projects addressed the efficient calculation of the Hartree potential (Bastug *et al.* 1995; Varga *et al.* 2000b), the interatomic forces (Varga *et al.* 2001) and the necessary multicentre integrals (Heitmann *et al.* 2001). As a result, the RDVM now allows theoretical studies of rather complex systems, as clusters (Bastug *et al.* 1997b) and complexes or compounds containing superheavy and transactinide elements (Fricke *et al.* 1997; Varga *et al.* 2000a) (compare Chapter 6). It can also be applied to the investigation of problems in surface physics, e.g. adsorption processes of adatoms on surfaces (Geschke *et al.* 2000).

4.2 Foundations

4.2.1 Existence theorem

The appropriate starting point for the discussion of the foundations of RDFT is QED. Although relativistic quantum field theories like QED do not provide a Schrödinger-like wave equation for the relativistic many-body problem, there nevertheless exists a well-defined procedure for the derivation of the Hamiltonian of a stationary system. It emerges as one component of the energy–momentum tensor, which is most conveniently established within the framework of Noether's theorem. For the standard QED Lagrangian of interacting electrons coupled to some classical, stationary C-number potential V^μ we obtain

$$\left.\begin{aligned}\hat{H} &= \hat{H}_{\rm e}(x^0) + \hat{H}_\gamma(x^0) + \hat{H}_{\rm int}(x^0) + \hat{H}_{\rm ext}(x^0),\\ \hat{H}_{\rm e}(x^0) &= \frac{1}{2}\int {\rm d}^3r\,[\hat{\psi}^\dagger(x), (-{\rm i}c\boldsymbol{\alpha}\cdot\nabla + \beta mc^2)\hat{\psi}(x)],\\ \hat{H}_\gamma(x^0) &= -\frac{1}{8\pi}\int {\rm d}^3r\,\{\partial^0\hat{A}_\nu(x)\partial^0\hat{A}^\nu(x) + \nabla\hat{A}_\nu(x)\cdot\nabla\hat{A}^\nu(x)\},\\ \hat{H}_{\rm int}(x^0) &= \frac{e}{c}\int {\rm d}^3r\,\hat{j}^\mu(x)\hat{A}_\mu(x),\\ \hat{H}_{\rm ext}(x^0) &= \frac{e}{c}\int {\rm d}^3r\,\hat{j}^\mu(x)V_\mu(\boldsymbol{r}).\end{aligned}\right\} \tag{4.1}$$

Here $\hat{\psi}(x)$ denotes the fermion field operator of the interacting, inhomogeneous system characterized by $\hat{H}$ (in the Heisenberg picture), $\hat{j}^\mu$ is the corresponding fermion four-current operator,

$$\hat{j}^\mu(x) = \tfrac{1}{2}c[\hat{\psi}^\dagger(x), \alpha^\mu\hat{\psi}(x)],$$

and $\hat{A}_\mu(x)$ represents the field operator of the photons, for which the covariant quantization scheme and Feynman gauge are used ($x^\mu = (ct, \boldsymbol{r})$, $\alpha^\mu = \gamma^0\gamma^\mu$). Both $\hat{H}$ and $\hat{j}^\mu$ have been formulated in the commutator form, which ensures the correct behaviour under charge conjugation (Källén 1958), although we will not dwell on this point in the following. The Hamiltonian commutes with the charge operator

$$\hat{Q} = \frac{e}{c}\int {\rm d}^3r\,\hat{j}^0(x).$$

This allows a classification of all many-electron states with respect to their total charge.

As is well known, the expectation values of $\hat{H}$ and $\hat{j}^\mu$ diverge if taken directly without some additional prescription. This is most easily seen for noninteracting electrons experiencing an external potential V^μ. The existence of the negative energy continuum states requires the redefinition of the energy scale in order to take into account the nonvanishing energy of the vacuum. Furthermore, the fact that V^μ can create virtual electron–positron pairs makes a renormalization of the four-current necessary. The situation is even more involved for interacting electrons. Within the standard perturbative approach the coupling between electrons and photons leads to large classes of divergent contributions which have to be first regularized and then renormalized by a suitable redefinition of the fundamental parameters of QED. For the ground state $|\Phi\rangle$ of the Fock space sector with charge Ne,

$$\hat{H}|\Phi\rangle = E|\Phi\rangle,$$

the total binding energy E_R of the electrons is thus given by the energy difference between $|\Phi\rangle$ and the ground state of the zero-charge sector, i.e. the interacting vacuum $|0\rangle$, augmented by the counterterm contributions (CTCs) required to keep E_R finite,

$$E_{\mathrm{R}} \equiv E_{\mathrm{tot}} = \langle\Phi|\hat{H}|\Phi\rangle - \langle 0|\hat{H}|0\rangle + \mathrm{CTCs}. \tag{4.2}$$

An analogous renormalization is necessary for the ground-state four-current,

$$j_{\mathrm{R}}^{\mu}(\boldsymbol{r}) = \langle\Phi|\hat{j}^{\mu}(x)|\Phi\rangle + \mathrm{CTCs}. \tag{4.3}$$

Of course, RDFT, whose central ingredients are the ground-state energy and four-current, must reflect this structure of the underlying quantum field theory. The need for renormalization thus shows up in the formulation of the basic existence theorem, in the single-particle equations and in the derivation of explicit functionals for the xc-energy. A detailed discussion of the various issues involved has been given in Engel and Dreizler (1996) and Engel *et al.* (1995b, 1998a), to which we refer the interested reader. In the following we always assume the quantities involved to be properly renormalized; all counterterms as well as the corresponding index 'R' will, however, be suppressed for brevity.

On this basis, we can summarize the existence theorem of RDFT (Engel and Dreizler 1996; Engel *et al.* 1995b; MacDonald and Vosko 1979; Rajagopal 1978; Rajagopal and Callaway 1973) as follows. There exists a one-to-one correspondence between the class of ground states which result from external potentials just differing by gauge transformations and the ground-state four-current,

$$\{|\Phi\rangle \mid |\Phi\rangle \text{ from } V_\nu + \partial_\nu \Lambda\} \Longleftrightarrow j^\nu(\boldsymbol{x}) \tag{4.4}$$

(here and in the following $|\Phi\rangle$ is always assumed to be nondegenerate). In other words, the class of physically equivalent realizations of the ground state is uniquely determined by j^ν . Choosing a suitable representative of each class, i.e. fixing the gauge for the complete set of V_ν, we can understand this representative $|\Phi\rangle$ as a unique functional of j^ν, $|\Phi[j^\nu]\rangle$. If we insert a specific four-current j_0^ν into this functional, we obtain the ground state $|\Phi_0\rangle$ of the corresponding system, $|\Phi_0\rangle = |\Phi[j_0^\nu]\rangle$. No information beyond j_0^ν is needed, i.e. the same functional applies to atoms, molecules and solids ($|\Phi[j^\nu]\rangle$ is universal).

The existence of a unique relation between the ground state and j^ν immediately leads to the statement that all ground-state observables are unique functionals of the four-current, most notably the ground-state energy,

$$E_{\mathrm{tot}}[j^\nu] = \langle\Phi[j^\nu]|\hat{H}|\Phi[j^\nu]\rangle.$$

In view of its field-theoretical basis, this energy functional not only accounts for the relativistic kinematics of both electrons and photons, but, in principle, also for all radiative corrections. With the Ritz principle,[1] avoiding the question of interacting v-representability (Dreizler and Gross 1990), we may then formulate the basic

---

[1] We are not aware of any rigorous minimum principle for the renormalized ground-state energies (4.2). There are, however, a number of arguments which can be given in favour of such a minimum principle.

variational equation of RDFT,

$$\frac{\delta}{\delta j^{\nu}(\boldsymbol{r})}\left\{E_{\text{tot}}[j^{\nu}] - \frac{\mu}{c}\int \mathrm{d}^3x\, j^0(\boldsymbol{x})\right\}\bigg|_{j=j_0} = 0, \tag{4.5}$$

where the subsidiary condition ensures charge conservation. Given the functional $E_{\text{tot}}[j^{\nu}]$, Equation (4.5) allows the determination of the ground-state four-current j_0 corresponding to the external potential V^{μ} in the Hamiltonian (4.1) and, by subsequent insertion of j_0 into $E_{\text{tot}}[j^{\nu}]$, the ground-state energy.

The proof for this existence theorem of RDFT (Engel and Dreizler 1996; Engel *et al.* 1995b) proceeds in a similar way as the original argument of Hohenberg and Kohn (HK) (1964), whose basic ingredients are the multiplicative nature of the external potential and the Ritz variational principle. This proof involves inequalities between ground-state expectation values of different Hamiltonians, so that we necessarily have to rely on their renormalized form (4.2). Fortunately, we can show that the HK-type proof is compatible with the QED renormalization scheme if we utilize the fact that the counterterms are unique functionals of the four-current (Engel *et al.* 1995b).

We may nevertheless ask whether it is possible to base RDFT on an approximate relativistic many-body approach, as, for example, the Dirac–Coulomb (DC) Hamiltonian,

$$\hat{H}^{\text{DC}} = \hat{H}_{\text{e}}(0) + \hat{H}_{\text{ext}}(0) + \hat{H}_{\text{e–e}},$$

$$\hat{H}_{\text{e–e}} = \tfrac{1}{2}e^2\int \mathrm{d}^3r\mathrm{d}^3r'\,\frac{\hat{\bar\psi}^{\dagger}(0,\boldsymbol{r})\hat{\bar\psi}^{\dagger}(0,\boldsymbol{r}')\hat{\psi}(0,\boldsymbol{r}')\hat{\psi}(0,\boldsymbol{r})}{|\boldsymbol{r}-\boldsymbol{r}'|},$$

or its Dirac–Coulomb–Breit (DCB) extension, so that we avoid the discussion of renormalization. In this case, the no-pair (np) approximation plays the role of the renormalization scheme,

$$\hat{H}^{\text{DC}}_{\text{np}} = \hat{\Lambda}_{+}\hat{H}^{\text{DC}}\hat{\Lambda}_{+}, \qquad \hat{j}^{\mu}_{\text{np}} = \hat{\Lambda}_{+}\hat{j}^{\mu}\hat{\Lambda}_{+},$$

where $\hat{\Lambda}_{+}$ is a projection operator onto positive-energy states. However, the no-pair approximation can be unambiguously specified only within some well-defined single-particle scheme. Even in this case $\hat{\Lambda}_{+}$ depends on the actual single-particle potential and thus on the external potential, $\hat{\Lambda}_{+}[V^{\mu}]$. As a consequence, $\hat{H}^{\text{DC}}_{\text{np}}$ is a nonlinear functional of V^{μ}, which does not allow the usual *reductio ad absurdum* of the HK-proof. In addition, the no-pair approximation introduces a gauge dependence into the

First of all, with increasing speed of light, i.e. in the nonrelativistic limit ($v/c \longrightarrow 0$), the energies (4.2) continuously approach values which do satisfy the Ritz principle. There seems to be no reason to assume that the minimum principle is restricted to the isolated value $c = \infty$. Secondly, we can explicitly verify that there exists a minimum principle for the renormalized ground-state energy of noninteracting fermions in an arbitrary four potential V^{μ} (within the Furry picture, compare Rafelski *et al.* (1978)). Finally, real atoms and molecules are stable (indicating that there exists a lower bound for energies) and QED has proved to be the most accurate theory available to date for describing these systems (note that, as a matter of principle, we do not have to rely on a perturbative treatment of QED systems, so that the asymptotic character of this perturbation expansion does not contradict this argument).

ground-state energy (Engel *et al.* 1998a), so that an unambiguous comparison of two ground-state energies is only possible if we neglect the Breit interaction and restrict ourselves to a purely electrostatic external potential. It thus seems that the existence theorem of RDFT can only be based on the field-theoretical Hamiltonian (4.1) together with the standard QED renormalization scheme. The no-pair approximation, which is used in most applications, is much more easily introduced at a later stage, i.e. in the context of the single-particle equations of RDFT.

From this discussion it is obvious that the two central approximations of the DC or DCB approach, the no-pair approximation and the neglect of either the complete Breit interaction (DC) or the retardation corrections to it (DCB), play a different role in RDFT. In fact, while the no-pair approximation is also a standard in RDFT, there exists no fundamental conceptual problem with including the full electron–electron interaction. In order to understand this we have to recall that RDFT is not based on a Schrödinger-like single-time single-field wave equation, but only requires a suitable approximation for $E_{\text{tot}}[j^\nu]$. The building blocks for the evaluation of (4.2), (4.3) are the propagators of the noninteracting particles, as the free-photon propagator,

$$D^{0,\mu\nu}(x-y) = -\mathrm{i}\frac{e^2}{c}\langle 0_\gamma | T\hat{A}_0^\mu(x)\hat{A}_0^\nu(y)|0_\gamma\rangle,$$

where $\hat{A}_0^\nu$ denotes the noninteracting photon field operator and $|0_\gamma\rangle$ is the corresponding vacuum state. The derivation of explicit approximations to $E_{\text{tot}}[j^\nu]$ via the usual field-theoretical methods thus automatically leads to the inclusion of the full transverse interaction. In fact, this is true not only in principle, but also in practice (Engel and Facco Bonetti 2000; Engel *et al.* 1998a; MacDonald and Vosko 1979; Rajagopal 1978; Ramana and Rajagopal 1981b).

On the other hand, there is also no fundamental problem with restricting RDFT to the Coulomb or Coulomb–Breit level. Choosing the Feynman gauge as used for the Hamiltonian (4.1), the full $D^0_{\mu\nu}$ is explicitly given by

$$\left.\begin{aligned} D^{0,\mathrm{F}}_{\mu\nu}(x-y) &= \int \frac{\mathrm{d}^4 q}{(2\pi)^4}\mathrm{e}^{-\mathrm{i}q(x-y)}D^{\mathrm{F}}_{\mu\nu}(q), \\ D^{\mathrm{F}}_{\mu\nu}(q) &= D(q^2)g_{\mu\nu}, \\ D(q^2) &= \frac{-4\pi e^2}{q^2+\mathrm{i}\eta}, \end{aligned}\right\} \tag{4.6}$$

which is easily reduced to the weakly relativistic (Coulomb–Breit),

$$D^{\mathrm{F,CB}}_{\mu\nu}(q) = D(-\boldsymbol{q}^2)\begin{pmatrix} 1+(q_0)^2/\boldsymbol{q}^2 & 0 \\ 0 & g_{ij}\end{pmatrix}, \tag{4.7}$$

or the nonrelativistic (Coulomb) level,

$$D^{\mathrm{F,C}}_{\mu\nu}(q) = D(-\boldsymbol{q}^2)g_{\mu 0}g_{\nu 0}.$$

However, the propagators (4.6) and (4.7) are equivalent to their Coulomb gauge counterparts (usually applied in quantum chemistry),

$$\left.\begin{aligned} D^{\mathrm{F}}_{\mu\nu}(q) &\longrightarrow D^{\mathrm{C}}_{\mu\nu}(q) = \begin{pmatrix} D(-\boldsymbol{q}^2) & 0 \\ 0 & D(q^2)(g_{ij} + q_i q_j/\boldsymbol{q}^2) \end{pmatrix}, \\ D^{\mathrm{F,CB}}_{\mu\nu}(q) &\longrightarrow D^{\mathrm{C,CB}}_{\mu\nu}(q) = D(-\boldsymbol{q}^2)\begin{pmatrix} 1 & 0 \\ 0 & g_{ij} + q_i q_j/\boldsymbol{q}^2 \end{pmatrix}, \end{aligned}\right\} \qquad (4.8)$$

only in gauge-invariant expressions. In general, gauge invariance can only be ensured by the inclusion of the negative continuum in all intermediate sums over states (Engel *et al.* 1998a) (with one important exception—see Section 4.3.1). As soon as the no-pair approximation is applied a gauge dependence is introduced, so that a consistent comparison of RDFT and DCB data should be based on the same gauge. The absolute size of this gauge dependence has been examined for the correlation energy of the relativistic homogeneous electron gas (RHEG) by comparison of the gauge-invariant standard form with its no-pair counterpart, evaluated for different gauges (Facco Bonetti *et al.* 1998). It was found that for high densities the error resulting from the combination of the no-pair approximation with a specific gauge can be substantial. On the other hand, the effect of this gauge dependence on atomic correlation energies is rather limited. Utilization of the no-pair form within the LDA showed that the gauge error is small compared with the error introduced by use of the LDA, in particular for the Coulomb gauge. Thus, assuming this result to also be characteristic of the gauge dependence of DCB data, gauge questions do not seem to be relevant for comparisons of DCB or experimental data with RLDA or RGGA results (or of different xc-functionals).

In most applications the external magnetic field

$$\boldsymbol{B}_{\mathrm{ext}}(\boldsymbol{r}) = \nabla \times \boldsymbol{V}(\boldsymbol{r})$$

vanishes. In this case, the external Hamiltonian reduces to

$$\hat{H}_{\mathrm{ext}} \longrightarrow \hat{H}'_{\mathrm{ext}} = \int \mathrm{d}^3 r\, \hat{n}(x) v_{\mathrm{ext}}(\boldsymbol{r}), \qquad v_{\mathrm{ext}}(\boldsymbol{r}) = eV^0(\boldsymbol{r}), \qquad (4.9)$$

where we have introduced the more familiar n for the density,

$$\hat{n}(x) = \tfrac{1}{2}[\hat{\bar{\psi}}^{\dagger}(x), \hat{\psi}(x)] \quad \longleftrightarrow \quad j^{\nu}(\boldsymbol{r}) = (cn(\boldsymbol{r}), \boldsymbol{j}(\boldsymbol{r})).$$

Following the standard HK scheme we can then prove that there exists a one-to-one mapping between the zeroth component of the external potential, the ground state and the ground state density (MacDonald and Vosko 1979),

$$\{v_{\mathrm{ext}} \mid v_{\mathrm{ext}} + \mathrm{const.}\} \Longleftrightarrow \{|\Phi\rangle \mid |\Phi\rangle \text{ from } v_{\mathrm{ext}} + \mathrm{const.}\} \Longleftrightarrow n(\boldsymbol{r}).$$

In this case, we can thus understand the ground state to be a functional of the density alone, $|\Phi[n]\rangle$. The same is then true for the ground-state observables as the energy,

$E_{\text{tot}}[n]$. As a consequence, there is only one single variational equation,

$$\frac{\delta}{\delta n(\boldsymbol{r})}\left\{E_{\text{tot}}[n]-\mu\int \mathrm{d}^3x\, n(\boldsymbol{x})\right\}\bigg|_{n=n_0}=0. \tag{4.10}$$

It must be emphasized that the restriction to an external potential of the type $V^\mu = (V^0, \boldsymbol{0})$ does not imply that the system cannot have some magnetic moment. Rather, the spatial components of the four-current must be viewed as functionals of the density, $\boldsymbol{j}[n] = \langle\Phi[n]|\hat{\boldsymbol{j}}|\Phi[n]\rangle \neq 0$.

Thus, in principle, the density is sufficient for an exact RDFT treatment of magnetic systems, similar to the situation in nonrelativistic DFT. In practice, on the other hand, spin-density functional theory proved to be necessary for the description of spin-polarized ground states in the nonrelativistic context. Its spin-density dependent energy functional allows a distinction between the spin-up and spin-down channels, which is also important if $\boldsymbol{B}_{\text{ext}} = \boldsymbol{0}$. As a matter of principle, the explicit inclusion of magnetic effects is possible via the four-current version of RDFT. However, the standard energy functionals of (R)DFT are based on the (relativistic) homogeneous electron gas (see Section 4.4.1), for which $\boldsymbol{j}$ vanishes. Consequently, explicitly $\boldsymbol{j}$-dependent approximations for the energy cannot be derived from the RHEG. Thus a direct relativistic extension of spin-density functional theory is desirable, whose basic variables are suitably generalized spin-densities.

The starting point for this generalization is the Gordon decomposition, in which the total current is split into the paramagnetic (orbital) component $\boldsymbol{j}_{\text{p}}$, a gauge term proportional to the scalar density ρ_{s}, and the curl of the magnetization density $\boldsymbol{m}$,

$$\begin{aligned}
\boldsymbol{j}(\boldsymbol{r}) &= \boldsymbol{j}_{\text{p}}(\boldsymbol{r}) - \frac{e}{mc}\boldsymbol{V}(\boldsymbol{r})\rho_{\text{s}}(\boldsymbol{r}) - \frac{c}{e}\nabla\times\boldsymbol{m}(\boldsymbol{r}),\\
\boldsymbol{j}_{\text{p}}(\boldsymbol{r}) &= -\frac{\mathrm{i}\hbar}{2m}\langle\Phi|\hat{\psi}^\dagger(x)\beta[\nabla\hat{\psi}(x)] - [\nabla\hat{\psi}^\dagger(x)]\beta\hat{\psi}(x)|\Phi\rangle,\\
\rho_{\text{s}}(\boldsymbol{r}) &= \langle\Phi|\hat{\psi}^\dagger(x)\beta\hat{\psi}(x)|\Phi\rangle,\\
\boldsymbol{m}(\boldsymbol{r}) &= -\mu_{\text{B}}\langle\Phi|\hat{\psi}^\dagger(x)\beta\boldsymbol{\Sigma}\hat{\psi}(x)|\Phi\rangle,
\end{aligned}$$

with

$$\mu_{\text{B}} = \frac{e\hbar}{2mc}, \qquad \boldsymbol{\Sigma} = \begin{pmatrix}\boldsymbol{\sigma} & 0\\ 0 & \boldsymbol{\sigma}\end{pmatrix}.$$

Neglect, for the sake of argument, the coupling of $\boldsymbol{V}$ to the orbital current,

$$\hat{H}_{\text{ext}} \longrightarrow \hat{H}''_{\text{ext}} = \int \mathrm{d}^3r\,\{\hat{n}(\boldsymbol{r})v_{\text{ext}}(\boldsymbol{r}) + \hat{\boldsymbol{m}}(\boldsymbol{r})\cdot\boldsymbol{B}_{\text{ext}}(\boldsymbol{r})\}.$$

The resulting Hamiltonian

$$\hat{H}'' = \hat{H}_{\text{e}} + \hat{H}_\gamma + \hat{H}_{\text{int}} + \hat{H}''_{\text{ext}} \tag{4.11}$$

is completely legitimate if we aim at the description of systems *not* subject to magnetic fields. In this case, we formulate a density functional approach for a more general class of systems than physically required. In this more general class the systems of interest are included as a subset which is obtained in the limit $\boldsymbol{B}_{\mathrm{ext}} \longrightarrow \mathbf{0}$: $\hat{H}''$ only serves to identify the fundamental variables of an RDFT scheme for $\boldsymbol{B}_{\mathrm{ext}} = \mathbf{0}$. On the basis of $\hat{H}''$ we can establish an existence theorem, connecting the ground state with the ground-state charge and magnetization densities (MacDonald and Vosko 1979),

$$|\Phi\rangle \Longleftrightarrow (n, \boldsymbol{m}). \tag{4.12}$$

In other words: $|\Phi\rangle$ is a unique and universal functional of $(n, \boldsymbol{m})$, $|\Phi[n, \boldsymbol{m}]\rangle$. Again the minimum principle for the ground-state energy provides a set of variational equations,

$$\frac{\delta}{\delta n(\boldsymbol{r})}\left\{E_{\mathrm{tot}}[n, \boldsymbol{m}] - \mu \int \mathrm{d}^3x\, n(\boldsymbol{x})\right\} = 0, \qquad \frac{\delta}{\delta \boldsymbol{m}(\boldsymbol{r})} E_{\mathrm{tot}}[n, \boldsymbol{m}] = 0. \tag{4.13}$$

RDFT in the form (4.12), (4.13) can be considered exact for $\boldsymbol{B}_{\mathrm{ext}} \longrightarrow \mathbf{0}$. Due to its universality $|\Phi[n, \boldsymbol{m}]\rangle$ remains unchanged in this limit. The same statement then holds for all components of $E_{\mathrm{tot}}[n, \boldsymbol{m}]$ for which the associated part of the Hamiltonian does not depend on $\boldsymbol{B}_{\mathrm{ext}}$. Thus, the only point at which the limit $\boldsymbol{B}_{\mathrm{ext}} \longrightarrow \mathbf{0}$ actually shows up is the explicit coupling term $\int \mathrm{d}^3r\, \boldsymbol{m} \cdot \boldsymbol{B}_{\mathrm{ext}}$. However, these statements should be taken with a grain of salt. In view of the prominent role of Ward identities and gauge invariance for the success of the QED renormalization scheme it is obvious that the arguments leading to (4.12), (4.13) are built on somewhat less solid ground than those underlying the four-current version of RDFT, Equations (4.4), (4.5). A detailed investigation of this issue is not yet available. It seems worthwhile remarking that for $\boldsymbol{B}_{\mathrm{ext}} \neq \mathbf{0}$ the Hamiltonian (4.11) and thus the RDFT variant (4.12), (4.13) represent approximations whose usefulness depends on the absolute size of $\boldsymbol{B}_{\mathrm{ext}}$.

In the form (4.12), (4.13) RDFT is perfectly suited for dealing with systems in which the direction of $\boldsymbol{m}$ varies with $\boldsymbol{r}$. On the other hand, if the noncollinearity of $\boldsymbol{m}$ is not an important feature for the system, we may restrict the artificial coupling between the electrons and the magnetic field to its z-component,

$$\hat{H}_{\mathrm{ext}}''' = \int \mathrm{d}^3r\, \{\hat{n}(\boldsymbol{r}) v_{\mathrm{ext}}(\boldsymbol{r}) + \hat{m}_z(\boldsymbol{r}) B_{\mathrm{ext},z}(\boldsymbol{r})\},$$

so that the one-to-one mapping reduces to

$$|\Phi\rangle \Longleftrightarrow (n, m_z). \tag{4.14}$$

In this case, the ground state is uniquely determined by n and m_z, $|\Phi[n, m_z]\rangle$, and the variational equations (4.13) reduce accordingly.

4.2.2 Single-particle equations

The next task is to derive an alternative form, more useful in practice, of the fundamental variational equations of Section 4.2.1. The basic idea is to represent the elementary density variables of RDFT in terms of auxiliary single-particle four spinors

ϕ_k (assuming that this is possible for arbitrary external fields, i.e. assuming noninteracting v-representability). Such a representation in general also includes all vacuum corrections to the ground-state four-current and energy (Engel and Dreizler 1996; Engel *et al.* 1998a). For most RDFT applications, however, these field-theoretical effects are irrelevant. In the following we thus restrict ourselves to summarizing the no-pair form for brevity.

In the four-current version of RDFT the auxiliary spinors are chosen to reproduce the complete j^μ,

$$j^\mu(\boldsymbol{r}) = c\sum_k \Theta_k \phi_k^\dagger(\boldsymbol{r})\alpha^\mu \phi_k(\boldsymbol{r}), \tag{4.15}$$

where

$$\Theta_k = \begin{cases} 0 & \text{for } \epsilon_k \leqslant -mc^2, \\ 1 & \text{for } -mc^2 < \epsilon_k \leqslant \epsilon_{\mathrm{F}}, \\ 0 & \text{for } \epsilon_{\mathrm{F}} < \epsilon_k, \end{cases} \tag{4.16}$$

in the no-pair approximation and ϵ_{F} is the Fermi energy. Equation (4.15) then induces a decomposition of E_{tot}, in which the manageable single-particle components are separated from the more complicated many-body contributions,

$$E_{\mathrm{tot}} = T_{\mathrm{s}} + E_{\mathrm{ext}} + E_{\mathrm{H}} + E_{\mathrm{xc}}. \tag{4.17}$$

Here T_{s} denotes the kinetic energy of the 'auxiliary particles',

$$T_{\mathrm{s}} = \int \mathrm{d}^3r \sum_k \Theta_k \phi_k^\dagger(\boldsymbol{r})[-\mathrm{i}c\boldsymbol{\alpha}\cdot\nabla + (\beta - 1)mc^2]\phi_k(\boldsymbol{r}), \tag{4.18}$$

and E_{ext} represents the coupling between the electrons and the external fields,

$$E_{\mathrm{ext}} = \frac{e}{c}\int \mathrm{d}^3r\, j_\mu(\boldsymbol{r})V^\mu(\boldsymbol{r}).$$

E_{H} is the 'covariant' form of the Hartree energy , which can be split into the Coulomb contribution $E_{\mathrm{H}}^{\mathrm{C}}$ and a transverse part $E_{\mathrm{H}}^{\mathrm{T}}$,

$$\left.\begin{aligned} E_{\mathrm{H}} &= \frac{1}{2c^2}\int \mathrm{d}^3x \int \mathrm{d}^4y\, D_{\mu\nu}^0(\boldsymbol{x}-\boldsymbol{y}, y^0)j^\mu(\boldsymbol{x})j^\nu(\boldsymbol{y}) = E_{\mathrm{H}}^{\mathrm{C}} + E_{\mathrm{H}}^{\mathrm{T}}, \\ E_{\mathrm{H}}^{\mathrm{C}} &= \tfrac{1}{2}e^2 \int \mathrm{d}^3r \int \mathrm{d}^3r'\, \frac{n(\boldsymbol{r})n(\boldsymbol{r}')}{|\boldsymbol{r}-\boldsymbol{r}'|}, \\ E_{\mathrm{H}}^{\mathrm{T}} &= -\frac{e^2}{2c^2}\int \mathrm{d}^3r \int \mathrm{d}^3r'\, \frac{\boldsymbol{j}(\boldsymbol{r})\cdot\boldsymbol{j}(\boldsymbol{r}')}{|\boldsymbol{r}-\boldsymbol{r}'|}. \end{aligned}\right\} \tag{4.19}$$

Finally, the xc-energy E_{xc}, in which all many-body aspects beyond the Pauli principle are absorbed, is defined by (4.17) (the rest mass of the electrons has been subtracted from E_{tot}). As the existence theorem (4.4) is equally valid for noninteracting particles, not only E_{tot} but also T_{s} is a unique functional of j^μ. Consequently, E_{xc} must also be a unique functional of j^μ, $E_{\text{xc}}[j^\mu]$. Minimization of E_{tot} with respect to the ϕ_k rather than j^μ thus leads to effective single-particle equations (MacDonald and Vosko 1979; Rajagopal 1978),

$$\{-\mathrm{i}c\boldsymbol{\alpha}\cdot\nabla + (\beta - 1)mc^2 + \alpha_\mu v_{\text{s}}^\mu(\boldsymbol{x})\}\phi_k(\boldsymbol{x}) = \epsilon_k\phi_k(\boldsymbol{x}), \tag{4.20}$$

with the multiplicative KS potential v_{s}^μ consisting of the sum of V^μ, the Hartree potential v_{H}^μ and the xc-potential v_{xc}^μ,

$$v_{\text{s}}^\mu(\boldsymbol{r}) = eV^\mu(\boldsymbol{r}) + v_{\text{H}}^\mu(\boldsymbol{r}) + v_{\text{xc}}^\mu(\boldsymbol{r}), \tag{4.21}$$

$$v_{\text{H}}^\mu(\boldsymbol{r}) = \frac{e^2}{c}\int \mathrm{d}^3r'\,\frac{j^\mu(\boldsymbol{r}')}{|\boldsymbol{r}-\boldsymbol{r}'|}, \tag{4.22}$$

$$v_{\text{xc}}^\mu(\boldsymbol{r}) = c\frac{\delta E_{\text{xc}}[j]}{\delta j_\mu(\boldsymbol{r})}. \tag{4.23}$$

Equations (4.15), (4.20)–(4.23) have to be solved self-consistently. With the exact $E_{\text{xc}}[j^\mu]$ their solution leads to the exact ground-state four-current, which, upon insertion into (4.17), yields the exact ground-state energy. On the other hand, no statement is made about the true many-body ground state $|\Phi\rangle$. Moreover, as a matter of principle, the eigenvalues ϵ_k have no physical meaning in the case of interacting particles. The only exception is the eigenvalue of the highest occupied KS state, which, in nonrelativistic DFT, can be shown to be identical with the ionization potential (Almbladh and von Barth 1985) (for finite systems, no rigorous proof of this statement is known in the field-theoretical situation).

At this point it is convenient, though not necessary, to define the exchange component E_{x} of E_{xc}. As in the nonrelativistic context (Langreth and Mehl 1983; Sahni *et al.* 1982; Sham 1985) we identify E_{x} with the first-order contribution to E_{xc} resulting from perturbation theory on the basis of the KS auxiliary Hamiltonian (Engel *et al.* 1998a). Within the no-pair approximation this leads to

$$E_{\text{x}} = -\tfrac{1}{2}e^2\sum_{k,l}\Theta_k\Theta_l\int \mathrm{d}^3r\int \mathrm{d}^3r'\,\frac{\cos(\omega_{kl}|\boldsymbol{r}-\boldsymbol{r}'|)}{|\boldsymbol{r}-\boldsymbol{r}'|}\phi_k^\dagger(\boldsymbol{r})\alpha_\mu\phi_l(\boldsymbol{r})\phi_l^\dagger(\boldsymbol{r}')\alpha^\mu\phi_k(\boldsymbol{r}'), \tag{4.24}$$

($\omega_{kl} = |\varepsilon_k - \varepsilon_l|/c$) which can be easily decomposed into a Coulomb part

$$E_{\text{x}}^{\text{C}} = -\tfrac{1}{2}e^2\sum_{k,l}\Theta_k\Theta_l\int \mathrm{d}^3r\int \mathrm{d}^3r'\,\frac{\phi_k^\dagger(\boldsymbol{r})\phi_l(\boldsymbol{r})\phi_l^\dagger(\boldsymbol{r}')\phi_k(\boldsymbol{r}')}{|\boldsymbol{r}-\boldsymbol{r}'|}, \tag{4.25}$$

and a transverse remainder $E_{\text{x}}^{\text{T}} = E_{\text{x}} - E_{\text{x}}^{\text{C}}$. It is worthwhile emphasizing that neither E_{x} nor E_{x}^{C} are identical with their Dirac–Fock (DF) counterparts. The KS spinors ϕ_k used for the evaluation of (4.24) or (4.25) are solutions of (4.20) with its *multiplicative* total potential, in contrast to the DF orbitals which experience the nonlocal DF exchange potential. In fact, the multiplicative nature of v_{s}^μ also ensures the gauge invariance of (4.24) (Engel *et al.* 1998a), which is lost as soon as the ϕ_k correspond to a nonlocal single-particle potential. So, while (4.24) is most easily derived in the Feynman gauge, the Coulomb gauge propagator (4.8) finally leads to the same $E_{\rm x}$. The correlation energy $E_{\rm c}$ of RDFT is then given by $E_{\rm c} = E_{\rm xc} - E_{\rm x}$.

Variants of the single-particle equations (4.20) are obtained for the other versions of RDFT. Starting from the zeroth component of (4.15),

$$n(\boldsymbol{r}) = \sum_k \Theta_k \phi_k^\dagger(\boldsymbol{r})\phi_k(\boldsymbol{r}), \tag{4.26}$$

the self-consistent equations of the purely n-dependent formalism (4.10) have the same form as the time-like component of Equation (4.20),

$$\{-\mathrm{i}c\boldsymbol{\alpha}\cdot\nabla + (\beta - 1)mc^2 + v_{\rm s}(\boldsymbol{r})\}\phi_k(\boldsymbol{r}) = \epsilon_k\phi_k(\boldsymbol{r}), \tag{4.27}$$

$$v_{\rm s}(\boldsymbol{r}) = v_{\rm ext}(\boldsymbol{r}) + v_{\rm H}(\boldsymbol{r}) + \frac{\delta E_{\rm xc}[n]}{\delta n(\boldsymbol{r})} + \frac{\delta E_{\rm H}^{\rm T}[\boldsymbol{j}[n]]}{\delta n(\boldsymbol{r})}, \tag{4.28}$$

with $v_{\rm H} \equiv v_{\rm H}^0$ (MacDonald and Vosko 1979). In Equations (4.27), (4.28) we have used the fact that $\boldsymbol{j}$ can be understood as a functional of n, $\boldsymbol{j}[n] = \langle\Phi[n]|\hat{\boldsymbol{j}}|\Phi[n]\rangle$. This not only allows the exact inclusion of $E_{\rm H}^{\rm T}$, but also relates the j-dependent $E_{\rm xc}$ of Equation (4.16) to the purely n-dependent $E_{\rm xc}$ of the present RDFT variant, $E_{\rm xc}[n] \equiv E_{\rm xc}[n, \boldsymbol{j}[n]]$. In practice, however, the functional $\boldsymbol{j}[n]$ is not at all known, so we usually simply neglect $E_{\rm H}^{\rm T}$ at this point. Note that for the large class of time-reversal invariant systems (closed shells), $\boldsymbol{j}$ vanishes, so that $E_{\rm H}^{\rm T}$ does not contribute anyway.

For the Hamiltonian (4.11) we start with a single-particle representation of the charge and the magnetization density rather than of the full j^μ,

$$n(\boldsymbol{r}) = \sum_k \Theta_k \phi_k^\dagger(\boldsymbol{r})\phi_k(\boldsymbol{r}), \qquad \boldsymbol{m}(\boldsymbol{r}) = -\mu_{\rm B}\sum_k \Theta_k \phi_k^\dagger(\boldsymbol{r})\beta\boldsymbol{\Sigma}\phi_k(\boldsymbol{r}). \tag{4.29}$$

The total energy can then be decomposed as in (4.17) with $E_{\rm ext}$ replaced by

$$E'_{\rm ext} = \int \mathrm{d}^3r\,\{n(\boldsymbol{r})v_{\rm ext}(\boldsymbol{r}) + \boldsymbol{m}(\boldsymbol{r})\cdot\boldsymbol{B}_{\rm ext}(\boldsymbol{r})\}. \tag{4.30}$$

In order to simplify the resulting single-particle equations, we next absorb $E_{\rm H}^{\rm T}$ into $E_{\rm xc}$ (MacDonald and Vosko 1979), relying on the fact that $\boldsymbol{j}$ is a unique functional of n, $\boldsymbol{m}$. However, as for Equations (4.27), (4.28) this usually implies the neglect of $E_{\rm H}^{\rm T}$. With this redefinition/approximation, Equation (4.13) leads to (Eschrig *et al.* 1985; Ramana and Rajagopal 1981a)

$$\{-\mathrm{i}c\boldsymbol{\alpha}\cdot\nabla + (\beta - 1)mc^2 + v_{\rm s} - \mu_{\rm B}\beta\boldsymbol{\Sigma}\cdot\boldsymbol{B}_{\rm s}\}\phi_k(\boldsymbol{r}) = \epsilon_k\phi_k(\boldsymbol{r}), \tag{4.31}$$

$$v_{\rm s}(\boldsymbol{r}) = v_{\rm ext}(\boldsymbol{r}) + v_{\rm H}(\boldsymbol{r}) + \frac{\delta E_{\rm xc}[n, \boldsymbol{m}]}{\delta n(\boldsymbol{r})}, \tag{4.32}$$

$$\boldsymbol{B}_{\rm s}(\boldsymbol{r}) = \boldsymbol{B}_{\rm ext}(\boldsymbol{r}) + \frac{\delta E_{\rm xc}[n, \boldsymbol{m}]}{\delta \boldsymbol{m}(\boldsymbol{r})}. \tag{4.33}$$

We can now easily take the limit $\boldsymbol{B}_{\text{ext}} = \boldsymbol{0}$. Equations (4.29)–(4.33) provide the appropriate starting point for density functional studies of magnetic systems. They have nevertheless not yet found widespread use due to their rather complex structure. Only recently, the first calculations with truly noncollinear $\boldsymbol{m}$ have been performed (Eschrig and Servedio 1999; Nordström and Singh 1996). While ground states with noncollinear $\boldsymbol{m}$ were found for a number of solids (see, for example, Sandratskii 1998), noncollinearity turned out to be only of limited importance for open-shell atoms (Eschrig and Servedio 1999). Moreover, presently only the exchange contribution to the RLDA for $E_{\text{xc}}[n, \boldsymbol{m}]$ is available (MacDonald 1983; Ramana and Rajagopal 1979, 1981a; Xu *et al.* 1984).

We are thus led to consider the RDFT formalism for collinear $\boldsymbol{m}$, Equation (4.14), which serves as a standard tool for the discussion of magnetic systems. The corresponding single-particle equations follow from Equations (4.29)–(4.33) by restriction to the z-component of $\boldsymbol{m}$. A particularly useful form of the equations for collinear $\boldsymbol{m}$ is found in terms of the generalized spin-densities $n_\pm$,

$$n_\pm(\boldsymbol{r}) = \frac{1}{2}\left[n(\boldsymbol{r}) \mp \frac{1}{\mu_{\text{B}}} m_z(\boldsymbol{r})\right] = \sum_k \Theta_k \phi_k^\dagger(\boldsymbol{r}) \frac{1 \pm \beta\Sigma_z}{2} \phi_k(\boldsymbol{r}). \tag{4.34}$$

Setting $B_{\text{ext},z} = 0$, we obtain

$$\{-ic\boldsymbol{\alpha} \cdot \nabla + (\beta - 1)mc^2 + \tfrac{1}{2}(1 + \beta\Sigma_z)v_{\text{s}+} + \tfrac{1}{2}(1 - \beta\Sigma_z)v_{\text{s}-}\}\phi_k = \epsilon_k \phi_k, \tag{4.35}$$

$$v_{\text{s},\sigma}(\boldsymbol{r}) = v_{\text{ext}}(\boldsymbol{r}) + v_{\text{H}}(\boldsymbol{r}) + v_{\text{xc},\sigma}(\boldsymbol{r}), \tag{4.36}$$

$$v_{\text{xc},\sigma}(\boldsymbol{r}) = \frac{\delta E_{\text{xc}}[n_+, n_-]}{\delta n_\sigma(\boldsymbol{r})} = \frac{\delta E_{\text{xc}}[n, m_z]}{\delta n(\boldsymbol{r})} - \operatorname{sgn}(\sigma)\mu_{\text{B}} \frac{\delta E_{\text{xc}}[n, m_z]}{\delta m_z(\boldsymbol{r})}. \tag{4.37}$$

Given the explicit form of the projection matrices in Equations (4.34), (4.35),

$$\tfrac{1}{2}(1 + \beta\Sigma_z) = \begin{pmatrix} 1 & 0 & 0 & 0 \\ 0 & 0 & 0 & 0 \\ 0 & 0 & 0 & 0 \\ 0 & 0 & 0 & 1 \end{pmatrix}, \qquad \tfrac{1}{2}(1 - \beta\Sigma_z) = \begin{pmatrix} 0 & 0 & 0 & 0 \\ 0 & 1 & 0 & 0 \\ 0 & 0 & 1 & 0 \\ 0 & 0 & 0 & 0 \end{pmatrix},$$

Equations (4.34)–(4.37) are immediately identified as the relativistic extension of the standard form of nonrelativistic spin-density functional theory.

Clearly, the choice of the appropriate variant of RDFT depends on the system under consideration. As already indicated, however, the availability of suitable approximations for E_{xc} is similarly important. Even within the purely n-dependent form of RDFT the relativistic $E_{\text{xc}}[n]$ is not identical with the nonrelativistic xc-functional. In $E_{\text{xc}}[n]$, relativity not only enters via the relativistic form of n, but also shows up in the functional dependence of E_{xc} on n. In applications, however, these corrections are often neglected, e.g. by the use of nonrelativistic spin-density functionals with the relativistic $n_\pm$ in Equation (4.37). Of course, the large variety of nonrelativistic forms for $E_{\text{xc}}[n_\uparrow, n_\downarrow]$ which have been suggested in the literature cannot be reviewed here; the interested reader is referred to Dreizler and Gross (1990). An overview of the few available relativistic forms for E_{xc} is given in the next two sections.

4.3 Implicit Density Functionals

4.3.1 Optimized potential method

The exact exchange (4.24) immediately raises the question whether orbital-dependent xc-functionals can be utilized in practice. As in the case of T_s, Equation (4.18), we can use the fact that, via the relativistic HK theorem for noninteracting particles, the KS orbitals are unique functionals of j^μ, $\phi_k[j^\mu]$. This allows the replacement of the functional derivative of E_{xc} with respect to j^μ required for the evaluation of v_{xc}^μ by functional derivatives to the ϕ_k and the corresponding KS eigenvalues ϵ_k, which leads to the ROPM integral equation of RDFT (Engel *et al.* 1998a) (in the no-pair approximation),

$$\int d^3r' \, \chi_0^{\mu\nu}(\boldsymbol{r}, \boldsymbol{r}') v_{xc,\nu}(\boldsymbol{r}') = \Lambda_{xc}^\mu(\boldsymbol{r}), \tag{4.38}$$

with χ_0 denoting the static response function of the KS system,

$$\chi_0^{\mu\nu}(\boldsymbol{r}, \boldsymbol{r}') = -\sum_k \Theta_k \phi_k^\dagger(\boldsymbol{r}) \alpha^\mu G_k(\boldsymbol{r}, \boldsymbol{r}') \alpha^\nu \phi_k(\boldsymbol{r}') + \text{c.c.}, \tag{4.39}$$

$$\Lambda_{xc}^\mu(\boldsymbol{r}) = -\sum_k \int d^3r' \left[\phi_k^\dagger(\boldsymbol{r}) \alpha^\mu G_k(\boldsymbol{r}, \boldsymbol{r}') \frac{\delta E_{xc}}{\delta \phi_k^\dagger(\boldsymbol{r}')} + \text{c.c.} \right] + \sum_k j_k^\mu(\boldsymbol{r}) \frac{\partial E_{xc}}{\partial \epsilon_k}, \tag{4.40}$$

$$G_k(\boldsymbol{r}, \boldsymbol{r}') = \sum_{l \neq k} \frac{\phi_l(\boldsymbol{r}) \, \phi_l^\dagger(\boldsymbol{r}')}{\epsilon_l - \epsilon_k}, \qquad j_k^\mu(\boldsymbol{r}) = \phi_k^\dagger(\boldsymbol{r}) \alpha^\mu \phi_k(\boldsymbol{r}). \tag{4.41}$$

Equations (4.38)–(4.41) are easily reduced to the purely density-dependent form of RDFT (Engel *et al.* 1995a; Shadwick *et al.* 1989). Recently, the ROPM equation for the $(n, \boldsymbol{m})$-version of RDFT has also been formulated (Auth 1999).

The numerical solution of Equations (4.38)–(4.41) is rather involved, due to the fact that the evaluation of G_k requires a summation over the complete KS spectrum. In the nonrelativistic case the semi-analytical Krieger–Li–Iafrate (KLI) approximation (Krieger *et al.* 1990) for (4.38)–(4.41) proved to be very accurate for atoms (Krieger *et al.* 1992), molecules (Engel *et al.* 2000b) and solids (Schmid 2000). The KLI approximation can be most easily extended into the relativistic domain by use of a closure approximation, $\epsilon_l - \epsilon_k \approx \Delta\bar{\epsilon}$, for G_k (Engel *et al.* 1998a),

$$\begin{aligned} v_{xc}^\mu(\boldsymbol{r}) = {} & \frac{1}{2n(\boldsymbol{r})} \left\{ -g_\nu^\mu + g^{\mu 0} g_{\nu 0} + \frac{j^\mu(\boldsymbol{r}) j_\nu(\boldsymbol{r})}{j^\rho(\boldsymbol{r}) j_\rho(\boldsymbol{r})} \right\} \\ & \times \left\{ \sum_k \left[\phi_k^\dagger(\boldsymbol{r}) \alpha^\nu \frac{\delta E_{xc}}{\delta \phi_k^\dagger(\boldsymbol{r})} - j_k^\nu(\boldsymbol{r}) \bar{e}_k + \text{c.c.} \right] + 2 \sum_k \Theta_k j_k^\nu(\boldsymbol{r}) \bar{v}_k \right\}, \end{aligned} \tag{4.42}$$

$$\bar{e}_k = \int d^3r \, \phi_k^\dagger(\boldsymbol{r}) \frac{\delta E_{xc}}{\delta \phi_k^\dagger(\boldsymbol{r})}, \qquad \bar{v}_k = \int d^3r \, j_{k,\lambda}(\boldsymbol{r}) v_{xc}^\lambda(\boldsymbol{r}), \tag{4.43}$$

Table 4.1 Exchange-only ground-state energies from ROPM and RHF calculations for noble gas atoms: Coulomb (C) and Coulomb–Breit (C + B) limit in comparison with complete transverse exchange (C + T) (Engel *et al.* 1998a). For the RHF approximation the energy difference with respect to the ROPM is given, $\Delta E = E_{\rm tot}({\rm RHF}) - E_{\rm tot}({\rm ROPM})$, providing results from (a) finite-differences calculations (Dyall *et al.* 1989) and (b) a basis-set expansion (Ishikawa and Koc 1994). All energies in mHartree. $v_{\rm ext}$ and c as in Ishikawa and Koc (1994).

Atom	$-E_{\rm tot}^{\rm C}$ ROPM	$\Delta E^{\rm C}$ RHF[a]	$\Delta E^{\rm C}$ RHF[b]	$-E_{\rm tot}^{\rm C+B}$ ROPM	$\Delta E^{\rm C+B}$ RHF[b]	$-E_{\rm tot}^{\rm C+T}$ ROPM
He	2 862	0	0	2 862	0	2 862
Ne	128 690	−2	−2	128 674	−2	128 674
Ar	528 678	−5	−5	528 546	−5	528 546
Kr	2 788 849	−13	−12	2 787 423	−12	2 787 431
Xe	7 446 882	−19	−6	7 441 115	−3	7 441 179
Rn	23 601 947	−35	−19	23 572 625	11	23 573 332

where, consistent with the closure approximation, the $\partial E_{\rm xc}/\partial\epsilon_k$-contribution to (4.40) has been neglected. Alternatively, Equations (4.42), (4.43) may be derived from a rearranged form of (4.38)–(4.41) (Kreibich *et al.* 1998).

4.3.2 Results for the exact exchange

The x-only ground-state energies of noble gas atoms obtained by solution of Equations (4.38)–(4.41) for different forms of the electron–electron interaction are listed in Table 4.1.

In the Coulomb limit a direct comparison with fully numerical RHF calculations is possible. Due to the multiplicative nature of $v_{\rm x}^{\rm OPM}$, the OPM energies are higher than the RHF data. The actual differences, however, are extremely small. As a consequence, basis-set limitations easily dominate over these conceptual differences, as can be seen from the Coulomb–Breit energies in Table 4.1. Finally, the comparison of the Coulomb–Breit values with those found by inclusion of the complete $E_{\rm x}^{\rm T}$, demonstrates the size of the retardation corrections to the Breit interaction. It is obvious that these corrections are only relevant for truly heavy atoms.

The importance of a self-consistent treatment of the transverse interaction is examined in Table 4.2.

The fully self-consistent handling is compared with a perturbative evaluation of only the beyond-Breit terms and a perturbative treatment of the complete $E_{\rm x}^{\rm T}$. Even for the heaviest atoms the perturbative evaluation of the retardation corrections to the Breit term seems to be sufficient. On the other hand, use of first-order perturbation theory for the complete $E_{\rm x}^{\rm T}$ leads to errors of the order of 1 eV for heavy atoms. An accurate description of inner shell transitions in these systems requires the inclusion of second-order Breit corrections.

Table 4.2 Exchange-only ground-state energies from ROPM, RHF, RKLI, RLDA and RGGA(PW91) calculations for noble gas atoms. In the case of the ROPM the self-consistent (SC) treatment of the complete $E_{\mathrm{x}}^{\mathrm{T}}$ (T) is compared with a self-consistent inclusion of only its Breit (B) limit (together with a first-order perturbative (PT) calculation of the beyond Breit contributions (T−B)) as well as a fully perturbative evaluation of $E_{\mathrm{x}}^{\mathrm{T}}$. The latter procedure has also been used for the RHF calculations (Dyall *et al.* 1989). All energies are in mHartree (Engel *et al.* 1998a). Reproduced with permission from Engel *et al.* (1998a) © 1998 Kluwer Academic/Plenum Publishers.

	$-E_{\mathrm{tot}}^{\mathrm{C+T}}$	$E_{\mathrm{tot}}^{\mathrm{C+T}} - E_{\mathrm{tot}}^{\mathrm{C+T}}$[ROPM: s(C + T)]						
	ROPM	ROPM		RHF	RKLI	RLDA	RGGA	GGA
SC:	C + T	C + B	C	C	C + T	C + T	C + T	C
PT:	—	T–B	T	T	—	—	—	—
He	2 862	0	0	0	0	138	6	6
Ne	128 674	0	0	−2	1	1 080	−24	−43
Ar	528 546	0	0	−5	2	2 458	41	−111
Kr	2 787 429	0	2	−12	3	6 543	−22	−1 683
Xe	7 441 173	1	10	−11	6	13 161	83	−6 705
Rn	23 573 354	8	68	29	9	35 207	−9	−35 145

Table 4.2 also demonstrates the accuracy of the KLI approximation in the relativistic situation. In fact, for heavy elements the differences between the KLI and the full OPM energies are smaller than those resulting from a perturbative treatment of the transverse interaction.

4.3.3 Correlation

Given the possibility of using the exact exchange in actual applications, we need a correlation functional which can be combined with the exact E_{x}. The most appropriate form of such an E_{c} is an open question even in the nonrelativistic case (Engel and Facco Bonetti 2000; Facco Bonetti *et al.* 2001; Görling and Levy 1994; Grabo and Gross 1995; Kotani 1998; Seidl *et al.* 2000). In most applications the exact E_{x} has thus been augmented by the LDA or GGA for correlation (Bylander and Kleinman 1995a,b, 1996, 1997; Chen *et al.* 1996; Engel and Dreizler 1999; Kim *et al.* 1999; Kotani 1994, 1995; Kotani and Akai 1996; Städele *et al.* 1997). However, this strategy does not lead to a consistent improvement over x-only results (Engel and Dreizler 1999; Kim *et al.* 1999). Error cancellation between exchange and correlation plays an important role for the success of the LDA and is also relevant in the case of the GGA. Conceptually, a fully nonlocal, orbital-dependent approximation for E_{c} appears to be most adequate. Presently, the most promising scheme for the derivation of such a functional is perturbation theory on the basis of the auxiliary KS Hamiltonian (Görling and Levy 1994; Sham 1985). This approach can be directly extended into the relativistic domain (Engel *et al.* 1998a), including all transverse and vacuum corrections.

Table 4.3 Coulomb correlation energies of the helium isoelectronic series: $E_c^{(2)}$ (OPM) versus CS (Colle and Salvetti 1975), LDA (Vosko *et al.* 1980), GGA (Perdew *et al.* 1992), MP2 (Ishikawa and Koc 1994) and exact results (Davidson *et al.* 1991). Both the nonrelativistic values E_c^{NR} and the difference between the relativistic numbers and E_c^{NR} are given (all energies in mHartree). Reproduced with permission from Engel and Dreizler (1999). © 1999 Academic Press.

	$-E_c^{NR}$						$E_c^{NR} - E_c^{R}$	
Ion	LDA	GGA	CS	OPM	MP2	exact	ROPM	RMP2
He	112.8	45.9	41.6	48.21	37.14	42.04	0.00	0.00
Ne^{8+}	203.0	61.7	40.6	46.81	44.37	45.69	−0.07	−0.07
Zn^{28+}	267.2	71.3	33.2	46.67	45.71	46.34	−0.19	−0.19
Sn^{48+}	297.7	76.0	30.0	46.65	45.98	46.47	0.72	0.69
Yb^{68+}	318.0	79.3	28.2	46.63		46.53	3.71	
Th^{88+}	333.2	81.7	27.0	46.62		46.56	11.00	

In principle, not only low-order perturbative E_c can be obtained in this way, but also resummed forms like the RPA (Engel and Facco Bonetti 2000). In practice, however, the resulting functionals are computationally much more demanding than the exact E_x, so that until now only the lowest-order contribution has been applied. Within the no-pair approximation and neglecting the transverse interaction, this second-order term reads

$$\left.\begin{aligned} E_c^{(2)} &= E_c^{MP2} + E_c^{\Delta HF}, \\ E_c^{MP2} &= \frac{1}{2} \sum_{ijkl;\epsilon_F<\epsilon_k,\epsilon_l} \Theta_i \Theta_j \frac{(ij||kl)[(kl||ij) - (kl||ji)]}{\epsilon_i + \epsilon_j - \epsilon_k - \epsilon_l}, \\ E_c^{\Delta HF} &= \sum_{il;\epsilon_F<\epsilon_l} \frac{\Theta_i}{\epsilon_i - \epsilon_l} \left| \langle i|\alpha_\mu v_x^\mu|l\rangle + e^2 \sum_j \Theta_j (ij||jl) \right|^2, \\ (ij||kl) &= \int d^3r_1 \int d^3r_2 \frac{\phi_i^\dagger(r_1)\phi_k(r_1)\phi_j^\dagger(r_2)\phi_l(r_2)}{|r_1 - r_2|}, \end{aligned}\right\} \tag{4.44}$$

where $\langle i|\alpha_\mu v_x^\mu|l\rangle = \int d^3r\, \phi_i^\dagger(r)\alpha_\mu \phi_l(r) v_x^{C,\mu}(r)$.

Illustrative results obtained by a perturbative evaluation of this functional on the basis of a self-consistent calculation with the exact E_x^C are given in Table 4.3 and Figures 4.1 and 4.2. Table 4.3 lists the correlation energies of the helium isoelectronic series, separating the nonrelativistic from the relativistic contribution.

As is well known, the LDA overestimates the exact E_c of neutral atoms by roughly a factor of 2. This error increases to a factor of 5 or more for highly charged ions. Moreover, while the PW91-GGA is rather close to the exact E_c for neutral helium, the error increases to a factor of 2 for Fm^{98+}. Obviously, these explicit density functionals do not scale properly with Z. The same is true for the orbital-dependent Colle–

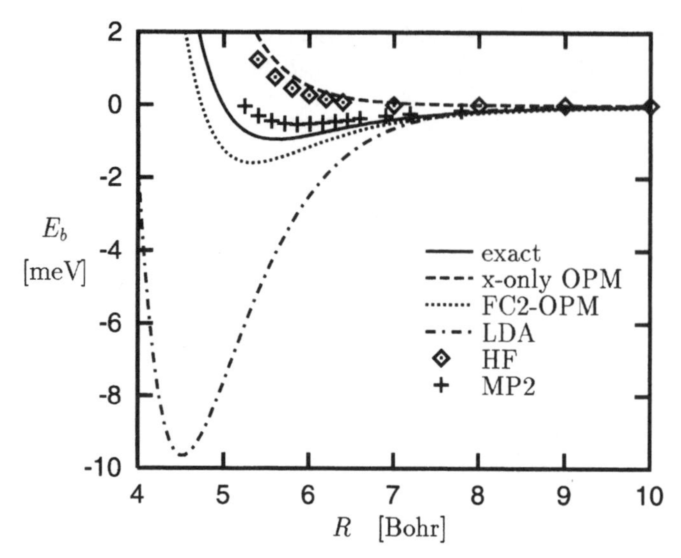

Figure 4.1 Energy surface of He_2: X-only and correlated OPM data (FC2 $\equiv E_{\rm x} + E_{\rm c}^{(2)}$) versus LDA, HF (Silver 1980), MP2 (Woon 1994) and exact (Aziz and Slaman 1991) results.

Salvetti (CS) functional, which, however, underestimates $E_{\rm c}$ by far. On the other hand, Equation (4.44) leads to very accurate $E_{\rm c}$ for the highly charged ions, reflecting its systematic origin. For neutral atoms $E_{\rm c}^{(2)}$ and the conventional MP2 energies (also listed in Table 4.3) bracket the exact values. Also, the relativistic corrections in $E_{\rm c}$ obtained with $E_{\rm c}^{(2)}$ are almost identical with their MP2 counterparts.

Table 4.3 shows that $E_{\rm c}^{(2)}$ is less accurate for neutral atoms than for positive ions. As an example of a negative ion, the most critical case, we consider Cs^-. For Cs^- $E_{\rm c}^{(2)}$ amounts to 3625 mHartree, which may be compared with the nonrelativistic value of 3593 mHartree. Lacking any information on the exact $E_{\rm c}$, we resort to an analysis of the corresponding electron affinities (EA), for which we obtain 1.31 eV (rel.) and 1.17 eV (nonrel.). These values are much larger than the experimental EA of 0.47 eV, which indicates the importance of higher-order correlation terms. In fact, the inclusion of the Epstein–Nesbet type diagrams in (4.44) reduces the EA to 0.46 eV. Nevertheless, even on the level of $E_{\rm c}^{(2)}$ this example demonstrates (i) the mere existence of negative ions within the OPM (which is due to the complete elimination of the electronic self-interaction by the exact $E_{\rm x}$), and (ii) the effect of relativity on the EA.

One can show that $E_{\rm c}^{(2)}$ includes the leading component of the dispersion force between two atoms (Engel *et al.* 1998a). Thus, from a fundamental point of view, $E_{\rm c}^{(2)}$ is the first xc-functional which allows a seamless description of van der Waals bond molecules. As an example we show the energy surface of He_2 in Figure 4.1 (Engel *et al.* 2000a).

As is well known, the LDA is not able to reproduce the van der Waals interaction, due to its local density-dependence. Figure 4.1 reflects this fact: in the LDA the He

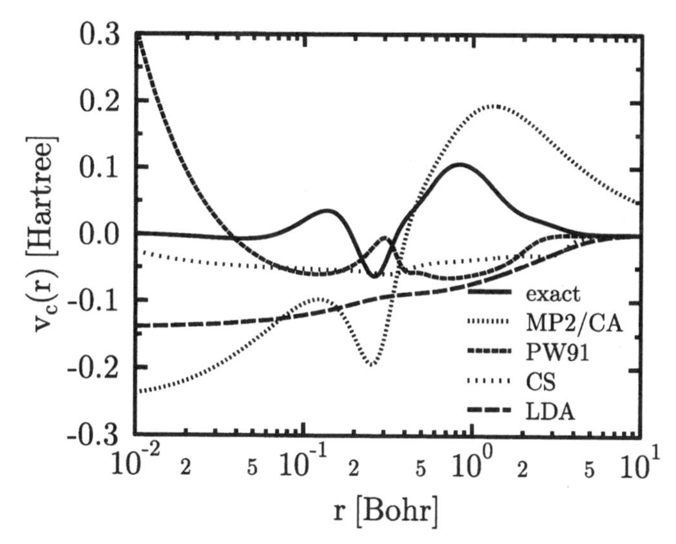

Figure 4.2 Correlation potential of Ne: exact v_c (Umrigar and Gonze 1994) versus LDA, PW91-GGA (Perdew *et al.* 1992) and CS (Colle and Salvetti 1975) results as well as closure approximation for $v_c^{(2)}$ (MP2/CA). Reproduced with permission from Facco Bonetti *et al.* (2001). © 2001 American Physical Society.

dimer is contracted until the individual atomic densities overlap substantially. On the other hand, the x-only OPM predicts He_2 to be unbound, consistent with the HF result. As for the behaviour of E_c for neutral atoms, the correlated OPM and the conventional MP2 results bracket the exact variational energy surface. Thus, while the need for higher-order correlation is again obvious, Figure 4.1 verifies the basic ability of functionals of the type (4.44) to describe dispersion forces.

All $E_c^{(2)}$-results discussed so far have been obtained by a perturbative evaluation of $E_c^{(2)}$ on the basis of an x-only OPM calculation. We may thus ask to what extent, for example, the comparatively poor EA found for Cs^- is due to the lack of the correlation component in v_s. A straightforward self-consistent application of $E_c^{(2)}$, however, is not only extremely involved, it also leads to a diverging v_c in the asymptotic regime (Facco Bonetti *et al.* 2001). This divergence originates from the dependence of (4.44) on unoccupied states, so that this problem can be avoided by use of a closure approximation in the OPM process (without affecting E_c) (Facco Bonetti *et al.* 2001). The resulting v_c for Ne is plotted in Figure 4.2.

Figure 4.2 exhibits the complete failure of the LDA and PW91-GGA for atomic v_c. The CS potential also has little in common with the exact v_c. The closure approximated potential corresponding to (4.44) is the first DFT potential which at least qualitatively follows the exact v_c. The need for the inclusion of higher-order contributions is apparent from the overestimation of the 'shell structure' of v_c. In addition, the asymptotic $1/r^4$-behaviour of the exact v_c is not reproduced. Clearly, perturbative correlation functionals like $E_c^{(2)}$ cannot be the final answer to the question of which E_c should be combined with the exact E_x.

4.4 Explicit Density Functionals

4.4.1 Local density approximation

In view of the computational demands of OPM calculations and the open question for a suitable orbital-dependent E_c explicitly density-dependent xc-functionals, i.e. the LDA or GGA, will remain the standard in RDFT applications in the nearer future. The RLDA is obtained from the xc-energy density $e_{xc}^{RHEG}(n_0)$ of the relativistic homogeneous electron gas (MacDonald and Vosko 1979; Rajagopal 1978; Ramana and Rajagopal 1981b) by substitution of the gas density n_0 by the local $n(\boldsymbol{r})$. Decomposing e_{xc}^{RHEG} into exchange and correlation we have

$$E_{xc}^{RDLA}[n] = \int d^3r \, [e_x^{HEG}(n(\boldsymbol{r}))\Phi_{x,0}(\beta) + e_c^{HEG}(n(\boldsymbol{r}))\Phi_{c,0}(\beta)], \tag{4.45}$$

where β is defined by

$$\beta = \frac{[3\pi^2 n(\boldsymbol{r})]^{1/3}}{mc}, \tag{4.46}$$

and both the exchange and correlation energy densities of the RHEG have been factorized into their respective nonrelativistic limits and relativistic corrections factors $\Phi_{x/c,0}(\beta)$. While $\Phi_{x,0}$ has been known for quite some time (Akhiezer and Peletminskii 1960),

$$\Phi_{x,0}(\beta) = 1 - \frac{3}{2}\left[\frac{\sqrt{1+\beta^2}}{\beta} - \frac{\text{Arsh}(\beta)}{\beta^2}\right], \tag{4.47}$$

$\Phi_{c,0}$ has only been evaluated within the RPA (Engel *et al.* 1995a; Ramana and Rajagopal 1981b). The resulting correction factor can be accurately parametrized in the form (Schmid *et al.* 1998)

$$\Phi_{c,0}^{RPA}(\beta) = \frac{1 + a_1\beta^3\ln(\beta) + a_2\beta^4 + a_3(1+\beta^2)^2\beta^4}{1 + b_1\beta^3\ln(\beta) + b_2\beta^4 + b_3[A\ln(\beta) + B]\beta^7}, \tag{4.48}$$

which incorporates the analytically known high-density limits of both the relativistic RPA as well as its nonrelativistic limit. Both $\Phi_{x,0}$ and $\Phi_{c,0}^{RPA}$ include the full transverse interaction, and, to some extent, vacuum corrections beyond the no-pair approximation (Engel and Dreizler 1996). As the relativistic form of all nonRPA contributions to e_c^{RHEG} is not known, the corresponding relativistic corrections must either be neglected (Engel *et al.* 1995a), using

$$E_c^{RDLA} = E_c^{RRPA} - E_c^{RPA} + E_c^{LDA},$$

or be taken into account in an empirical form, combining $\Phi_{c,0}^{RPA}$ with a parametrization for e_c^{HEG} which goes beyond the RPA (Schmid *et al.* 1998).

Applications of (4.45)–(4.48) to atoms, however, showed that the accuracy of the RLDA is not very satisfying (Das *et al.* 1980; Engel *et al.* 1995a; Ramana *et al.* 1982). This is illustrated in Table 4.2 and Figure 4.3.

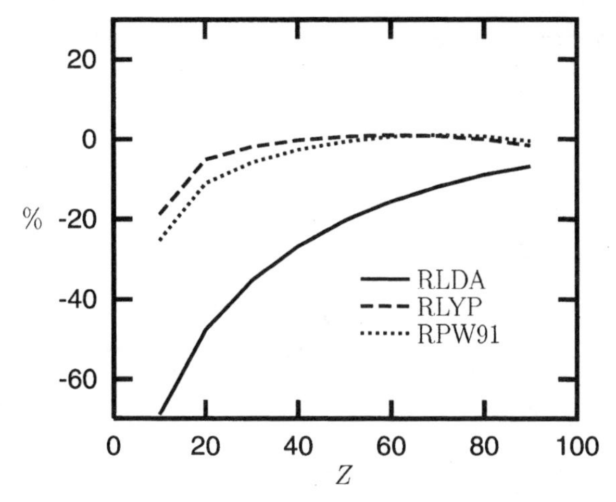

Figure 4.3 Relativistic contribution to E_{C}: percentage deviation of RLDA and RGGA results from relativistic MP2-data for Ne isoelectronic series. Reproduced with permission from Engel *et al.* (1998a) © 1998 Kluwer Academic/Plenum Publishers.

A rigorous reference standard for the analysis of the RLDA exchange is provided by x-only ROPM results. For heavy atoms the x-only RLDA ground-state energies are off by more than 10 Hartree, which is mainly due to the incomplete elimination of the K-shell self-interaction (see Table 4.2 (compare also MacDonald *et al.* 1981, 1982); for a relativistic version of the Perdew–Zunger self-interaction correction scheme (Perdew and Zunger 1981) see Rieger and Vogl (1995)). The picture is similar for the correlation energy. As discussed earlier, the nonrelativistic LDA overestimates atomic E_{c} by a factor of 2. At this point only the relativistic contribution to E_{c}, i.e. the performance of $\Phi_{\mathrm{c},0}^{\mathrm{RPA}}$, is of interest. In Figure 4.3 we plot the deviation of the relativistic correction evaluated from Equations (4.45), (4.48) for the Ne isoelectronic series from the corresponding MP2 data (Ishikawa and Koc 1994) (which can serve as reference values for the present purpose, in spite of the fact that they have been evaluated with the Coulomb–Breit interaction in Coulomb gauge and the no-pair approximation (compare Facco Bonetti *et al.* 1998)). As Figure 4.3 shows, the relativistic correction in E_{c} is overestimated by the RLDA, the error being as large as 70% for neutral atoms.

While Equations (4.45)–(4.48) correspond to a spin-saturated RHEG, the exchange energy has also been evaluated for a polarized RHEG (MacDonald 1983; Ramana and Rajagopal 1979, 1981a; Xu *et al.* 1984), which yields the input $E_{\mathrm{x}}[n, \boldsymbol{m}]$ to Equations (4.32), (4.33) or (4.37). As its $\boldsymbol{m}$-dependence is only implicitly given, this functional has only rarely been used (Cortona 1989; Cortona *et al.* 1985; Eschrig and Servedio 1999). For magnetic systems usually the nonrelativistic LDA is utilized with (4.37) (see, for example, Huhne *et al.* 1998). However, even for open-subshell atoms, the solution of (4.35) under the assumption of spherical symmetry is nontrivial. Several variants for handling the resulting intricate set of four coupled radial differential equa-

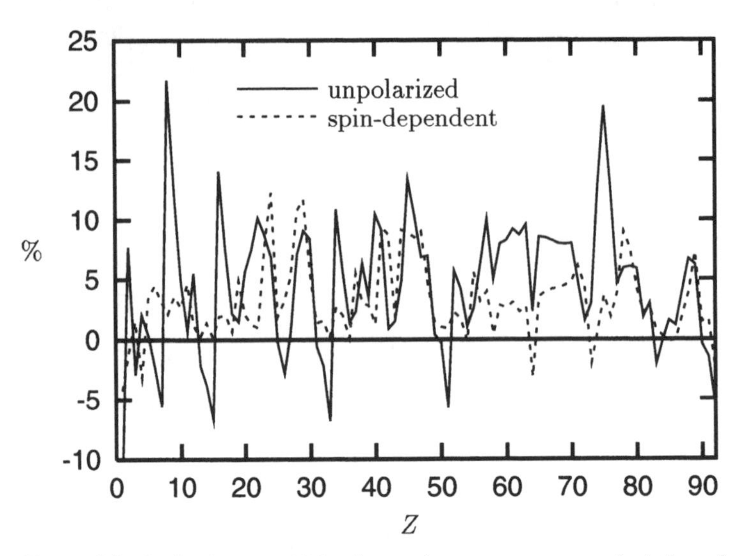

Figure 4.4 Ionization potentials of neutral atoms: percentage deviation of spin-dependent and unpolarized LDA from experiment (Lide 1996).

tions have been suggested in the literature (Cortona 1989; Cortona *et al.* 1985; Ebert 1989; Forstreuter *et al.* 1997; Yamagami *et al.* 1997). Their solution is particularly difficult for neutral atoms, so that the available results are mainly for ions. A new, more stable scheme has been developed recently (Engel *et al.* 2001a). This scheme relies on a careful analysis of the appropriate boundary conditions which allows the identification of a suitable criterion ('quantum number') to distinguish the 'spin-up' and 'spin-down' solutions. Its practical success is illustrated in Figure 4.4, in which the percentage deviation of the resulting ionization potentials (IPs) of neutral atoms from experiment is shown for the complete periodic table.

The spin-dependent treatment via (4.35) is compared with the solution of (4.27) (on the basis of the LDA). The error of the spin-dependent IP in general is considerably smaller than that of its unpolarized counterpart, most notably for the light atoms and the lanthanides. On the other hand, the error is still substantial for the 3d and 4d elements, so that the question of gradient corrections has to be raised.

4.4.2 Generalized gradient approximation

For a large variety of applications in quantum chemistry and condensed matter theory, the inclusion of gradient corrections to the LDA in the form of the generalized gradient approximation (GGA) (Becke 1988a; Perdew *et al.* 1992) turned out to be advantageous (Bagno *et al.* 1989; Becke 1992; Johnson *et al.* 1993a). The direct use of nonrelativistic GGAs in RDFT calculations, however, leads to substantial errors in total energies, as can be seen from Table 4.2. Consequently, a relativistic extension of the GGA (RGGA) is required. The most appropriate form of an RGGA for E_{x} is given by

$$E_{\mathrm{x}}^{\mathrm{RGGA}} = \int \mathrm{d}^3 r\, \mathrm{e}_{\mathrm{x}}^{\mathrm{HEG}}(n)[\Phi_{\mathrm{x},0}(\beta) + g(\xi)\Phi_{\mathrm{x},2}(\beta)], \tag{4.49}$$

with $\xi = [\nabla n/(2(3\pi^2 n)^{1/3} n)]^2$ and $g(\xi)$ being the gradient part of a nonrelativistic GGA (Engel *et al.* 1996). The correction factor $\Phi_{\mathrm{x},2}$ can, in principle, be calculated from the first-order response function of the RHEG. However, as this approach proved to be already very involved in the nonrelativistic limit (Antoniewicz and Kleinman 1985; Chevary and Vosko 1990; Engel and Vosko 1990) and as the gradient function $g(\xi)$ contains some semi-empirical information anyway, a semi-empirical approach to $\Phi_{\mathrm{x},2}$ should be sufficient. Following the strategy behind the Becke GGA (Becke 1988a), a reasonably accurate $\Phi_{\mathrm{x},2}$ may be obtained by making a sufficiently flexible ansatz (in the form of a Padé approximant) and fitting its coefficients to the exact relativistic E_{x} of a number of closed-subshell atoms (keeping the form of $g(\xi)$ fixed) (Engel *et al.* 1996, 1998b). Corresponding fits have been performed for the two most frequently used forms of GGAs (Becke 1988a; Perdew *et al.* 1992), the resulting $\Phi_{\mathrm{x},2}$ being very similar. As is demonstrated in Table 4.2 for the PW91-version, the RGGA leads to much more accurate atomic E_{x} (and v_{x}) than both the RLDA and the GGA.

The correlation functional requires a slightly different scheme, as, on the one hand, the RLDA is not known completely, and, on the other, some GGAs for E_{c} (Lee *et al.* 1988) are not based on the LDA. Therefore, only one overall correction factor for the complete correlation part of the GGA has been used,

$$E_{\mathrm{c}}^{\mathrm{RGGA}}[n] = \int \mathrm{d}^3 r\, \mathrm{e}_{\mathrm{c}}^{\mathrm{GGA}}(n, (\nabla n)^2, \dots)\Phi_{\mathrm{c}}^{\mathrm{GGA}}(\beta),$$

keeping the nonrelativistic form $e_{\mathrm{c}}^{\mathrm{GGA}}(n, (\nabla n)^2, \dots)$ fixed (Engel *et al.* 1998b). In view of the fact that the relativistic corrections to atomic E_{c} are much smaller than those to atomic E_{x} this less sophisticated approach should be sufficient. Again a Padé approximant has been used as ansatz for $\Phi_{\mathrm{c}}^{\mathrm{GGA}}$. Its coefficients have been fitted to the most systematic set of relativistic E_{c} available (MP2 results for the Ne isoelectronic series on the basis of the DCB Hamiltonian (Ishikawa and Koc 1994)), starting from two different $e_{\mathrm{c}}^{\mathrm{GGA}}$ (Lee *et al.* 1988; Perdew *et al.* 1992). As Figure 4.3 shows, atomic E_{c} are clearly improved by this RGGA.

In contrast to the rather inaccurate RLDA, the RGGA allows an examination of the importance of relativistic corrections to $E_{\mathrm{xc}}[n]$ for the properties of molecules and solids. This question has been investigated both for noble-metal compounds (Mayer *et al.* 1996; Varga *et al.* 1999) and for metallic gold and platinum (Schmid *et al.* 1998) within the framework of LAPW calculations (Blaha *et al.* 1995). Prototype results for Cu_2 and Au_2 are given in Table 4.4.

It turns out that even for Au, which usually exhibits the effects of relativity most clearly (Pyykkö 1988), the impact of the correction factors $\Phi_{\mathrm{x/c}}$ on the molecular binding properties is marginal, i.e. smaller than usual differences between two basis sets. It seems worthwhile pointing out that in the case of the dissociation energy the

Table 4.4 Spectroscopic parameters of noble-metal dimers: PP (Engel *et al.* 2001b) versus AE results for both LDA and BP86-GGA.

		Cu_2			Au_2		
Mode	E_{xc}	R_e (Bohr)	D_e (eV)	ω_e (cm^{-1})	R_e (Bohr)	D_e (eV)	ω_e (cm^{-1})
PP	LDA	4.04	2.83	304	4.61	3.02	198
AE	LDA[a,b]	4.05	2.86	307	4.64	3.00	196
PP	GGA				4.72	2.38	179
AE	GGA[a]	4.16	2.28	287	4.75	2.30	179
AE	RGGA[a]	4.17	2.27	285	4.76	2.27	177
AE	GGA[c]	4.20	3.27	283	4.78	(3.19)	188
AE	RGGA[c]	4.21	3.27	282	4.79	(3.17)	187
Exp.		4.20	2.05	265	4.67	2.30	191

[a]Mayer *et al.* (1996); [b]Liu and van Wüllen (2000); [c]Varga *et al.* (2000b), the D_e in parentheses correspond to unpolarized atomic ground-state energies.

Table 4.5 Lattice constant a_0 and cohesive energy E_{coh} of Au and Pt obtained from LAPW calculations with various xc-functionals in comparison to experiment (Brewer 1977; Khein *et al.* 1995). Reproduced with permission from Schmid *et al.* (1999). © 1999 Academic Press.

	Au		Pt	
	a_0 (Bohr)	$-E_{coh}$ (eV)	a_0 (Bohr)	$-E_{coh}$ (eV)
LDA	7.68	4.12	7.36	6.76
RLDA	7.68	4.09	7.37	6.73
GGA	7.87	2.91	7.51	5.34
RGGA	7.88	2.89	7.52	5.30
Exp.	7.67	3.78	7.40	5.85

similarity of GGA and RGGA results is due to the cancellation of the large relativistic corrections to the individual ground-state energies of the molecule and its constituents. The same observation is made for solids, as can be gleaned from Table 4.5.

On the other hand, a comparison of the (R)GGA results in Tables 4.4 and 4.5 with experiment reveals the limitations of GGAs for heavy elements. For both Au_2 and the Au metal the gradient terms overcorrect the errors of the LDA. It seems worthwhile noting that the particularly large deviations in the case of the metal are not due to the usual neglect of the spin–orbit coupling for the valence electrons within the LAPW

Table 4.6 Cohesive properties of 5d metals from LDA-LAPW calculations with (+SO) and without (−SO) spin–orbit coupling of the valence electrons.

		a_0 (Bohr)	$-E_{\text{coh}}$ (eV)	B_0 (GPa)
	−SO	5.93	10.47	335
W	+SO	5.94	10.72	316
	Exp.	5.98	8.90	310
	−SO	7.22	9.45	419
Ir	+SO	7.22	9.58	377
	Exp.	7.26	6.93	355
	−SO	7.67	4.29	214
Au	+SO	7.65	4.23	217
	Exp.	7.67	3.78	171

approach. Table 4.6 provides a comparison of LAPW results with and without spin–orbit coupling for some 5d metals (on the basis of the LDA).

While spin–orbit coupling contributes significantly to the cohesive energy, its effect is too small to explain the differences between GGA and experimental data in Table 4.5. Thus, on the one hand, the results in Tables 4.4 and 4.5 illustrate the role of error cancellation, in particular for the LDA. On the other hand, they indicate the need for fundamentally new concepts for $E_{\text{xc}}[n]$ (such as implicit functionals) in the relativistic regime.

4.5 Norm-Conserving Pseudopotentials

A particularly efficient method for the inclusion of relativity in electronic structure calculations is the pseudopotential (PP) approach. In the framework of DFT usually norm-conserving PPs (Bachelet *et al.* 1982; Hamann *et al.* 1979; Troullier and Martins 1991) are applied for this purpose. The standard form of norm-conserving PPs is given by

$$\langle \boldsymbol{r}|\hat{v}_{\text{ps}}|\boldsymbol{r}'\rangle = v_{\text{loc}}(r)\delta^{(3)}(\boldsymbol{r}-\boldsymbol{r}') + \frac{\delta(r-r')}{r^2} \times \sum_{l=0}^{L}\sum_{j=l\pm 1/2}\left[\frac{2j+1}{4l+2}v_{\text{ps},lj}(r) - v_{\text{loc}}(r)\right]\sum_{m=-l}^{l} Y_{lm}(\Omega)Y_{lm}^{*}(\Omega'), \tag{4.50}$$

where the individual components $v_{\text{ps},lj}$ correspond to the various relevant valence states of the atom (and as usual a local component v_{loc} has been extracted from the $v_{\text{ps},lj}$). In (4.50) a j-average is used to generate j-independent PPs as DFT PP calculations are typically based on nonrelativistic (spin-) density functional theory. However, the $v_{ps,lj}$ could equally well be utilized in relativistic PP calculations, similar to energy-adjusted PPs (Seth *et al.* 1997). The construction of the $v_{ps,lj}$ proceeds in three steps. First an all-electron (AE) calculation is performed for the atom of interest, utilizing the n-dependent version of RDFT. Next, a screened PP $v^{sc}_{ps,lj}$ is generated from the results of the AE calculation, requiring the lowest eigenstate obtained from $v^{sc}_{ps,lj}$, the pseudo-orbital (PO), to be identical with the corresponding AE orbital for r larger than a suitably chosen cut-off radius $r_{c,l}$. Finally, the interaction among the valence electrons is eliminated from $v^{sc}_{ps,lj}$. The simplest form for this unscreening reads

$$v_{ps,lj}(r) = v^{sc}_{ps,lj}(r) - v_H([n_{v,ps}];r) - v_{xc}([n_{v,ps}];r), \tag{4.51}$$

where $n_{v,ps}$ is the valence density obtained from the POs. In (4.51) the nonlinear contributions to the core–valence interaction, resulting from the nonlinearity of $E_{xc}[n]$, are neglected. While this is an acceptable approximation for first- and second-row atoms, the nonlinearity of the core–valence interaction cannot be ignored in LDA or GGA calculations for many others, most notably the transition-metal elements. In this case, the inclusion of nonlinear core corrections (NLCCs) is necessary (Louie *et al.* 1982).

Among the various schemes for the construction of $v^{sc}_{ps,lj}$ presently the Troullier–Martins (TM) form (Troullier and Martins 1991) seems to be most widely used. However, in contrast to the original approach (Bachelet *et al.* 1982; Hamann *et al.* 1979) the TM scheme has been formulated for nonrelativistic situations. A consistent relativistic extension (Engel *et al.* 2001b) is presented in the next section.

4.5.1 Relativistic Troullier–Martins scheme

For atoms Equation (4.27) reduces to two coupled radial equations for the large and small components of the orbitals,

$$c\left(\partial_r + \frac{\kappa}{r}\right)a_{nlj}(r) = (2mc^2 - v_s(r) + \epsilon_{nlj})b_{nlj}(r), \tag{4.52}$$

$$c\left(\partial_r - \frac{\kappa}{r}\right)b_{nlj}(r) = (v_s(r) - \epsilon_{nlj})a_{nlj}(r) \tag{4.53}$$

(Equations (4.52), (4.53) imply a spherical average in the case of open subshells, $\kappa = -2(j-l)(j+\frac{1}{2})$.) The corresponding components of the PO satisfy Equations (4.52), (4.53) with v_s replaced by $v^{sc}_{ps,lj}$. Given the AE solutions of Equations (4.52), (4.53) the explicit construction of $v^{sc}_{ps,lj}$ starts with an ansatz for the large component of the PO,

$$a_{ps,lj}(r) = \begin{cases} a_{nlj}(r) & \text{for } r > r_{c,l}, \\ r^{l+1}\exp[p(r)] & \text{for } r \leqslant r_{c,l}, \end{cases} \qquad p(r) = \sum_{i=0}^{6} c_{2i} r^{2i}.$$

For valence states for which the associated screened PP has a depth of no more than 100 Hartree, $v^{\rm sc}_{{\rm ps},lj}$ can then be extracted from (4.52), (4.53), using a weakly relativistic expansion,

$$v^{\rm sc}_{{\rm ps},lj} = \begin{cases} v_{\rm s} & \text{for } r > r_{{\rm c},l}, \\ v^{\rm sc,nr}_{{\rm ps},lj} + \delta v_{lj} & \text{for } r \leqslant r_{{\rm c},l}, \end{cases}$$

$$v^{\rm sc,nr}_{{\rm ps},lj} = \epsilon_{nlj} + \frac{l+1}{r}\frac{p'}{m} + \frac{p'' + (p')^2}{2m},$$

$$\delta v_{lj} = \frac{(v^{\rm sc,nr}_{{\rm ps},lj} - \epsilon_{nlj})^2}{2mc^2} + \frac{(v^{\rm sc,nr}_{{\rm ps},lj})'}{4m^2c^2}\left(\frac{a'_{{\rm ps},lj}}{a_{{\rm ps},lj}} + \frac{\kappa}{r}\right),$$

where the primes denote derivatives with respect to r. The corresponding small component $b_{{\rm ps},lj}$ of the PO then follows from (4.52). Finally, the coefficients c_{2i} are determined by requiring continuity of $a_{{\rm ps},lj}$ and its first four derivatives at $r_{{\rm c},l}$, proper normalization as well as a smooth PP at the origin, $(v^{\rm sc,nr}_{{\rm ps},lj})''(0) = 0$.

Prototype LDA and GGA results obtained with these PPs are given in Tables 4.4 and 4.7. As Table 4.4 demonstrates, the PP calculations reproduce the AE results for both Cu_2 (Mayer *et al.* 1996) and Au_2 (Liu and van Wüllen 2000) very accurately, both on the LDA and on the GGA level. Similar agreement is found for the transition-metal compounds listed in Table 4.7, for which, however, only nonrelativistic AE reference values (Castro and Salahub 1994; Engel *et al.* 2001b) are available.

Nevertheless, the relativistic corrections are not negligible even for these 3d elements. In fact, in the case of FeO relativity reduces the excitation energy from the $^5\Delta$ ground state to the first excited state ($^5\Sigma$) from the nonrelativistic value of 0.4 eV to 0.2 eV. On the other hand, the comparison of the LDA results with experiment clearly shows the need for nonlocal corrections. The GGA results are consistently closer to the experimental data, in particular for $R_{\rm e}$. The GGA values for $D_{\rm e}$ are nonetheless not completely satisfying, which underlines the importance of the truly nonlocal contributions to $E_{\rm xc}$.

4.5.2 Results for the exact exchange

The use of PPs is particularly attractive in the case of implicit xc-functionals. Even an x-only calculation within the KLI approximation (4.42), (4.43) is computationally more demanding than corresponding LDA or GGA calculations as an evaluation of the Slater integrals in (4.24) or (4.25) is required. PPs for the exact $E_{\rm x}$ have been introduced both on the KLI level (Bylander and Kleinman 1995a) and for the full OPM (Höck and Engel 1998; Moukara *et al.* 2000). However, it was noted very early on that standard norm-conserving PPs for the exact $E_{\rm x}$ suffer from a spurious long-range exchange component in the ionic PP (Bylander and Kleinman 1995a). This xc-tail originates from the nonlocality of $E_{\rm x}$, which leads to a contribution of the core states to the exchange potential in the valence regime. This feature of the

Table 4.7 Spectroscopic parameters of transition-metal compounds: PP versus AE (Castro and Salahub 1994; Engel *et al.* 2001b) results and experimental data (Cheung *et al.* 1981; Moskovits and DiLella 1980; Moskovits *et al.* 1984; Murad 1980; Purdum *et al.* 1982). For 3d elements the valence space includes the complete M shell. NLCCs have been used.

			nonrelativistic			relativistic		
	Mode	E_{xc}	R_e (Bohr)	D_e (eV)	ω_e (cm^{-1})	R_e (Bohr)	D_e (eV)	ω_e (cm^{-1})
Fe_2	AE	LDA	3.68	4.38	497			
$^7\Delta$	PP	LDA	3.68	4.31	440	3.66	3.95	451
	PP	GGA				3.77	2.67	414
	Exp.					3.82	1.30	300
FeO	AE	LDA	3.01	7.06	957			
$^5\Delta$	PP	LDA	2.99	7.00	968	2.97	6.80	984
	PP	GGA				3.04	5.36	913
	Exp.					3.06	4.06	881
FeO	AE	LDA	3.06	6.70	942			
$^5\Sigma$	PP	LDA	3.04	6.60	947	3.01	6.59	969

Table 4.8 Equilibrium lattice constant a, cohesive energy E_{coh} and bulk modulus B of FCC Al: Exact exchange in comparison with LDA results.

E_{xc}					
x	c	Mode	a (Bohr)	E_{coh} (eV/atom)	B (GPa)
Exact	—	OPM-PP	7.10	3.98	135
Exact	—	SC-OPM-PP	7.79	1.37	71
LDA	LDA	LDA-PP+NLCC	7.48	4.05	88
		Exp.	7.65	3.39	77

core–valence interaction cannot be eliminated by (4.51). In applications to molecules or solids the xc-tail leads to a spurious ionic force which prevents accurate structural optimizations, in particular for solids.

As an example we show FCC aluminium in Table 4.8. The original OPM-PP for the exact E_x yields a lattice constant which is much too small, although the bond length of Al_2 is overestimated by an AE-OPM calculation (Engel *et al.* 2001c).

In order to eliminate the spurious xc-tail, a self-consistent scheme for the generation of norm-conserving PP has been developed (Engel *et al.* 2001c). In this parameter-free scheme the screened PP is iterated until its asymptotic structure matches that of

Table 4.9 s–p and p–p transfer energies of first-row atoms: PP versus AE x-only OPM results versus LDA data (all energies in eV).

	$2s_\downarrow \longrightarrow 2p_\uparrow$				$2p_\uparrow \longrightarrow 2p_\downarrow$		
	Mode	LDA	OPM		Mode	LDA	OPM
Be	AE	2.47	1.67	C	AE	1.20	1.35
	PP	2.37	1.67		PP	1.34	1.35
	PP+NLCC	2.46			PP+NLCC	1.21	
B	AE	3.23	2.13	N	AE	2.70	3.24
	PP	3.00	2.16		PP	2.98	3.26
	PP+NLCC	3.21			PP+NLCC	2.73	
C	AE	4.05	2.46	O	AE	1.48	1.93
	PP	3.68	2.50		PP	1.62	1.95
	PP+NLCC	4.01			PP+NLCC	1.49	

$v_{xc}([n_{v,ps}]; r)$ in (4.51), so that the final unscreened PP $v_{ps,lj}$ has the correct ionic behaviour. The resulting PPs (SC-OPM-PP) substantially improve OPM results for molecules and solids, as can be seen from Table 4.8. While the cohesive energy reflects the missing correlation energy, both the lattice constant and the bulk modulus are clearly closer to the experimental values than the original OPM-PP data.

It seems worthwhile emphasizing that, apart from the fact that they lead to the spurious xc-tail, the nonlinear contributions to the core–valence interaction seem to be less important for the exact E_x than for the LDA. This is illustrated in Table 4.9, which lists atomic excitation energies associated with the transfer of an electron from one spin channel into the other. While in the case of the LDA, NLCCs are required for the accurate reproduction of these transfer energies, linear unscreening is sufficient for the exact E_x.

As already noted in Section 4.4, the exact E_x provides a standard that can be used to analyse conventional density functionals. As an example we show in Figure 4.5 the v_x of bulk aluminium obtained within the PP approach.

The solution of (4.38) (in a plane-wave basis) is compared with the KLI approximation as well as the corresponding LDA and PW91-GGA potentials (Schmid 2000). While the LDA is reasonably close to the exact v_x (at least if we exclude the vicinity of the atomic core), the GGA is much less accurate, in spite of the improved cohesive properties found with the GGA (Fuchs *et al.* 1998). Even in the interstitial regime the gradient contributions to the GGA potential have the wrong sign. This demonstrates once more that the performance of GGAs is to some extent based on the cancellation of local errors (Engel and Vosko 1993). The KLI approximation, on the other hand, just underestimates the shell structure in the exact v_x.

As a final example of the application of the exact E_x, the band structure of FeO in AFM-II structure is plotted in Figure 4.6.

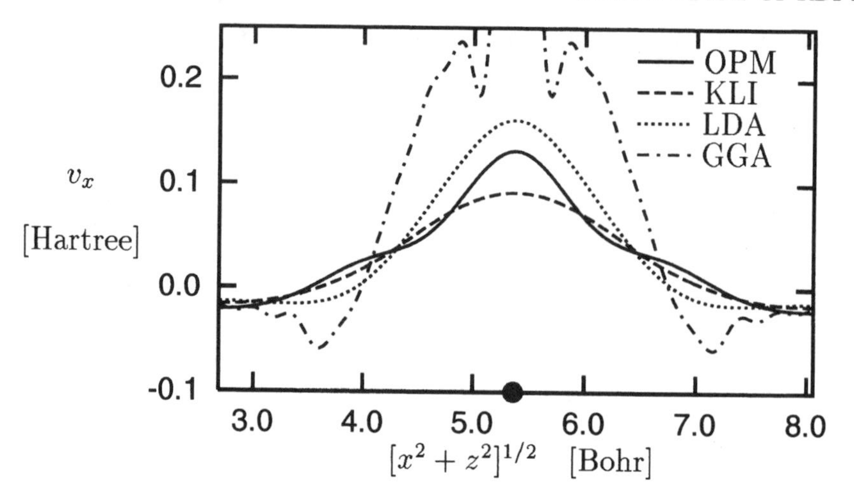

Figure 4.5 Exchange potential of FCC Al in [110] direction: full OPM versus KLI approximation, LDA and PW91-GGA ($E_{\text{cut}} = 100$ Ryd, 44 special k-points, 750 states (per k-point) in G_k; •, the position of the atom).

For this PP calculation the exact E_{x} has been combined with the LDA for E_{c}. In spite of the use of the exact exchange functional, FeO is predicted to be a metal, in contrast to experiment. At present, it is not clear whether this failure to reproduce the insulating ground state of FeO originates from the use of the LDA for E_{c} or from the technical limitations of the PP calculation (KLI approximation for v_{x}, only three special k-points for the integration over the Brillouin zone, 3s electrons in the core). It must be emphasized, however, that the band structure shown in Figure 4.6 is rather different from its LDA counterpart (Dufek *et al.* 1994), which emphasizes the importance of the exact v_{x} for this system.

4.6 Applications of RDFT using the Relativistic Discrete Variational Method

In this section a summary of four-component molecular density functional calculations within the RDVM (Rosén and Ellis 1975) is given. In the RDVM we restrict ourselves to the no-pair limit of RDFT, since QED effects are irrelevant for chemical bonding. On the other hand, all relevant relativistic effects are fully included *a priori*. The starting point for the molecular calculations is the purely n-dependent version of RDFT, Equations (4.26)–(4.28), with neglect of E_{H}^{T}. Accordingly, the total energy consists of the kinetic energy (4.18), the external energy (4.9) with inclusion of the associated ionic interaction energy,

$$E_{\text{ext}}[n] = \int \mathrm{d}^3 r\, n(\boldsymbol{r}) v_{\text{ext}}(\boldsymbol{r}) + \frac{1}{2} \sum_{\alpha} \sum_{\beta \neq \alpha} \frac{Z_\alpha Z_\beta e^2}{|\boldsymbol{R}_\alpha - \boldsymbol{R}_\beta|}, \tag{4.54}$$

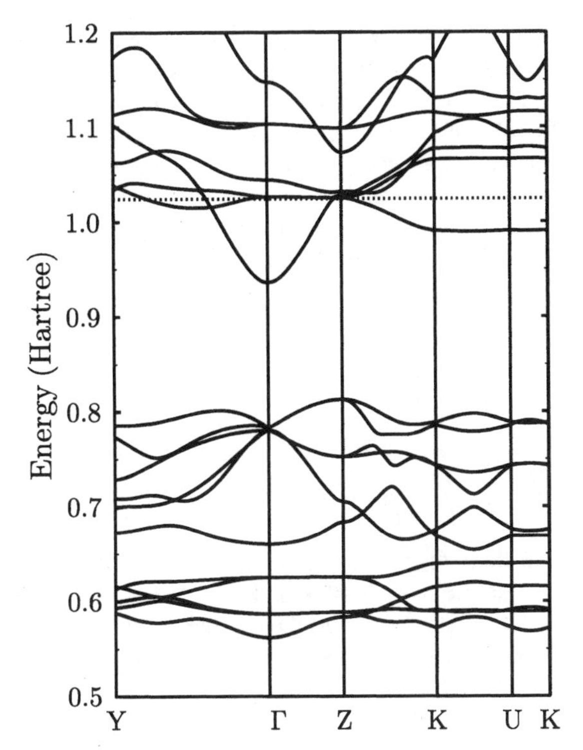

Figure 4.6 Band structure of FeO from OPM-PP calculation with exact E_{X} and LDA correlation. The dashed line represents ϵ_{F} (KLI approximation, 3p, 3d, 4s states of Fe in valence space, $E_{\mathrm{cut}} = 250$ Ryd, three special k-points).

where

$$v_{\mathrm{ext}}(\boldsymbol{r}) = -\sum_{\alpha} \frac{Z_{\alpha} e^2}{|\boldsymbol{R}_{\alpha} - \boldsymbol{r}|},$$

the Coulomb contribution to the Hartree energy (4.19) and the xc-energy. The molecular wave functions ϕ_k in the RDVM are linear combinations of numerical atom-centred Dirac spinors ξ_{μ},

$$\phi_k(\boldsymbol{r}) = \sum_{\mu} c_{k\mu} \xi_{\mu}(\boldsymbol{r}), \qquad \xi_{\mu}(\boldsymbol{r}) = \frac{1}{r} \begin{pmatrix} f_i^{n\kappa}(r) \mathcal{Y}_{\kappa}^{m}(\Omega) \\ i g_i^{n\kappa}(r) \mathcal{Y}_{-\kappa}^{m}(\Omega) \end{pmatrix}$$

($\mu \equiv (in\kappa m)$). An important computational simplification is obtained by the decomposition of the molecular orbital basis into a symmetry-adapted basis according to the irreducible representations of the molecular point (double) group (Meyer *et al.* 1996). Equation (4.27) is thus recast as an algebraic eigenvalue problem. The corresponding matrix elements are evaluated with the highly accurate multicentre integration scheme of Boerrigter *et al.* (1988). It is based on a clever partitioning of the whole space into so-called Voronoi cells around each atom. These cells have the shape of

Wigner–Seitz-type cells. The singularities caused by the cusps of relativistic wave functions at the nuclear sites are eliminated by suitable transformations of the sample points, which leads to an improved numerical representation of the wave functions (Bastug *et al.* 1995). With this method, a total of approximately 1400 sample points is needed to achieve a relative accuracy of 10^{-8} in calculations for diatomic molecules.

The most difficult step in this molecular approach is the evaluation of the Hartree potential $v_H = v_H^0$, Equation (4.22) and the Hartree energy (4.19). In the present version of the RDVM v_H is obtained by utilizing an auxiliary charge density $\tilde{n}(\boldsymbol{r})$, which is related to the full density (4.26) by

$$n(\boldsymbol{r}) = \tilde{n}(\boldsymbol{r}) + \Delta n(\boldsymbol{r}).$$

$\tilde{n}(\boldsymbol{r})$ is expanded as

$$\tilde{n}(\boldsymbol{r}) = \sum_{i}^{N_{\text{atom}}} \sum_{j}^{M_i} \sum_{l=0}^{L_j} \sum_{m=-l}^{l} Q_{jm}^{il} \, |f_i^j(r_i)|^2 \, Y_{lm}(\Omega_i), \tag{4.55}$$

where i runs over all atoms (symmetry equivalent centres), f_i^j represents the jth basis function at centre i and j runs over all orbitals of interest. The coefficients Q_{jm}^{il} are determined by a least-squares fit of $\tilde{n}$ to the complete n in such a way that the total electronic charge is conserved. In (4.28) v_H is then approximated by the electrostatic potential corresponding to (4.55),

$$\tilde{v}_H(\boldsymbol{r}) = e^2 \sum_{ijlm} Q_{jm}^{il} \frac{4\pi}{2l+1} \frac{1}{r_i^{l+1}} Y_{lm}(\Omega_i) \times \left[\int_0^{r_i} dr' \, r'^l \, |f_j^i(r'_i)|^2 + r_i^{2l+1} \int_{r_i}^{\infty} dr' \frac{1}{r'^{l+1}} |f_j^i(r')|^2 \right], \tag{4.56}$$

while E_H is approximated by

$$E_H \approx \int d^3r \, n(\boldsymbol{r}) \tilde{v}_H(\boldsymbol{r}) - \frac{1}{2} \int d^3r \, \tilde{n}(\boldsymbol{r}) \tilde{v}_H(\boldsymbol{r}),$$

so that only terms of second order in Δn are neglected. It has been shown that this procedure not only provides an efficient computational scheme, but also yields a variationally consistent total energy (Bastug *et al.* 1995).

4.6.1 Results

In this section we give some exemplary results from RDVM calculations using the Slater expression for E_{xc} (with $X_\alpha = 0.7$). In Table 4.10 the spectroscopic parameters of some heavy diatomic molecules obtained by the RDVM (Bastug *et al.* 1997a) are compared with relativistic configuration interaction (RCI) data on the basis of relativistic effective core potentials (RECPs). An interesting result of this investigation is the observation that the binding energies from relativistic calculations are smaller

Table 4.10 Spectroscopic parameters of some heavy dimers: relativistic and nonrelativistic DVM results versus RECP-RCI data.

molecule	method	R_e (Bohr)	D_e (eV)	ω_e (cm^{-1})
Tl_2	RECP-RCI[a]	6.69	0.16	39
	AE-RDVM	6.13	0.63	66
	AE-DVM	6.24	1.41	76
	Exp.[d]	5.67	0.43	80
Pb_2	RECP-RCI[b]	5.61	0.88	103
	AE-RDVM	5.68	1.33	110
	AE-DVM	5.72	3.35	119
	Exp.[e]		1.02	108
Bi_2	RECP-RCI[c]	5.27	2.30	170
	AE-RDVM	5.16	2.88	175
	AE-DVM	5.23	5.51	189
	Exp.[f]	5.03	2.04	173

[a]Christiansen (1983); [b]Balasubramanian and Pitzer (1983); [c]Christiansen (1984); [d]Froben *et al.* (1983); [e]Stranz and Khanna (1981); [f]Huber and Herzberg (1950).

than their nonrelativistic counterparts (AE-DVM). This can be understood by an analysis of the molecular valence orbitals: for Tl_2 and Pb_2 they are of $\frac{1}{2}(\sigma)$-type, for Bi_2 there are four $\frac{1}{2}(\sigma)$-type and two $\frac{3}{2}(\pi)$-type orbitals. These orbitals are mainly linear combinations of atomic $6p_{1/2}$ and $6p_{3/2}$ wave functions of the constituent atoms. As a result of spin–orbit splitting, the $6p_{(3/2)}$ state is less bound, which leads to a weaker bonding in the relativistic case.

In another study the geometric and electronic structures of highly symmetric neutral and multiply charged C_{60}^{x+} ($x = 0$–7) fullerenes were investigated (Bastug *et al.* 1997b). From these calculations we can estimate that stable C_{60}^{x+} clusters should exist up to $x = 13$. This value can be compared with a value of at least $x = 9$ obtained from slow ion impact experiments (Jin *et al.* 1996). Another theoretical investigation predicted a critical charge of $x = 16$ (Seifert *et al.* 1996).

The binding energies for the molecules calculated so far systematically overestimate the experimental data, which is a hint that a better description of the xc-energy is required. For that purpose, the RDVM was extended using the xc-functionals discussed in Section 4.4. Results for some diatomic molecules have already been presented in Table 4.4.

4.6.2 Geometry optimization

To obtain equilibrium geometries for small molecules and clusters we have implemented a variable metric method which is based on a quasi-Newton scheme and is widely used in optimization theory (Lipkowitz and Boyd 1993; Schlegel 1987). In this method the energy surface $E(\boldsymbol{R})$ is expanded around the present set of internuclear distances $\boldsymbol{R}_k$ up to second order. An improved geometry $\boldsymbol{R}_{k+1}$ can then be obtained by

$$\boldsymbol{R}_{k+1} = \boldsymbol{R}_k - \lambda_k H_k^{-1} \boldsymbol{g}_k, \tag{4.57}$$

where H denotes the matrix of the second derivatives (Hessian), $\boldsymbol{g} = \mathrm{d}E/\mathrm{d}\boldsymbol{R}$ and λ is a step-length parameter. Due to the huge computational effort required, H cannot be calculated directly. It can, however, be approximated by the displacements and the gradients of previous steps. Iterative use of (4.57) finally leads to the equilibrium geometry.

For the calculation of the energy gradient we follow the computational scheme applied in nonrelativistic codes (Fournier *et al.* 1989; Satoko 1981, 1984; Versluis and Ziegler 1988). The only contribution to the total energy which explicitly depends on the nuclear coordinates is the energy (4.54). The force on nucleus α can thus be obtained by calculating the first derivative of (4.54) with respect to the nuclear coordinates $\boldsymbol{R}_\alpha$,

$$\left(\frac{\mathrm{d}E}{\mathrm{d}\boldsymbol{R}_\alpha}\right)_{\mathrm{HFF}} = \int \mathrm{d}^3 r \, \frac{Z_\alpha e^2 n(\boldsymbol{r})}{|\boldsymbol{R}_\alpha - \boldsymbol{r}|^3}(\boldsymbol{R}_\alpha - \boldsymbol{r}) - \sum_{\alpha \neq \beta} \frac{Z_\alpha Z_\beta e^2}{|\boldsymbol{R}_\alpha - \boldsymbol{R}_\beta|^3}(\boldsymbol{R}_\alpha - \boldsymbol{R}_\beta). \tag{4.58}$$

This so-called Hellmann–Feynman force (HFF) represents the electrostatic interaction between the negatively charged electrons and the nuclei as well as the interaction among the nuclei. Equation (4.58) would describe the forces correctly for an exact solution of the Dirac equation. However, in practical calculations we have to introduce approximations which have a rather large influence on the forces. There are two such 'artificial' forces resulting from the following.

A. Finite-basis sets. The molecular wave functions are represented as a finite sum of atom-centred four-component basis functions, which causes a spurious force often called orbital basis correction (OBC) (also known as Pulay force (Pulay 1983)). For an atom centred at $\boldsymbol{R}_\alpha$ it reads

$$\left(\frac{\mathrm{d}E}{\mathrm{d}\boldsymbol{R}_\alpha}\right)_{\mathrm{OBC}} = \sum_i \Theta_i \sum_{\mu\nu} c_{i\mu}^* c_{i\nu} \int \mathrm{d}^3 r \frac{\partial \xi_\mu^\dagger}{\partial \boldsymbol{R}_\alpha}([-\mathrm{i}c\boldsymbol{\alpha}\cdot\nabla + (\beta - 1)mc^2] + v_{\mathrm{ext}} + \tilde{v}_{\mathrm{H}} + v_{\mathrm{xc}} - \epsilon_i)\xi_\nu + \mathrm{c.c.}$$

Here, μ runs over the basis functions centred at position $\boldsymbol{R}_\alpha$, while ν runs over all basis functions.

B. Charge density fit. The approximation of v_{H} by (4.56) leads to the density fit correction (DFC) (again for an atom at $\boldsymbol{R}_\alpha$),

$$\left(\frac{\mathrm{d}E}{\mathrm{d}\boldsymbol{R}_\alpha}\right)_{\mathrm{DFC}} = \int \mathrm{d}^3 r \, (n(\boldsymbol{r}) - \tilde{n}(\boldsymbol{r}))\frac{\partial \tilde{v}_{\mathrm{H}}}{\partial \boldsymbol{R}_\alpha}.$$

Table 4.11 Analytical versus numerical gradient for Au_2.

R (Bohr)	D_e (eV)	force (Hartree/Bohr) numerical	analytical	HFF	OBC	DFC
4.5	−2.977	−0.038	−0.037	1.954	−1.905	−0.087
4.6	−3.051	−0.018	−0.018	2.091	−2.028	−0.081
4.65	−3.069	−0.009	−0.009	2.147	−2.078	−0.078
4.7	−3.076	−0.001	−0.001	2.194	−2.122	−0.075
4.75	−3.074	0.005	0.004	2.236	−2.159	−0.072
4.8	−3.064	0.011	0.011	2.270	−2.190	−0.069
4.9	−3.023	0.020	0.020	2.320	−2.237	−0.063

Table 4.12 Estimated M–F bond lengths (in Bohrs) of MF_6 compounds (M = U, Np, Pu). The calculations were performed in O_h-symmetry. In the RECP calculations the Becke–Lee–Yang–Parr GGA has been applied.

Method	UF_6	NpF_6	PuF_6
AE-RLDA	3.79	3.76	3.76
RECP-LDA[a]	3.87		3.81
RECP-GGA[a]	3.97		
RECP-GGA[b]	3.86	3.87	3.82
Exp.[c]	3.77	3.74	3.72

[a]Gagliardi *et al.* (1998); [b]Hay and Martin (1998); [c]Weinstock and Goodman (1965).

The total energy gradient used in (4.57) is thus given by

$$\frac{\mathrm{d}E}{\mathrm{d}\boldsymbol{R}_\alpha} = \left(\frac{\mathrm{d}E}{\mathrm{d}\boldsymbol{R}_\alpha}\right)_{\mathrm{HFF}} + \left(\frac{\mathrm{d}E}{\mathrm{d}\boldsymbol{R}_\alpha}\right)_{\mathrm{OBC}} + \left(\frac{\mathrm{d}E}{\mathrm{d}\boldsymbol{R}_\alpha}\right)_{\mathrm{DFC}}. \tag{4.59}$$

To check the accuracy of the analytical energy gradient (4.59) we have calculated both (4.59) and the numerical gradient for Au_2 (using the RLDA). In this case, the numerical force can be obtained easily by numerical differentiation of $E(\boldsymbol{R})$ via a two-point formula. As can be seen from Table 4.11, the analytical gradient is in very good agreement with the numerical gradient. Moreover, the force vanishes at the minimum of the binding energy at about 4.71 Bohr. It is worth noting that the artificial gradient terms are of the same order of magnitude as the physical Hellmann–Feynman force. They exactly compensate the errors introduced by the use of finite atom-centred basis sets and the approximate treatment of v_{H}.

In Table 4.12 our results of geometry optimizations for the hexafluoride actinide compounds UF_6, NpF_6 and PuF_6 are compared with results of RECP calculations, using different xc-functionals. The optimized bond lengths obtained with the RDVM on the level of the RLDA are in very good agreement with the experimental data. Moreover, the RLDA result for UF_6 is very close to the value of 3.77 Bohr obtained by a DCB calculation (Jong and Nieuwpoort 1996). The RECP values, on the other hand, overestimate the bond lengths substantially. In addition, the bond-length reduction from UF_6 to NpF_6 is not reproduced by the RECP calculations.

4.6.3 Adsorption on surfaces

Another investigation was dedicated to the question whether the RDVM is able to describe adsorption processes on surfaces. In a first conceptual study we have calculated the binding energy of a single barium adatom on a barium(110) surface, varying the distance perpendicular to the surface (Geschke *et al.* 2000) (within the Slater-X_α approximation for E_{xc}). A first investigation showed that the adatom is preferably adsorbed at a hollow position, as shown in Figure 4.7. The Ba surface was modelled by an atomic cluster, consisting of four first- and five second-layer atoms in the most restricted case. The internuclear distances in the cluster were kept fixed at the bulk values, i.e. relaxation effects were not considered. In order to check the sensitivity of the results to the size of the cluster the number of atoms in the first layer was further increased laterally up to 12 atoms. As can be seen from Figure 4.7, which shows the binding energy of the adatom as a function of the distance to the surface, the cluster approach is justified at least for the system chosen here. The differences of the potential energy curves around the equilibrium distance between the simulation of the surface by a 13-atom cluster and that by a 17-atom cluster are marginal.

We have also investigated the adsorption of a carbon oxide molecule on a platinum (111)-surface, comparing different adsorption sites. Due to its catalytic properties there exists an enormous interest in the chemistry of Pt surfaces. In particular, the adsorption of CO on Pt(111) has been extensively studied both experimentally (Ertl *et al.* 1977; Froitzheim *et al.* 1977; Seebauer *et al.* 1982) and theoretically (Brako and Şokçević 1998; Hammer *et al.* 1996; Kopalj and Causà 1999; Lynch and Hu 2000). Our results are consistent with the experimental observation that the CO molecule is preferably adsorbed in an on-top position, in which the internuclear axis of CO is perpendicular to the surface and the carbon atom sits directly above a Pt surface atom (Seebauer *et al.* 1982). This result emphasizes the importance of relativity, as a corresponding nonrelativistic calculation predicts a bridge position to be energetically preferred (Philipsen *et al.* 1997). Detailed data are given for the on-top position in Table 4.13. In our calculations for this geometry the surface was modelled by an atomic cluster with seven atoms in a first layer and six atoms representing a second layer. The RLDA value for the distance between the C atom and the closest Pt atom in the surface is in very good agreement with experiment. On the other hand, the binding energy is overestimated by the RLDA, even though, in view of the influence of the coverage, the recent experimental value of 1.89 ± 0.20 eV (Yeo *et al.* 1997) obtained for low coverages has to be taken with a grain of salt. This deviation is significantly reduced by use of the RGGA (employing the relativistic form (4.49) of the Becke GGA (Becke 1988a) for E_x and the Perdew GGA for E_c (Perdew 1986a)).

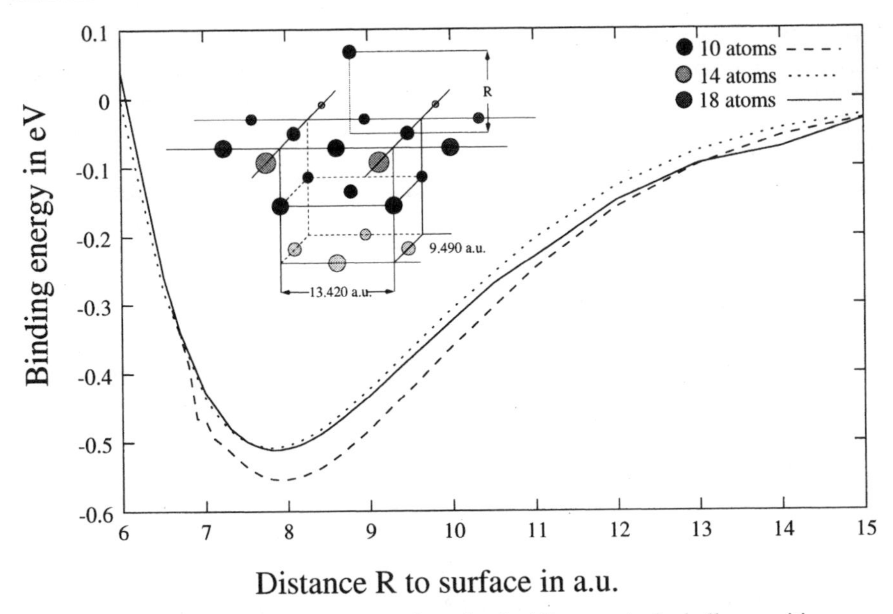

Figure 4.7 Potential energy curves of an adsorbed Ba atom in the hollow position on a Ba(110) surface for different cluster sizes.

Table 4.13 Distance of C atom to closest Pt atom in the surface ($R_{\text{Pt–C}}$) and binding energy (E_{b}) for CO adsorbed on Pt(111) in the on-top position: RDVM results versus complete active space self-consistent-field data (CASSCF) (Roszak and Balasubramanian 1995) and GGA results within the zeroth-order regular approximation (ZORA) (Philipsen *et al.* 1997). Exp. Ogletree *et al.* (1986).

method	$R_{\text{Pt–C}}$ (Bohr)	E_{b} (eV)	model
RDVM RLDA	3.50	3.25	two-layer cluster
RDVM RGGA	3.57	2.30	two-layer cluster
CASSCF	3.84	1.44	
ZORA GGA	3.59	1.41	two-layer slab
Exp.	3.49 ± 0.19	1.89 ± 0.20	

From the results given in this paragraph we can conclude that the RDVM gives realistic data within a reasonable computing time even for such complex systems.

4.6.4 Improved numerical integration scheme

The most time-consuming step in RDVM calculations is the numerical evaluation of multicentre matrix elements. Although the Baerends integration scheme used so far (te Velde and Baerends 1992) is rather accurate and efficient due to the partitioning

Table 4.14 Comparison of the Baerends integration scheme with new method: total charge and energy of neutral $RfCl_4$ (172 electrons) for different numbers of sample points n_{tot} (δ_B, relative accuracy in the Baerends method; δ_Ω, relative accuracy in the angular integration of the new scheme). The values in parentheses denote the number of radial points used in the new scheme.

	δ	n_{tot}		Q	$-E_{tot}$ (Hartree)
	1.0×10^{-6}	4 442		172.000 76	40 491.4
δ_B	1.0×10^{-8}	8 753		172.000 002 4	40 490.991
	1.0×10^{-10}	15 716		172.000 000 47	40 490.930 52
	1.0×10^{-11}	25 934		172.000 000 40	40 490.930 497
	1.0×10^{-3}	589	(42)	171.999 10	40 490.88
	1.0×10^{-5}	2 646	(80)	172.000 016	40 490.929 4
δ_Ω	0.3×10^{-6}	8 776	(109)	172.000 001 1	40 490.930 435
	1.0×10^{-7}	14 125	(119)	172.000 000 38	40 490.930 433
	0.3×10^{-7}	22 247	(128)	172.000 000 019	40 490.930 414

of the complete integration area into smaller subregions, this scheme still suffers from the discontinuities between different subregions. In order to avoid numerical inaccuracies a large number of sample points is necessary in these critical regions. Hence, for very large systems or for molecules of low symmetry too many sample points are required.

We have implemented a new integration scheme which requires fewer grid points (Heitmann *et al.* 2001). We followed the basic concept of decomposing the whole space into smaller subregions via overlapping cell functions $z_A(\boldsymbol{r})$ (Becke 1988b; Becke and Dickinson 1988; Perez-Jorda *et al.* 1994) in order to reduce the multicentre integrals to integrals of lower dimensions (Ishikawa *et al.* 1999),

$$\int F(\boldsymbol{r})\,\mathrm{d}^3r = \sum_A \int \mathrm{d}^3r\, F(\boldsymbol{r}) z_A(\boldsymbol{r}) = \sum_A I_A, \qquad \sum_A z_A(\boldsymbol{r}) = 1.$$

The $z_A(\boldsymbol{r})$ are chosen so that they tend to 1 in the vicinity of atom A but drop to zero in the direction of all other nuclei. Thus, even for integrals involving atomic basis functions of two different atoms the integrand of each contribution I_A has no more than one singular point. The integration can be further simplified by suitable transformations to intrinsic coordinates, e.g. elliptic-hyperbolic coordinates for diatomic molecules or spherical coordinates for polyatomic systems.

In Table 4.14 the performance of this new integration scheme is compared with that of the Baerends method for the case of $RfCl_4$ (for fixed T_d symmetry), focusing on the most basic properties, the total charge and energy. Since Rf is the first transactinide element with nuclear charge $Z = 104$, this 'test system' offers the possibility to study the performance of the integration scheme in the high-Z regime. It can be seen that with the new method the same accuracy is reached for both Q and E_{tot} with half as many grid points as in the case of the Baerends scheme.

Acknowledgments

We thank all the collaborators, I. Andrejkovics, J. Anton, T. Auth, T. Baştuğ, P. Blaha, N. Chetty, A. Facco Bonetti, E. K. U. Gross, A. Heitmann, A. Höck, D. Geschke, S. Keller, T. Kreibich, C. Mosch, H. Müller, R. N. Schmid, K. Schulze, K. Schwarz and W.-D. Sepp, who contributed to the work summarized in this review. We are grateful to the Deutsche Forschungsgemeinschaft for the financial support of our projects by grants Dr 113/20-1, -2, -3 and Fr 637/8-1, -2, -3.

5 Magnetic Phenomena in Solids

Hubert Ebert
Ludwig-Maximilians-Universität, München
Eberhard K. U. Gross
Julius-Maximilians-Universität, Würzburg

5.1 Introduction

Relativistic influences on the electronic properties of solids have been known for quite a long time. One of the most prominent examples of these is the position of the optical absorption edge of Au. Compared with that of Ag, this is higher in energy giving rise to the characteristic yellow colour of Au (Christensen and Seraphin 1971). If the properties of a solid that are determined by the electronic structure at the Fermi level are considered, it is often the case that scalar relativistic effects are most important. For transition-metal systems these have a very strong impact on the relative position of the sp and d bands and therefore on the density of states at the Fermi energy. As a consequence the electronic specific heat and magnetic susceptibility may be influenced in a very pronounced way by relativistic effects (Liu *et al.* 1979). More examples of this are the quadrupolar and magnetic hyperfine interactions (Pyykkö *et al.* 1973). Because these take place in the vicinity of the nucleus, where relativistic influences on the electrons are most pronounced, these will show up even for relatively light elements (Tterlikkis *et al.* 1968).

As implied by the term scalar, the corresponding relativistic corrections do not influence the symmetry of a system. In particular, they do not involve the spin operator, leaving the spin quantum number as a good quantum number even for a magnetic solid. The spin–orbit coupling on the other hand, removes degeneracies because it couples the two spin subsystems. For magnetic solids this has many far-reaching consequences, which will be the main focus of this contribution. Compared with its nonmagnetic state, spin–orbit coupling obviously gives rise for a magnetic solid to a lowering of its symmetry that is reflected by an anisotropy for more or less any of its physical properties. Concerning the optical properties of a magnetic solid, this leads, for example, to the Faraday and Kerr effects, discoveries of the 19th century, and could be ascribed to the presence of spin–orbit coupling in the 1930s (Hulme 1932). Further examples are the anomalous Hall resistivity and magnetoresistance (McGuire and Potter 1975) as well as the magnetocrystalline anisotropy and the magnetostriction (Bruno 1993). These examples demonstrate that many physical phenomena caused

Relativistic Effects in Heavy-Element Chemistry and Physics. Edited by B. A. Hess

by spin–orbit coupling have very important applications in everyday technology. To some extent this explains the numerous theoretical research activities at present underway in this field. In addition, these are strongly motivated by the discovery of new phenomena found during the last two decades in electronic spectroscopy using synchrotron radiation of well-defined polarization. The most prominent of these effects is the magnetic dichroism in X-ray absorption in its linear (van der Laan *et al.* 1986) and circular (Schütz *et al.* 1987) versions. This magneto-optical effect is now an established tool for the investigation of the spin and orbital moments of an absorber atom on the basis of the so-called sum rules (Carra *et al.* 1993; Schütz *et al.* 1993; Thole *et al.* 1992; Wienke *et al.* 1991) and is also exploited for element-specific domain imaging (Fischer *et al.* 1999; Stöhr *et al.* 1999). Interestingly, circular dichroism has not only been intensively investigated for magnetic solids, but has been predicted to also occur for superconducting systems (Capelle *et al.* 1997).

The first band-structure calculations aiming to calculate spin–orbit-induced properties in magnetic solids have been done by Callaway and co-workers (Callaway and Wang 1973). These authors and later on many others (Antropov *et al.* 1995; Brooks and Kelly 1983; Eriksson *et al.* 1990b; Lim *et al.* 1991; Min and Jang 1991; Sticht and Kübler 1985; Temmerman and Sterne 1990; Wang *et al.* 1995) accounted for spin–orbit coupling in the variational step with the unperturbed Hamiltonian matrix describing a spin-polarized system. As an alternative to this approach it was suggested to perform band-structure calculations on the basis of the Dirac equation for a spin-dependent potential to deal with all relativistic effects and magnetic ordering on the same level (Doniach and Sommers 1981; Feder *et al.* 1983; Strange *et al.* 1984). During the last few years several conventional band-structure schemes have been adopted to fit into this framework (Ebert 1988; Krutzen and Springelkamp 1989; Solovyev *et al.* 1989). In particular, multiple scattering theory, which leads directly to the electronic Green function, has been generalized in an appropriate way (Schadler *et al.* 1987; Strange *et al.* 1989). Because of its great flexibility, this approach has been used for most of the investigations to be reviewed in this contribution. Accordingly, relativistic multiple scattering theory for magnetic solids will be briefly sketched in the next section.

Nearly all theoretical investigations of the consequences of spin–orbit coupling for magnetic solids that can be found in the literature are done within the framework of density functional theory. As discussed in Chapter 4, relativistic spin density functional theory supplies in principle a consistent formal platform for this type of investigations. In many applications, however, it was found that the calculations based on relativistic spin density functional theory could reproduce experimental findings only in a semi-quantitative way. This applies in particular if we study properties connected with the spin–orbit-induced orbital current density. To cure the shortcomings of plain relativistic spin density functional theory, several schemes have been suggested in the literature. A short review of current density functional theory (CDFT) (Vignale and Rasolt 1987) and on Brooks's orbital polarization (OP) formalism (Brooks 1985) will be given in the next section. In addition, the relativistic version of the Bogoliubov–

de Gennes equations will be introduced that supplies a sound formal platform for dealing with dichroic effects in superconductors.

5.2 Formalism

5.2.1 Relativistic density functional theory

When dealing with the electronic structure of magnetic solids, we usually neglect the influence of orbital magnetism on it. Accordingly, corresponding band-structure calculations are mostly done on the basis of spin density functional theory (SDFT) as derived by von Barth and Hedin (1972) and others in a nonrelativistic way. This framework still seems to be acceptable when relativistic effects are included by introducing corresponding corrections terms to the Schrödinger equation. If fully relativistic calculations are performed instead, in principle a corresponding basis should be adopted to deal with many-body effects. The first step in this direction was made by Rajagopal and Callaway (1973), who derived the SDFT starting from a relativistic level. These authors demonstrated in particular that quantum electrodynamics supplies the proper framework for a relativistically consistent density functional theory and derived the corresponding relativistic Kohn–Sham–Dirac equations:

$$\left[c\boldsymbol{\alpha}\cdot\left(\frac{\hbar}{\mathrm{i}}\nabla+\frac{e}{c}\boldsymbol{A}_{\mathrm{eff}}(\boldsymbol{r})\right)+\beta mc^2+V_{\mathrm{eff}}(\boldsymbol{r})\right]\Psi_i(\boldsymbol{r})=\epsilon_i\Psi_i(\boldsymbol{r}) \tag{5.1}$$

with

$$V_{\mathrm{eff}}(\boldsymbol{r})=-e\left[A^0_{\mathrm{ext}}(\boldsymbol{r})+\frac{1}{c}\int \mathrm{d}^3r'\frac{J^0(\boldsymbol{r}')}{|\boldsymbol{r}-\boldsymbol{r}'|}+c\frac{\delta E_{\mathrm{xc}}[J^\mu]}{\delta J^0(\boldsymbol{r})}\right], \tag{5.2}$$

$$\boldsymbol{A}_{\mathrm{eff}}(\boldsymbol{r})=-e\left[\boldsymbol{A}_{\mathrm{ext}}(\boldsymbol{r})+\frac{1}{c}\int \mathrm{d}^3r'\frac{\boldsymbol{J}(\boldsymbol{r}')}{|\boldsymbol{r}-\boldsymbol{r}'|}+c\frac{\delta E_{\mathrm{xc}}[J^\mu]}{\delta \boldsymbol{J}(\boldsymbol{r})}\right]. \tag{5.3}$$

A detailed derivation is given in Chapter 4. Other treatments can be found elsewhere (Dreizler and Gross 1990; Engel and Dreizler 1996; Eschrig 1996). In Equation (5.1) the $\Psi_i(\boldsymbol{r})$ are four-component (Dirac spinor) wave functions with corresponding single-particle energies ϵ_i. The matrices α_i and β are the standard 4×4 Dirac matrices (Rose 1961). The effective potentials, V_{eff} and $\boldsymbol{A}_{\mathrm{eff}}$, respectively, contain as a first term the corresponding external contributions. The second terms in Equations (5.2) and (5.3) are the familiar Hartree potentials, and the third terms arise from exchange and correlation. The corresponding exchange-correlation energy $E_{\mathrm{xc}}[J^\mu]$ is a functional of the electronic four-current J^μ, which is determined self-consistently via

$$J^0=-ec\sum_i \Psi_i^\dagger\Psi_i,$$

$$J^l=-ec\sum_i \Psi_i^\dagger\beta\alpha^l\Psi_i.$$

Here J^0/c is identical to the ordinary electronic charge density ρ, while the other three components represent the electronic current density $\boldsymbol{j}$. J^μ is the central quantity of relativistic density functional theory. All properties of the system are determined by J^μ.

Thus, in contrast to nonrelativistic SDFT, where the central quantities are the spin densities $n^{\uparrow(\downarrow)}$ or, more generally, the particle density n and spin magnetization $\boldsymbol{m}$, the relativistic formalism leads in a natural way to a current density functional theory (CDFT). In view of the complexity of this very general scheme, an approximate relativistic version of SDFT has been worked out by several authors (MacDonald 1983; MacDonald and Vosko 1979; Rajagopal 1978; Rajagopal and Callaway 1973; Ramana and Rajagopal 1979). The first step in this direction is the Gordon decomposition of the spatial current density into its orbital and spin parts (Eschrig 1996; Eschrig *et al.* 1985; Rajagopal and Callaway 1973),

$$\boldsymbol{j}_{\text{orb}} = \frac{1}{2m}\Psi^\dagger\beta\left[\frac{1}{\mathrm{i}}\overleftarrow{\nabla} - \frac{1}{\mathrm{i}}\nabla + 2e\boldsymbol{A}\right]\Psi + \frac{1}{2m}\nabla\times\Psi^\dagger\beta\boldsymbol{\sigma}\Psi, \tag{5.4}$$

where $\boldsymbol{\sigma}$ is the vector of 4×4 Pauli matrices (Rose 1961). The coupling of the spin part $\boldsymbol{j}_{\text{spin}}$ (the second term in Equation (5.4)) to the vector potential $\boldsymbol{A}_{\text{eff}}$ may alternatively be described by introducing the corresponding spin magnetization

$$\boldsymbol{m} = -\mu_{\text{B}}\sum_i \Psi_i^\dagger\beta\boldsymbol{\sigma}\Psi_i.$$

This leads to the coupling term

$$-\boldsymbol{m}\cdot\boldsymbol{B}_{\text{eff}},$$

with $\boldsymbol{B}_{\text{eff}}$ the effective magnetic field corresponding to $\boldsymbol{A}_{\text{eff}}$. Thus, ignoring the orbital current density contribution $\boldsymbol{j}_{\text{orb}}$ we arrive at a Kohn–Sham–Dirac equation completely analogous to the nonrelativistic SDFT Schrödinger equation (Eschrig 1996; Eschrig *et al.* 1985):

$$\left[\frac{\hbar}{\mathrm{i}}c\boldsymbol{\alpha}\cdot\nabla + \beta mc^2 + V_{\text{eff}}(\boldsymbol{r}) - \mu_{\text{B}}\beta\boldsymbol{\sigma}\cdot\boldsymbol{B}_{\text{eff}}(\boldsymbol{r})\right]\Psi_i(\boldsymbol{r}) = \epsilon_i\Psi_i(\boldsymbol{r}) \tag{5.5}$$

with

$$\boldsymbol{B}_{\text{eff}}(\boldsymbol{r}) = \boldsymbol{B}_{\text{ext}}(\boldsymbol{r}) + \frac{\delta E_{\text{xc}}[n,\boldsymbol{m}]}{\delta\boldsymbol{m}(\boldsymbol{r})}.$$

This approach was first suggested by MacDonald and Vosko (1979) and MacDonald (1983), who justified the simplification by introducing a fictitious magnetic field that couples only to the spin degree of freedom as reflected by Equation (5.5). This formal justification was criticized by Xu *et al.* (1984) because, by describing a relativistic electronic system in terms of the particle density n and spin magnetization density $\boldsymbol{m}$ alone, the magnetic interaction part connected with the electronic current density is not Lorentz invariant. This problem was circumvented by Rajagopal and co-workers (Ramana and Rajagopal 1979, 1981a; Xu *et al.* 1984) by first considering the problem in the rest frame of an electron—for which $\boldsymbol{j}_{\rm orb}$ vanishes—giving a consistent justification for the use of relativistic SDFT.

The orbital current density contribution to $E_{\rm xc}$—ignored within SDFT—was first considered by Vignale and Rasolt on a nonrelativistic level (Vignale 1995; Vignale and Rasolt 1987, 1988, 1989). As one of the central quantities, these authors introduce the paramagnetic orbital current density $\boldsymbol{j}_{\rm orb,p}$. Gauge invariance then implies that the exchange-correlation energy depends on $\boldsymbol{j}_{\rm orb,p}$ only through the so-called vorticity:

$$\boldsymbol{\nu} = \nabla \times \frac{\boldsymbol{j}_{\rm orb,p}(\boldsymbol{r})}{n(\boldsymbol{r})}.$$

This step in particular allows us to derive a local version of nonrelativistic CDFT. A corresponding explicit expression for the corresponding $E_{\rm xc}$ has been given for the first time by Vignale and Rasolt (1988),

$$E_{\rm xc}[n, \boldsymbol{\nu}] = E_{\rm xc}[n, 0] + \int {\rm d}x \left(\frac{9\pi}{4}\right)^{1/3} \frac{1}{24\pi^2 r_{\rm s}} \left(\frac{\chi_{\rm L}}{\chi_{\rm L}^0} - 1\right) |\boldsymbol{\nu}(x)|^2,$$

where $r_{\rm s} = (3/(4\pi n))^{1/3}$ and

$$\frac{\chi_{\rm L}}{\chi_{\rm L}^0} = 1 + 0.027\,64 r_{\rm s} \ln r_{\rm s} + 0.014\,07 r_{\rm s} + O(r_{\rm s}^2 \ln r_{\rm s})$$

is the ratio of the diamagnetic susceptibility for the interacting and noninteracting electron gas. Later, more sophisticated expressions for $E_{\rm xc}$ have been given (Capelle and Gross 1997a; Skudlarski and Vignale 1993).

The Vignale–Rasolt CDFT formalism can be obtained as the weakly relativistic limit of the fully relativistic Kohn–Sham–Dirac equation (5.1). This property has been exploited to set up a computational scheme that works in the framework of nonrelativistic CDFT and accounts for the spin–orbit coupling at the same time (Ebert *et al.* 1997a). This hybrid scheme deals with the kinematic part of the problem in a fully relativistic way, whereas the exchange-correlation potential terms are treated consistently to first order in $1/c$. In particular, the corresponding modified Dirac equation

$$\left[\frac{\hbar}{\rm i} c\boldsymbol{\alpha} \cdot \nabla + \beta mc^2 + V_{\rm eff}(\boldsymbol{r}) - \mu_{\rm B} \beta \boldsymbol{\sigma} \cdot \boldsymbol{B}_{\rm eff}(\boldsymbol{r}) + \sum_{\sigma} \beta H_{{\rm op},\sigma} P_{\sigma}\right] \Psi_i(\boldsymbol{r}) = \epsilon_i \Psi_i(\boldsymbol{r}) \tag{5.6}$$

incorporates a term

$$H_{{\rm op},\sigma} = -\frac{{\rm i}\hbar e}{2mc} [\boldsymbol{A}_{{\rm xc},\sigma}(\boldsymbol{r}), \nabla]_+$$

that explicitly represents the coupling of the orbital current and the exchange-correlation vector potential $\boldsymbol{A}_{{\rm xc},\sigma}$. Since in the CDFT formalism of Vignale and Rasolt $\boldsymbol{A}_{{\rm xc},\sigma}$ is defined in a spin-dependent way, the spin-projection operator $P_{\sigma} = \frac{1}{2}(1 \pm \beta\sigma_z)$ appears in addition to $H_{{\rm op},\sigma}$ in Equation (5.6).

Within the above approximate relativistic CDFT scheme, the Breit interaction has been ignored. This radiative correction accounts for the retardation of the Coulomb interaction and exchange of transversal photons. A more complete version than that included in Equation (5.1) is given by the Hamiltonian (Bethe and Salpeter 1957; Pyykkö 1978):

$$H_{\text{Breit}} = \frac{-e^2}{2R}\boldsymbol{\alpha}_1 \cdot \boldsymbol{\alpha}_2 + \frac{e^2}{2R}\left[\boldsymbol{\alpha}_1 \cdot \boldsymbol{\alpha}_2 - \boldsymbol{\alpha}_1 \cdot \hat{\boldsymbol{R}}\,\boldsymbol{\alpha}_2 \cdot \hat{\boldsymbol{R}}\right] \quad \text{with } \boldsymbol{R} = \boldsymbol{r}_2 - \boldsymbol{r}_1,$$

where the first part is the magnetic Gaunt part and the second is the retardation term. While inclusion of the Breit interaction within quantum-chemical calculations for atoms and molecules is nearly standard (Pyykkö 1978), so far only one model (Jansen 1988) and one fully relativistic (Ebert 1995) calculation have been done in the case of solids. This is quite astonishing, because the Breit interaction gives rise to the so-called shape anisotropy that contributes in general to the magnetocrystalline anisotropy energy to the same order of magnitude as the spin–orbit coupling (see below).

An alternative to the CDFT approach is the heuristic suggestion by Brooks and co-workers (Brooks 1985; Eriksson *et al.* 1989, 1990a) to use a $\boldsymbol{k}$-space method and to add a so-called orbital polarization (OP) term to the Hamiltonian matrix. This additional term has been borrowed from atomic theory and is meant to account for Hund's second rule, i.e. to maximize the orbital angular momentum. During the last few years, this approach has been applied with remarkable success to d- as well as f-electron systems and has been refined by various authors (Severin *et al.* 1993; Shick and Gubanov 1994). As can be shown (Ebert and Battocletti 1996), Brooks's OP term can be formulated in a way that can be incorporated into the Dirac equation allowing in that way for a corresponding extension of band-structure methods based on multiple scattering theory (Ebert and Battocletti 1996). For a d-electron system, i.e. for the case in which the orbital magnetism is primarily due to an open d-electron shell, Brooks's OP term takes the form $-B_{m_s}\langle\hat{l}_z\rangle_{m_s} m_l\, \delta_{l2}$. This term describes a shift in energy for an orbital with quantum numbers $l = 2$, m_l and m_s that is proportional to the average orbital angular momentum $\langle l_z\rangle_{m_s}$ for the m_s-spin subsystem and the so-called Racah parameters B_{m_s} (Racah 1942) that in turn can be represented by the Coulomb integrals $F^2_{m_s}$ and $F^4_{m_s}$. An operator that corresponds to this energy shift is given by

$$H^{\text{OP}}_{m_s} = -B^{\text{OP}}_{m_s}(r)\langle\hat{l}_z\rangle_{m_s}\hat{l}_z\, \delta_{l2},$$

with

$$B^{\text{OP}}_{m_s}(r) = \frac{2}{441}\int\left[9\frac{r^2_<}{r^3_>} - 5\frac{r^4_<}{r^5_>}\right]\rho_{\text{d}m_s}(r')4\pi r'^2\,\text{d}r',$$

where $\rho_{\text{d}m_s}$ describes the average charge density of a d electron with spin character m_s. Obviously, the operator $H^{\text{OP}}_{m_s}$ has the form expected within CDFT for rotational symmetry (Vignale and Rasolt 1988). This is emphasized by introducing the vector potential function $A^{\mathrm{OP}}_{m_s} = -B^{\mathrm{OP}}_{m_s}(r)\langle l_z\rangle$ that leads to the Dirac equation:

$$\left[\frac{\hbar}{\mathrm{i}}c\boldsymbol{\alpha}\cdot\nabla + \beta mc^2 + V_{\mathrm{eff}}(\boldsymbol{r}) - \mu_{\mathrm{B}}\beta\boldsymbol{\sigma}\cdot\boldsymbol{B}_{\mathrm{eff}}(\boldsymbol{r}) + A^{\mathrm{OP}}\beta l_z\right]\Psi_i(\boldsymbol{r}) = \epsilon_i\Psi_i(\boldsymbol{r}). \quad (5.7)$$

For further discussion of the connection of this equation with CDFT see below.

In addition to the OP formalism, several alternative schemes have been suggested in the past to account, within a relativistic band-structure calculation, for correlation effects not incorporated within the local approximation to SDFT (LSDA). For example, the LDA+U scheme has been applied to the compound CeSb (Antropov *et al.* 1995), a system that has a maximum Kerr-rotation angle of 90° (Pittini *et al.* 1996). Similar experience has been made for other f-electron systems. Nevertheless, we should point out that by applying the LDA+U scheme we leave the framework of DFT. This does not apply to the SIC (self-interaction correction) formalism (Dreizler and Gross 1990), for which a proper relativistic formulation has been worked out recently (Forstreuter *et al.* 1997; Temmerman *et al.* 1997) and applied to magnetic solids (Temmerman *et al.* 1997).

From the above, it is obvious that relativistic effects influence the electronic structure in a twofold way. On the one hand, there is the influence of the electronic kinetics, which is accounted for by working with the Dirac-formalism. On the other hand, relativity leads to pronounced corrections in the exchange-correlation energy E_{xc} compared with its nonrelativistic counterpart. Explicit approximations for the relativistic exchange-correlation energy functional were derived and analysed in detail by MacDonald and Vosko (1979), MacDonald (1983), Ramana and Rajagopal (1979, 1981a), Xu *et al.* (1984) and Engel *et al.* (1998a). Until now, however, only a few investigations have been performed on the importance of these corrections (Koelling and MacDonald 1983; MacDonald *et al.* 1981; Schmid *et al.* 1999; Severin *et al.* 1997). Nevertheless, we may conclude from these few studies that the absolute magnitude of the total energies as well as the binding energies of tightly bound core states is affected in a rather appreciable way. However, for properties such as the equilibrium lattice parameter and even for many magnetic properties, no pronounced changes are expected. For this reason, the use, in Equation (5.5), of approximate exchange-correlation functionals derived within nonrelativistic SDFT seems to be well justified.

5.2.2 Relativistic Bogoliubov–de Gennes equations

In the nonrelativistic regime, the microscopic description of inhomogeneous singlet superconductors is based on the Bogoliubov–de Gennes (BdG) equations (de Gennes 1966):

$$[-\tfrac{1}{2}\nabla^2 + (V_{\mathrm{eff}}(\boldsymbol{r}) - \mu)]u_k(\boldsymbol{r}) + \int \Delta_{\mathrm{eff}}(\boldsymbol{r},\boldsymbol{r}')v_k(\boldsymbol{r}')\,\mathrm{d}^3r' = E_k u_k(\boldsymbol{r}), \quad (5.8)$$

$$\int \Delta^*_{\mathrm{eff}}(\boldsymbol{r},\boldsymbol{r}')u_k(\boldsymbol{r}')\,\mathrm{d}^3r' - [-\tfrac{1}{2}\nabla^2 + (V_{\mathrm{eff}}(\boldsymbol{r}) - \mu)]v_k(\boldsymbol{r}) = E_k v_k(\boldsymbol{r}). \quad (5.9)$$

Originally, the effective potential $V_{\text{eff}}(\boldsymbol{r})$ and the effective pairing potential $\Delta_{\text{eff}}(\boldsymbol{r}, \boldsymbol{r}')$ were either treated as given functions of $\boldsymbol{r}$ or they were approximated by the standard mean-field expressions, i.e. the Hartree potential for $V_{\text{eff}}(\boldsymbol{r})$ and the BCS term for $\Delta_{\text{eff}}(\boldsymbol{r}, \boldsymbol{r}')$. More recently, a density functional theory for superconductors was developed (Oliveira *et al.* 1988) where the effective potentials $V_{\text{eff}}(\boldsymbol{r})$ and $\Delta_{\text{eff}}(\boldsymbol{r}, \boldsymbol{r}')$ are expressed as functionals of the density,

$$n(\boldsymbol{r}) = \langle \hat{n}(\boldsymbol{r}) \rangle,$$
$$\hat{n}(\boldsymbol{r}) = \sum_{\sigma=\uparrow,\downarrow} \hat{\psi}_\sigma^\dagger(\boldsymbol{r}) \hat{\psi}_\sigma(\boldsymbol{r}),$$

and the superconducting order parameter,

$$\left.\begin{aligned} \chi(\boldsymbol{r}, \boldsymbol{r}') &= \langle \hat{\chi}_0(\boldsymbol{r}, \boldsymbol{r}') \rangle, \\ \hat{\chi}_0(\boldsymbol{r}, \boldsymbol{r}') &= \hat{\psi}_\uparrow(\boldsymbol{r}) \hat{\psi}_\downarrow(\boldsymbol{r}') - \hat{\psi}_\downarrow(\boldsymbol{r}) \hat{\psi}_\uparrow(\boldsymbol{r}'). \end{aligned}\right\} \tag{5.10}$$

Here, $\hat{\psi}_\sigma(\boldsymbol{r})$ and $\hat{\psi}_\sigma^\dagger(\boldsymbol{r})$ represent the ordinary nonrelativistic electron annihilation and creation operators. An LDA-type approximation has recently been derived for the exchange-correlation free energy $F_{\text{xc}}[n, \chi]$, leading to explicit expressions for the effective potentials $V_{\text{eff}}(\boldsymbol{r})$ and $\Delta_{\text{eff}}(\boldsymbol{r}, \boldsymbol{r}')$ (Kurth *et al.* 1999).

In second-quantized notation, the BdG Hamiltonian corresponding to Equations (5.8) and (5.9) reads

$$\begin{aligned} \hat{H} = \sum_{\sigma=\uparrow,\downarrow} \int \mathrm{d}^3 r \, \hat{\psi}_\sigma^\dagger(\boldsymbol{r}) [-\tfrac{1}{2}\nabla^2 + V_{\text{eff}}(\boldsymbol{r}) - \mu] \hat{\psi}_\sigma(\boldsymbol{r}) \\ - \int \mathrm{d}^3 r \int \mathrm{d}^3 r' \, (\Delta_{\text{eff}}^*(\boldsymbol{r}, \boldsymbol{r}') \hat{\chi}_0(\boldsymbol{r}, \boldsymbol{r}') + \Delta_{\text{eff}}(\boldsymbol{r}, \boldsymbol{r}') \hat{\chi}_0^\dagger(\boldsymbol{r}, \boldsymbol{r}')). \end{aligned} \tag{5.11}$$

In the relativistic generalization of this Hamiltonian, two separate issues have to be addressed: (i) the relativistic analogue of the nonrelativistic order parameter in Equation (5.10) needs to be constructed, and (ii) explicit expressions for the effective potentials have to be found. A relativistic density functional theory for superconductors, properly derived from quantum electrodynamics, has not yet been formulated. Ultimately, such a theory should provide explicit functionals for the effective potentials. For the time being, we shall consider these potentials as given functions of $\boldsymbol{r}$ modelling, for example, an SNS multilayer or a Josephson junction (Plehn *et al.* 1994). Even in this simplified case, it is by no means obvious how the second term on the right-hand side of Equation (5.11), i.e. the order parameter $\hat{\chi}_0(\boldsymbol{r}, \boldsymbol{r}')$, should be generalized to the relativistic domain. As we shall see below, this generalization leads to a number of highly interesting consequences.

Before proceeding with the construction of relativistic order parameters, we note that the singlet order parameter given in Equation (5.10) does not represent the only possible pair state in the nonrelativistic domain. Spin-triplet order parameters found, for example, in liquid ^{3}He, can be formed as well. A convenient way of describing arbitrary pairing potentials is given by

$$\hat{\Delta} = \int \mathrm{d}^3 r \int \mathrm{d}^3 r' (\hat{\psi}_\uparrow(\boldsymbol{r}), \hat{\psi}_\downarrow(\boldsymbol{r})) D_{2\times 2}(\boldsymbol{r}, \boldsymbol{r}') \begin{pmatrix} \hat{\psi}_\uparrow(\boldsymbol{r}') \\ \hat{\psi}_\downarrow(\boldsymbol{r}') \end{pmatrix}, \tag{5.12}$$

where $D_{2\times 2}(\boldsymbol{r}, \boldsymbol{r}')$ is a general 2×2 matrix of functions of $(\boldsymbol{r}, \boldsymbol{r}')$. Expanding this quantity in a suitable basis $\{M_i\}$ of 2×2 matrices

$$D_{2\times 2}(\boldsymbol{r}, \boldsymbol{r}') = \sum_{i=0}^{3} \Delta_i(\boldsymbol{r}, \boldsymbol{r}') M_i,$$

the expression in Equation (5.12) reduces to

$$\hat{\Delta} = \int \mathrm{d}^3 r \int \mathrm{d}^3 r' \left(\sum_{i=0}^{3} \Delta_i(\boldsymbol{r}, \boldsymbol{r}') \hat{\chi}_i(\boldsymbol{r}, \boldsymbol{r}') \right)$$

with the four order parameters

$$\hat{\chi}_i(\boldsymbol{r}, \boldsymbol{r}') = (\hat{\psi}_\uparrow(\boldsymbol{r}), \hat{\psi}_\downarrow(\boldsymbol{r})) M_i \begin{pmatrix} \hat{\psi}_\uparrow(\boldsymbol{r}') \\ \hat{\psi}_\downarrow(\boldsymbol{r}') \end{pmatrix}.$$

In principle, the basis $\{M_i\}$ can be chosen arbitrarily. However, if we want the order parameters $\hat{\chi}_i(\boldsymbol{r}, \boldsymbol{r}')$ to have well-defined transformation properties under the Galilei group we are led to the Balian–Werthamer matrices (Balian and Werthamer 1963):

$$M_0 = \mathrm{i}\sigma_\mathrm{y}, \qquad \begin{pmatrix} M_1 \\ M_2 \\ M_3 \end{pmatrix} = \boldsymbol{\sigma}(\mathrm{i}\sigma_\mathrm{y}).$$

With this choice, the singlet order parameter given in Equation (5.10) is represented as

$$\hat{\chi}_0(\boldsymbol{r}, \boldsymbol{r}') = (\hat{\psi}_\uparrow(\boldsymbol{r}), \hat{\psi}_\downarrow(\boldsymbol{r}))(\mathrm{i}\sigma_\mathrm{y}) \begin{pmatrix} \hat{\psi}_\uparrow(\boldsymbol{r}') \\ \hat{\psi}_\downarrow(\boldsymbol{r}') \end{pmatrix} \tag{5.13}$$

and can easily be shown to transform like a scalar, while the three triplet order parameters

$$\hat{\boldsymbol{\chi}}(\boldsymbol{r}, \boldsymbol{r}') = (\hat{\psi}_\uparrow(\boldsymbol{r}), \hat{\psi}_\downarrow(\boldsymbol{r}))\boldsymbol{\sigma}(\mathrm{i}\sigma_\mathrm{y}) \begin{pmatrix} \hat{\psi}_\uparrow(\boldsymbol{r}') \\ \hat{\psi}_\downarrow(\boldsymbol{r}') \end{pmatrix} \tag{5.14}$$

transform like a vector.

Normally, we would use the unit matrix $I_{2\times 2}$ and the vector of Pauli matrices $\boldsymbol{\sigma}$ as basis for this type of expansion, since by using this symmetry-adapted set of matrices one achieves a separation of scalar quantities $(\hat{\psi}^\dagger I_{2\times 2}\hat{\psi})$ from quantities transforming as a vector $(\hat{\psi}^\dagger \boldsymbol{\sigma} \hat{\psi})$, i.e. a classification with respect to irreducible representations of the rotation group. However, Cooper pairs are formed with two annihilation (or two creation) operators, and not with one creation and one annihilation operator, and a scalar-vector separation is only achieved in terms of the Balian–Werthamer matrices M_i. We note that a density functional theory in terms of the above set of singlet and triplet order parameters was recently formulated (Capelle and Gross 1997b).

To construct the relativistic analogue of the nonrelativistic order parameters in Equations (5.13) and (5.14), the two-component Pauli field operators

$$\begin{pmatrix} \hat{\psi}_{\uparrow}(\boldsymbol{r}) \\ \hat{\psi}_{\downarrow}(\boldsymbol{r}) \end{pmatrix}$$

have to be replaced by the four-component Dirac field operators $\hat{\Psi}(\boldsymbol{r})$. Hence, the most general relativistic pairing field is given by

$$\hat{\Delta} = \int d^3r \int d^3r' \, \hat{\Psi}^{\mathrm{T}}(\boldsymbol{r}) D_{4\times 4}(\boldsymbol{r}, \boldsymbol{r}') \hat{\Psi}(\boldsymbol{r}'). \tag{5.15}$$

Expanding the 4×4 matrix of pairing potentials in a basis $\{\eta_i\}$ of the space of 4×4 matrices

$$D_{4\times 4}(\boldsymbol{r}, \boldsymbol{r}') = \sum_{i=0}^{15} \Delta_i(\boldsymbol{r}, \boldsymbol{r}') \eta_i,$$

the expression in Equation (5.15) reduces to

$$\hat{\Delta} = \int d^3r \int d^3r' \left(\sum_{i=0}^{15} \Delta_i(\boldsymbol{r}, \boldsymbol{r}') \hat{\chi}_i^{\mathrm{rel}}(\boldsymbol{r}, \boldsymbol{r}') \right)$$

with the 16 relativistic order parameters

$$\hat{\chi}_i^{\mathrm{rel}}(\boldsymbol{r}, \boldsymbol{r}') = \hat{\Psi}^{\mathrm{T}}(\boldsymbol{r}) \eta_i \hat{\Psi}(\boldsymbol{r}'). \tag{5.16}$$

Requiring these order parameters to transform in a Lorentz-covariant way, we are led to a particular basis of 4×4 matrices $\{\eta_i\}$ which was recently derived in detail (Capelle and Gross 1999a). The resulting order parameters represent a Lorentz scalar (one component), a four vector (four components), a pseudo scalar (one component), an axial four vector (four components), and an antisymmetric tensor of rank two (six independent components). This set of 4×4 matrices is different from the usual Dirac γ matrices. The latter only lead to a Lorentz scalar, a four vector, etc., when combined with one creation and one annihilation operator, whereas the order parameter consists of two annihilation operators.

The 16 order parameters in Equation (5.16), with the matrices $\{\eta_i\}$ given in Capelle and Gross (1999a), exhaust the possible pairings which can be formed from two Dirac spinors. The most important among these 16 order parameters is the Lorentz scalar

$$\hat{\chi}_0^{\mathrm{rel}}(\boldsymbol{r}, \boldsymbol{r}') = \hat{\Psi}^{\mathrm{T}}(\boldsymbol{r}) \eta_0 \hat{\Psi}(\boldsymbol{r}'), \tag{5.17}$$

where η_0 can be written in terms of the standard γ matrices as

$$\eta_0 = \gamma^1\gamma^3.$$

Equation (5.17) represents the relativistic generalization of the nonrelativistic singlet order parameter given in Equation (5.10). Initially (Capelle and Gross 1995), the relativistic generalization $\hat{\chi}_0^{\text{rel}}$ of the nonrelativistic singlet order parameter in Equation (5.10) was constructed by replacing the nonrelativistic time-reversal matrix $t = \mathrm{i}\sigma_y$ in Equation (5.13) by the relativistic one, $T = \gamma^1\gamma^3$, i.e. it was postulated that the relativistic Cooper pairs should still consist of time-conjugate states. In the complete construction of all relativistic order parameters, which was derived later (Capelle and Gross 1999a), the matrix $\eta_0 = T$ naturally emerges as that of the 16 matrices $\{\eta_i\}$ leading to the order parameter which reduces to the singlet order parameter in Equation (5.10) in the nonrelativistic limit.

Having derived the relativistic order parameter in Equation (5.17), the relativistic generalization of the BdG equations (5.8) and (5.9) is straightforward (Capelle and Gross 1995, 1999a). We obtain

$$\gamma^0[c\boldsymbol{\gamma}\cdot\boldsymbol{p} + mc^2(1-\gamma^0) + q\gamma^\mu A_\mu]u_k(\boldsymbol{r}) \\ + \int \mathrm{d}^3r'\,\Delta(\boldsymbol{r},\boldsymbol{r}')\eta_0 v_k(\boldsymbol{r}') = E_k u_k(\boldsymbol{r}), \tag{5.18}$$

$$-\gamma^0[c\boldsymbol{\gamma}\cdot\boldsymbol{p} + mc^2(1-\gamma^0) + q\gamma^\mu A_\mu]^* v_k(\boldsymbol{r}) \\ - \int \mathrm{d}^3r'\,\Delta^*(\boldsymbol{r},\boldsymbol{r}')\eta_0 u_k(\boldsymbol{r}') = E_k v_k(\boldsymbol{r}). \tag{5.19}$$

Here both u_k and v_k are four-component (Dirac) spinors, representing the particle and hole amplitudes, while γ^μ is the four-vector of γ matrices and $\boldsymbol{\gamma}$ the corresponding three-vector, containing γ^1, γ^2 and γ^3. A_μ is the four-potential, and Δ the pair potential. The subscript 'eff', indicating that these potentials will ultimately be self-consistent effective potentials, has been dropped for notational simplicity. The chemical potential μ has been absorbed in the component A_0 of the four-potential. The Dirac–Bogoliubov–de Gennes (DBdG) equations (5.18) and (5.19) generalize the conventional BdG equations (5.8) and (5.9) to the relativistic domain in the same way as the Dirac equation generalizes the Schrödinger equation.

The DBdG equations (5.18) and (5.19) can be diagonalized analytically (Capelle and Gross 1995) for spatially uniform superconductors, employing a point-contact pair potential

$$\Delta(\boldsymbol{r},\boldsymbol{r}') = \Delta\delta(\boldsymbol{r}-\boldsymbol{r}')$$

and setting $A_\mu = (-\mu, 0, 0, 0)$. The resulting energy spectrum is given by

$$E_k = \pm\sqrt{(\epsilon_k \pm a)^2 + |\Delta|^2}, \tag{5.20}$$

where

$$a = mc^2 + \mu \quad \text{and} \quad \epsilon_k = +\sqrt{(\hbar k)^2c^2 + m^2c^4}.$$

The energy spectrum has four branches corresponding to the four possible choices of the signs. We can immediately work out two important limiting cases.

(i) In the nonsuperconducting limit ($\Delta \equiv 0$), Equation (5.20) reduces to $\pm(\epsilon_k \pm a)$. The two branches $\pm\epsilon_k - a$ are the usual Dirac spectrum shifted by a. The remaining two branches, $-(\pm\epsilon_k - a)$ being the negative of the first, represent the hole spectrum, as always for BdG-type equations.

(ii) In the nonrelativistic limit ($v/c \to 0$), Equation (5.20) reduces to the well-known BCS result $\pm\sqrt{(\hbar^2k^2/2m - \mu)^2 + |\Delta|^2}$.

In both the relativistic and nonrelativistic cases, the superconducting gap is $2|\Delta|$. The relativistic theory predicts, however, that the position of the gap is slightly shifted away from the Fermi wave vector k_F. We find

$$k_{\mathrm{gap}}^2 = k_\mathrm{F}^2[1 + \tfrac{1}{4}(v_\mathrm{F}/c)^2],$$

with the Fermi velocity v_F. The predicted shift is of the same order of magnitude as the experimentally confirmed relativistic correction to the Cooper-pair mass (Cabrera and Peskin 1989).

For ordinary matter, terms of higher than second order in (v/c) are often very small. Hence it is desirable to have a set of weakly relativistic (Pauli-type rather than Dirac-type) BdG equations. By systematic elimination of the lower components of the DBdG equations (i.e. those which are suppressed by factors of (v/c) in the weakly relativistic limit) we obtain from Equations (5.18) and (5.19) the two equations (Capelle and Gross 1995, 1999b):

$$[h(\boldsymbol{r}) + \delta h(\boldsymbol{r})]u_k(\boldsymbol{r}) + \int [\Delta(\boldsymbol{r}, \boldsymbol{r}')\mathrm{i}\sigma_y + \delta\Delta(\boldsymbol{r}, \boldsymbol{r}')]v_k(\boldsymbol{r}')\,\mathrm{d}^3r' = E_k u_k(\boldsymbol{r}), \tag{5.21}$$

$$-[h(\boldsymbol{r}) + \delta h(\boldsymbol{r})]^* v_k(\boldsymbol{r}) - \int [\Delta^*(\boldsymbol{r}', \boldsymbol{r})\mathrm{i}\sigma_y - \delta\Delta^\dagger(\boldsymbol{r}', \boldsymbol{r})]u_k(\boldsymbol{r}')\,\mathrm{d}^3r' = E_k v_k(\boldsymbol{r}), \tag{5.22}$$

where u_k and v_k are two-component (Pauli) spinors,

$$h(\boldsymbol{r}) = \frac{1}{2m}\left[\boldsymbol{p} - \frac{q}{c}\boldsymbol{A}(\boldsymbol{r})\right]^2 + (V(\boldsymbol{r}) - \mu) - \mu_\mathrm{B}\boldsymbol{\sigma}\boldsymbol{B}(\boldsymbol{r})$$

is the normal-state Hamiltonian, including the vector potential $\boldsymbol{A}$ and the magnetic field $\boldsymbol{B}$, and μ_B is the Bohr magneton. Neglecting the relativistic correction terms δh and $\delta\Delta$, we obtain from Equations (5.21) and (5.22) the traditional spin-dependent version of the BdG equations for spin-singlet superconductors (de Gennes 1966). The term

$$\delta h(\boldsymbol{r}) = \frac{1}{4m^2c^2}\left[\hbar\boldsymbol{\sigma}\cdot[\nabla V(\boldsymbol{r})] \times \boldsymbol{p} + \tfrac{1}{2}\hbar^2\nabla^2 V(\boldsymbol{r}) - \frac{p^4}{2m}\right]$$

represents the well-known second-order relativistic corrections, i.e. spin–orbit coupling, the Darwin term, and the mass–velocity correction. Analogously,

$$\delta\Delta(\boldsymbol{r},\boldsymbol{r}') = \frac{1}{4m^2c^2}[\hbar\boldsymbol{\sigma}\cdot[\nabla\Delta(\boldsymbol{r},\boldsymbol{r}')]\times\boldsymbol{p}' + \tfrac{1}{2}\hbar^2(\nabla+\nabla')^2\Delta(\boldsymbol{r},\boldsymbol{r}')]\mathrm{i}\sigma_{\mathrm{y}} \quad (5.23)$$

contains the second-order relativistic corrections involving the pair potential. These terms, evidently, appear only in the superconducting state. (The prime on $\boldsymbol{p}$ and ∇ denotes a derivative with respect to the primed coordinate.) Since these terms depend on the pair potential in a similar way to those of $\delta h(\boldsymbol{r})$ depend on the lattice potential, they will be referred to as the *anomalous* spin–orbit coupling and anomalous Darwin terms, respectively.

The spin–orbit, mass–velocity and Darwin corrections contained in δh already appear in the normal state and are thus not of superconducting origin. Nevertheless, their effect on observables in a superconductor can be dramatically modified by superconducting coherence. An example will be given in Section 5.3.5, where we show the dichroic response of a superconductor.

The anomalous spin–orbit and Darwin corrections $\delta\Delta(\boldsymbol{r},\boldsymbol{r}')$, given in Equation (5.23), are fundamentally different from $\delta h(\boldsymbol{r})$, since they depend explicitly on the pair potential and are nonzero only in the superconducting state of matter. These terms were derived for the first time only a few years ago (Capelle and Gross 1995). The anomalous spin–orbit coupling provides a contribution to dichroism in superconductors, which can be distinguished from that of conventional spin–orbit coupling due to their very different temperature dependence (Capelle *et al.* 1997, 1998). The first appearance of a spin–orbit term containing the pair potential dates back to 1985, when Ueda and Rice postulated such a term on group theoretical grounds in their phenomenological treatment of p-wave superconductivity (Ueda and Rice 1985). However, at that time it was not clear how such a term could be obtained microscopically, and what its detailed form was. These questions were answered only 10 years later on the basis of the theory outlined above (Capelle and Gross 1995). Both Darwin terms, the conventional and the anomalous one, have also been rederived phenomenologically (Capelle 2001). This rederivation showed that the anomalous Darwin term can be understood as a consequence of relativistic fluctuations of paired particles in the pair potential of the superconductor, in a similar way in which the conventional Darwin term can be understood as a consequence of fluctuations of charged particles in the electric (lattice) potential (Sakurai 1967).

More correction terms arise if, additionally, the relativistic generalizations of the nonrelativistic triplet order parameters in Equation (5.14) are considered. Some of these terms are highly unusual, containing, for example, products of the pair potential with the vector potential (Capelle *et al.* 2001; Marques *et al.* 1999).

5.2.3 Multiple scattering formalism

Most methods of band-structure calculation are based on the muffin-tin, atomic sphere approximation (ASA) or Wigner–Seitz construction for the electronic potential and charge density. These geometric schemes give in particular a subdivision of space into atom-centred spheres or cells, respectively. Accordingly, the first step of a relativistic band-structure calculation is to deal with the Dirac equation for an isolated atomic potential well. Based on the resulting solution to the single-site problem, the solution to the Dirac equation for the solid can be obtained by making use of the variational principle (Ebert 1988; Solovyev *et al.* 1989). This approach in general relies on three-dimensional periodicity and leads to a representation of the electronic structure in terms of Bloch states $|\Psi_{j\boldsymbol{k}}\rangle$ and the associated energy eigenvalues $E_{j\boldsymbol{k}}$. Alternatively, we may use multiple scattering theory as introduced by Korringa, Kohn and Rostoker (KKR) and extended by many others (Gonis 1992; Weinberger 1990) to solve the Dirac equation for a solid. In this case, the electronic structure is represented by the Green function leading to an extreme flexibility. In particular, we are able to deal with impurities, disordered alloys, surfaces and so on. Because most examples to be shown below have been obtained by the spin-polarized relativistic version of multiple scattering theory (SPR-KKR) this approach will be outlined briefly in the following.

To solve the single-site Dirac equation for a spin-dependent potential well, we start from the ansatz (Doniach and Sommers 1981; Feder *et al.* 1983; Strange *et al.* 1984),

$$\begin{aligned} \Phi_\nu(\boldsymbol{r}, E) &= \sum_\Lambda \Phi_{\Lambda\nu}(\boldsymbol{r}, E) \\ &= \begin{pmatrix} g_{\kappa\nu}(r, E)\, \chi_\Lambda(\hat{\boldsymbol{r}}) \\ \mathrm{i} f_{\kappa\nu}(r, E)\, \chi_{-\Lambda}(\hat{\boldsymbol{r}}) \end{pmatrix}, \end{aligned} \tag{5.24}$$

where the index ν numbers the various linearly independent solutions. The large and small components of the partial waves $\Phi_{\Lambda\nu}(\boldsymbol{r}, E)$ are composed of the radial wave functions $g_\kappa(r, E)$ and $f_\kappa(r, E)$ and the spin-angular functions,

$$\chi_\Lambda(\hat{\boldsymbol{r}}) = \sum_{m_s=\pm 1/2} C(l\tfrac{1}{2}j; \mu - m_s, m_s)\, Y_l^{\mu - m_s}(\hat{\boldsymbol{r}})\, \chi_{m_s},$$

with the Clebsch–Gordon coefficients $C(l\frac{1}{2}j; m_l, m_s)$, the complex spherical harmonics $Y_l^{m_l}$ and the Pauli spinors χ_{m_s} (Rose 1961). The spin–orbit and magnetic quantum numbers κ and μ, respectively, have been combined to $\Lambda = (\kappa, \mu)$ and $-\Lambda = (-\kappa, \mu)$, respectively.

For the case where the magnetic ordering of the system is accounted for by a spherically symmetric potential with a spin-dependent part $\beta B(r)\hat{\boldsymbol{e}}_z \cdot \boldsymbol{\sigma}$ that is set up within the framework of SDFT, the ansatz given in Equation (5.24) leads to the following set of radial differential equations,

$$P'_{\Lambda\nu} = -\frac{\kappa}{r} P_{\Lambda\nu} + \left[\frac{E - V}{c^2} + 1\right] Q_{\Lambda\nu} + \frac{B}{c^2} \sum_{\Lambda'} \langle \chi_{-\Lambda} | \sigma_z | \chi_{-\Lambda'} \rangle Q_{\Lambda'\nu}, \tag{5.25}$$

$$Q'_{\Lambda\nu} = \frac{\kappa}{r} Q_{\Lambda\nu} - [E - V] P_{\Lambda\nu} + B \sum_{\Lambda'} \langle \chi_\Lambda | \sigma_z | \chi_{\Lambda'} \rangle P_{\Lambda'\nu}, \tag{5.26}$$

where the usual notation $P_{\Lambda\nu} = rg_{\Lambda\nu}$ and $Q_{\Lambda\nu} = crf_{\Lambda\nu}$ has been used. The coupling coefficients occurring here are given by

$$\langle\chi_\Lambda|\sigma_z|\chi_{\Lambda'}\rangle = G(\kappa,\kappa',\mu)\,\delta_{\mu\mu'}$$

$$= \delta_{\mu\mu'}\begin{cases} -\dfrac{\mu}{(\kappa+1/2)} & \text{for } \kappa = \kappa', \\ -\sqrt{1-\left(\dfrac{\mu}{\kappa+1/2}\right)^2} & \text{for } \kappa = -\kappa'-1, \\ 0 & \text{otherwise.} \end{cases}$$

Because of the properties of $G(\kappa,\kappa',\mu)$ only partial waves for the same μ get coupled, i.e. μ is still a good quantum number. In addition, we can see that for the orbital angular momentum quantum numbers l and l' of two coupled partial waves we have the restriction $l - l' = 0, \pm 2, \ldots$, i.e. only waves of the same parity are coupled. Nevertheless, this still implies that an infinite number of partial waves are coupled. In practice, however, all coupling terms for which $l - l' = \pm 2$ are ignored. A justification for this restriction has been given by Feder *et al.* (1983) and Cortona *et al.* (1985). Results of numerical studies further justify this simplification (Ackermann 1985; Jenkins and Strange 1994). Altogether, this restricts the number of terms in Equations (5.25) and (5.26) to 2 if $|\mu| < j$. For the case $|\mu| = j$, there is no coupling at all, i.e. the solutions ψ_ν have pure spin-angular character Λ.

By requiring that the wave function $\Phi_\nu(\boldsymbol{r}, E)$ in Equation (5.24) has a unique spin-angular character Λ in the limit $r \to 0$, the index ν may be identified with Λ. The corresponding single-site t-matrix is then obtained by introducing the auxiliary matrices a and b (Ebert and Gyorffy 1988; Faulkner 1977):

$$a_{\Lambda\Lambda'}(E) = -\mathrm{i}pr^2[h^-_\Lambda(p\boldsymbol{r}), \Phi_{\Lambda\Lambda'}(\boldsymbol{r}, E)]_\mathrm{r},$$
$$b_{\Lambda\Lambda'}(E) = \mathrm{i}pr^2[h^+_\Lambda(p\boldsymbol{r}), \Phi_{\Lambda\Lambda'}(\boldsymbol{r}, E)]_\mathrm{r}.$$

Here $p = \sqrt{E(1+E/c^2)}$ is the momentum, the functions $h^\pm_\Lambda(p\boldsymbol{r})$ are the relativistic Hankel functions of the first and second kind (Rose 1961) and $[\cdots]_\mathrm{r}$ denotes the relativistic form of the Wronskian evaluated at $\boldsymbol{r}$ outside the potential well. Finally, the single-site t-matrix $t(E)$ is obtained from the expression (Ebert and Gyorffy 1988):

$$t(E) = \frac{\mathrm{i}}{2p}(a(E) - b(E))b^{-1}(E). \tag{5.27}$$

The scattering path operator τ (Gyorffy and Stott 1973) supplies a very powerful tool for dealing with the multiple scattering problem for an arbitrary array of scatterers. The corresponding matrix $\tau^{nn'}_{\Lambda\Lambda'}$ describes the transfer of a wave with spin-angular character Λ' coming in at site n' into a wave outgoing from site n with character Λ and with all possible scattering events that may take place in between accounted for.

According to this definition it has to fulfil the following self-consistency condition,

$$\underline{\tau}^{nn'} = \underline{t}^n \delta_{nn'} + \underline{t}^n \sum_{k \neq n} \underline{G}^{nk} \underline{\tau}^{kn'}, \tag{5.28}$$

where all quantities are energy dependent and the underline denotes matrices with their elements labelled by $\Lambda = (\kappa, \mu)$. Here the single-site t-matrix $\underline{t}^n$ is fixed by the solutions to the single-site Dirac equation for site n. Furthermore, $\underline{G}^{nn'}$ is the relativistic real-space Green function or structure-constants matrix that represents the propagation of a free electron between sites n and n' (Wang *et al.* 1992).

For a finite cluster of atoms, Equation (5.28) can be solved by inverting the corresponding real-space KKR matrix (Ebert *et al.* 1999),

$$\underline{\underline{\tau}} = [\underline{\underline{m}} - \underline{\underline{G}}]^{-1}, \tag{5.29}$$

where the double underline indicates supermatrices with the elements being labelled by the site indices of the cluster. The elements themselves are matrices labelled by Λ as for example $(\underline{\underline{G}})^{nn'} = \underline{G}^{nn'}$ with $(\underline{G}^{nn'})_{\Lambda\Lambda'} = G^{nn'}_{\Lambda\Lambda'}$. The matrix $\underline{\underline{m}}$ in Equation (5.29) is site-diagonal and has the inverse of the single-site t-matrix $\underline{t}^n$ as its diagonal elements, i.e. $(\underline{\underline{m}})^{nn'} = \underline{m}^n \delta_{nn'} = (\underline{t}^n)^{-1} \delta_{nn'}$.

For ordered infinite systems, Equation (5.28) can be solved exactly by means of Fourier transformation. For one atom per unit cell, the term $\tau^{nn'}_{\Lambda\Lambda'}(E)$ is obtained from the Brillouin-zone integral

$$\tau^{nn'}_{\Lambda\Lambda'}(E) = \frac{1}{\Omega_{\mathrm{BZ}}} \int_{\Omega_{\mathrm{BZ}}} \mathrm{d}^3 k \, [\underline{t}^{-1}(E) - \underline{G}(\boldsymbol{k}, E)]^{-1}_{\Lambda\Lambda'} \mathrm{e}^{\mathrm{i}\boldsymbol{k}(\boldsymbol{R}_n - \boldsymbol{R}_{n'})}. \tag{5.30}$$

Here $\boldsymbol{R}_{n(n')}$ denotes the lattice vector for site n (n') and $\underline{G}(\boldsymbol{k}, E)$ is the relativistic $\boldsymbol{k}$-dependent structure-constants matrix (Wang *et al.* 1992).

To deal with the electronic structure of surfaces within the framework of the spin-polarized relativistic KKR formalism, the standard layer techniques used for LEED and photoemission investigations (Pendry 1974) have been generalized by several authors (Fluchtmann *et al.* 1995; Scheunemann *et al.* 1994). As an alternative to this, Szunyogh and co-workers introduced the so-called screened version of the KKR method (Szunyogh *et al.* 1994, 1995). A firm basis for this approach has been supplied by the tight-binding (TB) KKR scheme introduced by Zeller *et al.* (1995). The corresponding spin-polarized relativistic version has been applied by various authors to multilayer and surface-layer systems (Nonas *et al.* 2001).

To simplify the evaluation of the Brillouin-zone integral given in Equation (5.30) on the basis of group theory requires the use of magnetic groups. This problem was first considered for the cubic case (Hörmandinger and Weinberger 1992; Zecha 1997). Recently, a general scheme has been presented to deal with the integral in Equation (5.30) and related schemes for arbitrary two- and three-dimensional periodic systems (Huhne and Ebert 2002).

Having solved the multiple scattering problem, the electronic Green function $G(\boldsymbol{r}, \boldsymbol{r}', E)$ for a solid can be set up in analogy to the nonrelativistic formalism

(Faulkner and Stocks 1980):

$$G(\boldsymbol{r}, \boldsymbol{r}', E) = \sum_{\Lambda\Lambda'} Z^n_\Lambda(\boldsymbol{r}, E)\tau^{nn'}_{\Lambda\Lambda'}(E)Z^{n'\times}_{\Lambda'}(\boldsymbol{r}', E)$$
$$- \sum_{\Lambda} [\, Z^n_\Lambda(\boldsymbol{r}, E)J^{n\times}_\Lambda(\boldsymbol{r}', E)\Theta(r' - r)$$
$$+ J^n_\Lambda(\boldsymbol{r}, E)Z^{n\times}_\Lambda(\boldsymbol{r}', E)\Theta(r - r')\,]\,\delta_{nn'}. \quad (5.31)$$

Here $\boldsymbol{r}$ ($\boldsymbol{r}'$) is confined to the atomic cell at site n (n') and the wave functions $Z^n_\Lambda(\boldsymbol{r}, E)$ and $J^n_\Lambda(\boldsymbol{r}, E)$ are the properly normalized regular and irregular solutions of the corresponding single-site problem for site n (see above). In Equation (5.31) the multiplication sign '$\times$' indicates that the wave functions $Z^{n\times}_\Lambda(\boldsymbol{r}, E)$ and $J^{n\times}_\Lambda(\boldsymbol{r}, E)$ are the left-hand side regular and irregular solutions of the corresponding modified Dirac equation (Tamura 1992). Fortunately, for most situations these are obtained from the same radial differential equations as the conventional right-hand side solutions $Z^n_\Lambda(\boldsymbol{r}, E)$ and $J^n_\Lambda(\boldsymbol{r}, E)$.

Representing the electronic structure in terms of the Green function has many appealing advantages compared with an approach using Bloch states. Systems that are locally distorted but otherwise ordered can straightforwardly be treated on the basis of the Dyson equation. Important examples for this situation are impurities that distort the host matrix only within a certain spatial range (Cabria *et al.* 2000) and adsorbates on surfaces (Nonas *et al.* 2001). The treatment of randomly disordered alloys can be achieved by adopting in addition an appropriate alloy theory. In particular, Soven's coherent potential approximation (CPA) (Soven 1967) has proved to be extremely successful in the past. Finally, we should mention that the Green function formalism supplies an outstanding theoretical platform for dealing with linear response functions (Banhart and Ebert 1995; Matsumoto *et al.* 1990) and spectroscopic properties (Fluchtmann *et al.* 1995).

The form $\beta B(r)\hat{\boldsymbol{e}}_z \cdot \boldsymbol{\sigma}$ for the spin-dependent exchange-correlation potential term corresponds to a magnetic moment pointing along the z-direction. This strongly facilitates the solution of the single-site Dirac equation because it keeps the number of coupling terms in Equations (5.25) and (5.26) to a minimum compared with any other orientation. In the case of a noncollinear spin structure it is therefore most efficient to solve the single-site Dirac equation with respect to a local frame of reference for which the magnetic moment points along the local z-axis. To deal with the multiple scattering problem, the resulting t-matrix is transferred to the global crystallographic frame of reference. Transformation of the resulting scattering path operator matrix τ to the local frame of reference finally allows the corresponding Green function to be set up according to Equation (5.31) (Ebert *et al.* 2000).

The use of a spherically symmetric potential is also not a necessary requirement for the SPR-KKR formalism. Representing the angular dependency of the potential terms $\bar{V}(\boldsymbol{r})$ and $\boldsymbol{B}(\boldsymbol{r})$ in terms of spherical harmonics leads to a coupled system of radial differential equations analogous to Equations (5.25) and (5.26). This implies that the details of the potential enter the calculation of the single-site wave functions

$Z^n_\Lambda(\boldsymbol{r}, E)$ and $J^n_\Lambda(\boldsymbol{r}, E)$ (see Equation (5.31)), while the scheme to set up the single-site t-matrix (Equation (5.27)) and the various ways for dealing with the multiple scattering problem (Equations (5.29) and (5.30)) remain unchanged. This important feature of the KKR formalism also applies to the use of more complex Hamiltonians, as demonstrated by the inclusion of the Breit interaction (Ebert 1995) and the OP term (Battocletti and Ebert 1996), as well as for the use of CDFT (Ebert *et al.* 1997a).

While a fully relativistic approach is very satisfying from a formal point of view, it is in general not very transparent. In particular, analysing the results for spin–orbit-induced properties is often quite difficult. In this situation it is very helpful to vary the speed of light (Banhart *et al.* 1996). Of course, this simple trick leads not only to a corresponding change for the spin–orbit coupling, but for all other relativistic corrections as well. To allow a manipulation of the spin–orbit coupling alone within an SPR-KKR calculation, an alternative scheme has been worked out by Ebert and co-workers. This consists of a change from a four- to a two-component formalism by the elimination of the minor component. Because one term of the resulting second-order differential equation for the wave function can unambiguously be associated with the spin–orbit coupling, it can be manipulated independently without affecting the other relativistic corrections (Ebert *et al.* 1996a). In addition, it allowed us to split the spin–orbit coupling into two terms that have rather different consequences (Ebert *et al.* 1997b). The first term (ξ_{zz}) commutes with the spin-dependent part of the exchange-correlation potential and for this reason it primarily leads to a removal of degeneracies. The second term (ξ_{xy}), on the other hand, leads to a mixing of the two spin subsystems.

5.3 Applications

5.3.1 Ground-state properties

While all relativistic corrections to the Schrödinger equation may influence the properties of a magnetic solid, the consequences of the spin–orbit coupling are most interesting, because it leads to effects that cannot be understood on the basis of a nonrelativistic or scalar relativistic description. The most obvious consequence of spin–orbit coupling for a magnetic solid is that it lowers its symmetry compared with the corresponding paramagnetic state. This can be demonstrated in the most detailed way on a microscopic level by investigation of the electronic structure. On the other hand, this reduction in symmetry also shows up on a macroscopic level, giving rise to the magnetocrystalline anisotropy, orbital magnetic moments and other important physical phenomena.

Electronic properties

The various consequences of the spin–orbit coupling for the electronic structure are demonstrated in Figure 5.1 for the dispersion relation $E_j(\boldsymbol{k})$ of Ni (Ebert *et al.* 1997b). Note that the spin–orbit coupling gives rise to a lifting of degeneracies (e.g. at A and

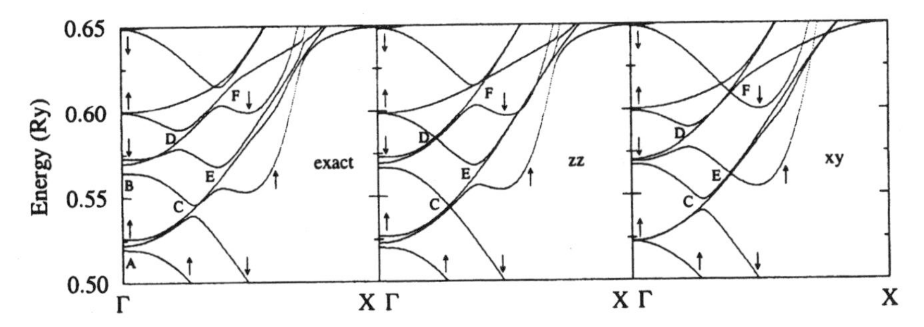

Figure 5.1 Dispersion relation $E_j(\boldsymbol{k})$ of FCC-Ni for the magnetization $\boldsymbol{M}$ and the wave vector $\boldsymbol{k}$ along the [001]- and [100]-axes, respectively. The panels show from left to right results based on the full Dirac equation and those obtained keeping only the zz- and xy-parts of the spin–orbit coupling.

B in Figure 5.1, left) and causes hybridization or mixing of bands (e.g. at C, D, E and F) that otherwise would cross.

In addition, we find that for Bloch states $|\Psi_{j\boldsymbol{k}}\rangle$ the expectation value $\langle\Psi_{j\boldsymbol{k}}|\sigma_z|\Psi_{j\boldsymbol{k}}\rangle$ is not restricted to ± 1 (Ackermann *et al.* 1984; Richter and Eschrig 1989), i.e. spin is no longer a good quantum number. However, remarkable deviations from the values ± 1 occur in general only in the region where bands cross if spin–orbit coupling is neglected. For this reason it is justified to attach the labels $\downarrow$ and $\uparrow$ to the bands to indicate their dominant spin character for a certain range of the wave vector $\boldsymbol{k}$.

Keeping only the ξ_{zz}-part of the spin–orbit interaction all electronic states have pure spin character that does not change along a band. However, this does not rule out the hybridization of bands induced by ξ_{zz} (E and F in the middle panel of Figure 5.1). On the other hand, no hybridization is found at C and D, where bands of different spin character now cross. Furthermore, note that the splitting of the bands, e.g. at A, B, E and F, caused by the ξ_{zz}-part is nearly the same as for the full spin–orbit interaction.

In contrast to ξ_{zz}, the ξ_{xy}-part gives rise to a pronounced hybridization of bands of different spin character (C and D in the right panel of Figure 5.1)—just as for the full spin–orbit interaction. While hybridization is also present at E and F, it is much less pronounced than for ξ_{zz}. Surprisingly, the splitting of the bands caused by ξ_{xy}, while being in general smaller than for ξ_{zz}, is still quite appreciable.

In the case of disordered alloys, the concept of the dispersion relation $E_j(\boldsymbol{k})$ is no more meaningful. Instead, we use the Bloch spectral function $A_{\mathrm{B}}(\boldsymbol{k}, E)$ that can be viewed as a $\boldsymbol{k}$-dependent density of states (DOS) function (Faulkner 1982). Due to the chemical disorder, $A_{\mathrm{B}}(\boldsymbol{k})$ for a given energy E is in general spread out in $\boldsymbol{k}$-space, implying that the wave vector $\boldsymbol{k}$ is not a good quantum number. By introducing the spin-projected Bloch-spectral function within the framework of the SPR-KKR-CPA approach, the mixing of the spin subsystems due to the spin–orbit coupling could also be demonstrated for disordered alloys (Ebert *et al.* 1997c). In addition, the anisotropy of the Fermi surface represented by $A_{\mathrm{B}}(\boldsymbol{k}, E_{\mathrm{F}})$ could be shown. These spin–orbit-

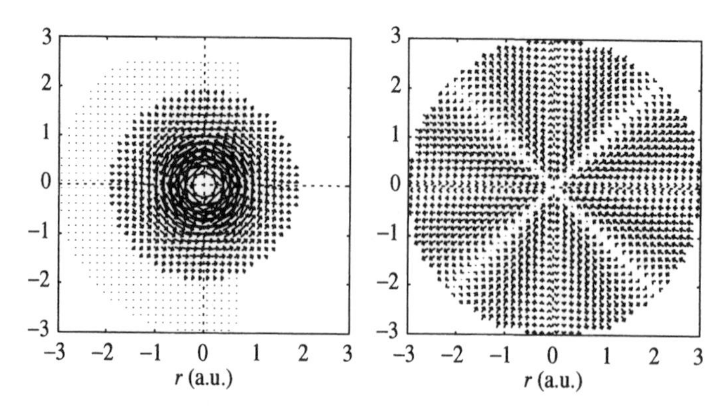

Figure 5.2 Orbital current density $\boldsymbol{j}_{\mathrm{p}}$ for BCC-Fe in the (001) plane (left). The right part gives the corresponding radial component scaled by a factor of around 350 with respect to the left part of the figure. For display $\boldsymbol{j}_{\mathrm{p}}$ has been weighted with r^2.

induced features of the electronic structure of disordered alloys have important consequences for their transport properties (see below).

For a paramagnetic solid, time-reversal symmetry implies that spin–orbit coupling viewed as a perturbation leads for states with quantum numbers (m_l, m_s) to the same changes as for $(-m_l, -m_s)$. As a consequence, the spatial symmetry of the charge distribution is not affected and no orbital current is induced. For a spin-polarized solid, on the other hand, this no longer holds, i.e. states with quantum numbers (m_l, m_s) and $(-m_l, -m_s)$ are affected by the inclusion of the spin–orbit coupling in a different manner because of the exchange splitting. As a consequence, the charge distribution will be rearranged according to the lowered symmetry of the system compared with its paramagnetic state. For a magnetic solid with a cubic lattice and the magnetization along the [001]-, [111]- or [110]-axes the effective symmetry is only tetragonal, trigonal or orthorhombic, respectively (Cracknell 1970). This could be demonstrated by a determination of the corresponding electric field gradient (EFG) tensor for substitutionally dissolved impurities in BCC-Fe by NMR measurements (Ebert 2000; Seewald *et al.* 1999, 1997).

A further consequence of the presence of the spin–orbit coupling for a spin-polarized solid is that its orbital angular momentum is quenched no more. This corresponds to the occurrence of a finite paramagnetic orbital current density $\boldsymbol{j}_{\mathrm{p}}$, which can be obtained from the expression

$$\boldsymbol{j}_{\mathrm{p}} = -\frac{1}{\pi} \operatorname{Tr} \operatorname{Im} \int^{E_{\mathrm{F}}} \mathrm{d}E \, \frac{1}{\mathrm{i}} [\overrightarrow{\nabla} - \overleftarrow{\nabla}] G(\boldsymbol{r}, \boldsymbol{r}', E)|_{\boldsymbol{r}=\boldsymbol{r}'}.$$

Corresponding results (Huhne *et al.* 1998) obtained for the current density $\boldsymbol{j}_{\mathrm{p}}$ in BCC-Fe are shown in Figure 5.2.

Here the direction and magnitude of $\boldsymbol{j}_{\mathrm{p}}$ are represented by arrows for the (001) plane with the z- and magnetization axes pointing upwards. At first sight the current density distribution seems to be rotational symmetric. However, a closer look reveals

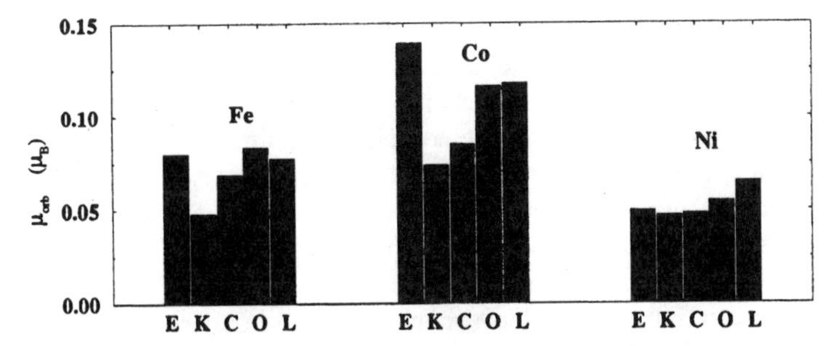

Figure 5.3 Orbital magnetic moments for BCC-Fe, FCC-Co and FCC-Ni. The various columns represent from left to right the experimental data (E) (Stearns 1987) and theoretical data obtained by the KKR method on the basis of plain SDFT (K), CDFT (C) (Ebert *et al.* 1997a) and using the OP-potential term (O) (Ebert and Battocletti 1996). The last column labelled with L gives results obtained using the LMTO with the OP term included (Eriksson *et al.* 1990b,c).

that it has in fact a lower symmetry. This is demonstrated in the right part of Figure 5.2 which gives the radial component of j_p within the (001) plane. Note that there is only a fourfold symmetry axis along the z-axis. For the paramagnetic state the x- and y-axes as well as the diagonal axes in between would be twofold symmetry axes. Obviously, the corresponding symmetry operation C_2 is missing here because of the ferromagnetic state and the spin–orbit coupling accounted for. However, we can also clearly see from the right part of Figure 5.2 that this symmetry operation combined with the time-reversal operator T results in the proper symmetry operation ($TC_{2\perp}$) for the ferromagnetic state (Cracknell 1970).

Orbital magnetic moments and hyperfine fields

With the spin–orbit-induced orbital current density in magnetic solids there is obviously a finite orbital angular momentum density associated. The corresponding orbital magnetic moment $\mu_{\rm orb}$ can be obtained from the expression (Ebert *et al.* 1988a):

$$\mu_{\rm orb} = -\frac{\mu_{\rm B}}{\pi} \,{\rm Tr}\,{\rm Im} \int^{E_{\rm F}} {\rm d}E \int {\rm d}^3 r\, \beta l_z G(\boldsymbol{r}, \boldsymbol{r}, E).$$

As Figure 5.3 shows, the spin–orbit-induced $\mu_{\rm orb}$ contributes 5–10% of the total magnetic moments of the elemental ferromagnets Fe, Co and Ni.

However, we also note from this figure that the results obtained on the basis of plain SDFT are much too small compared with experiment in the case of Fe and Co. An obvious way to cure this shortcoming of plain SDFT is to calculate the orbital magnetic moments on the basis of CDFT. Corresponding results for $\mu_{\rm orb}$ of Fe, Co and Ni, which have been obtained using the relativistic version of Vignale and Rasolt's CDFT formalism, are given in Figure 5.3 (Ebert *et al.* 1997a). Using CDFT instead of plain SDFT obviously leads to an appreciable enhancement of $\mu_{\rm orb}$ for Fe and Co. Although this effect is found to be too small, we can expect that the remaining deviation from experiment will be reduced with improved parametrizations for the exchange-correlation potentials available.

The basic CDFT Hamiltonian in Equation (5.6) does not rule out the existence of a finite orbital magnetic moment in the nonrelativistic limit. With the help of model calculations, it could be demonstrated that this is not the case for bulk Fe, Co and Ni (Ebert *et al.* 1997a). Starting an SCF calculation with a finite spin–orbit-induced orbital current density and switching off the spin–orbit coupling during the SCF-cycle the orbital magnetic moment vanished (see also next section).

An alternative to CDFT, which leads in most cases to a rather satisfying agreement with experiment and which is numerically not very demanding, has been suggested by Brooks. To deal with the orbital magnetism of f-electron systems, this author introduced the so-called OP formalism (Brooks 1985), which was originally restricted to $\boldsymbol{k}$-space band-structure methods. Using the real-space formulation given above, we can see that it effectively leads to a feedback of the spin–orbit-induced orbital current into the potential term of the Dirac equation (see Equation (5.7)). Based on the corresponding spin- and orbital-polarized relativistic KKR formalism (Ebert and Battocletti 1996), we find a strong enhancement of the orbital magnetic moment for Fe and Co leading to a rather satisfying agreement with experiment (see Figure 5.3). The spin magnetic moment, on the other hand, is hardly affected by inclusion of the OP term. Furthermore, calculations done in the full-potential mode (Huhne *et al.* 1998) clearly demonstrated that the OP term does not include aspherical potential terms that would be counted twice in a full-potential calculation, as sometimes suspected in the past.

Apart from minor numerical differences, the results obtained with the SPR-KKR are completely in line with those obtained before using the LMTO method (Eriksson *et al.* 1990b,c; Trygg *et al.* 1995). However, the latter approach accounts for spin–orbit coupling and the OP term only in the variational step, while for the SPR-KKR these are also included when calculating the wave functions and the corresponding single-site t-matrices. As a consequence, the SPR-KKR can straightforwardly be combined with the CPA to deal with disordered alloys. As an example of an application of the SPR-KKR-CPA, results for μ_{orb} of BCC-$\mathrm{Fe}_x\mathrm{Co}_{1-x}$ are shown in Figure 5.4.

As we can see in this figure, the enhancement of μ_{orb} for Fe and Co in BCC-$\mathrm{Fe}_x\mathrm{Co}_{1-x}$ is very similar to that found for the pure metals. Again this enhancement brings the average orbital magnetic moment for the alloy into very satisfying agreement with experimental data deduced from magnetomechanical and spectroscopic g-factor measurements.

Within a nonrelativistic calculation of the hyperfine fields in cubic solids, one gets only contributions from s electrons via the Fermi contact interaction. Accounting for the spin–orbit coupling, however, leads to contributions from non-s electrons as well. On the basis of the results for the orbital magnetic moments we may expect that these are primarily due to the orbital hyperfine interaction. Nevertheless, there might be a contribution via the spin–dipolar interaction as well. A most detailed investigation of this issue is achieved by using the proper relativistic expressions for the Fermi-contact (F), spin–dipolar (dip) and orbital (orb) hyperfine interaction operators (Battocletti

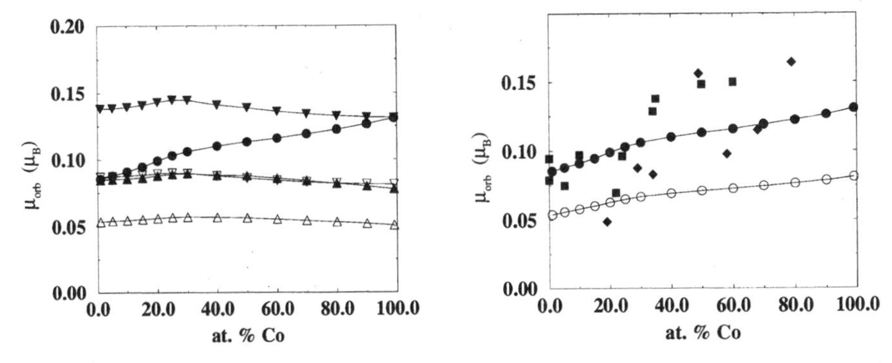

Figure 5.4 Orbital magnetic moments in BCC-Fe_xCo_{1-x}. The triangles pointing upwards and downwards represent the theoretical moments of Fe and Co, respectively, while the concentration weighted sum is given by circles. Full and open symbols stand for results obtained with and without the OP term included, respectively (Ebert and Battocletti 1996). Experimental data (Stearns 1987) for the average magnetic moment (right part) stemming from magnetomechanical and spectroscopic g-factors are given by full squares and diamonds.

1997; Pyper 1988):

$$H_{\mathrm{hf}} = e\boldsymbol{\alpha} \cdot \boldsymbol{A}_{\mathrm{n}}(\boldsymbol{r}) \tag{5.32}$$
$$= e\boldsymbol{\alpha} \cdot (\boldsymbol{\mu}_{\mathrm{n}} \times \boldsymbol{r}) A_{\mathrm{n}}(r),$$

$$H_{\mathrm{F}} = 2\mu_{\mathrm{B}}\beta\boldsymbol{\mu}_{\mathrm{n}} \cdot \boldsymbol{\sigma}\frac{1}{r_{\mathrm{n}}^3}\Theta(r_{\mathrm{n}} - r), \tag{5.33}$$

$$H_{\mathrm{dip}} = \mu_{\mathrm{B}}\beta\frac{1}{r^5}[3(\boldsymbol{\mu}_{\mathrm{n}} \cdot \boldsymbol{r})(\boldsymbol{\sigma} \cdot \boldsymbol{r}) - (\boldsymbol{\mu}_{\mathrm{n}} \cdot \boldsymbol{\sigma})r^2]\Theta(r - r_{\mathrm{n}}), \tag{5.34}$$

$$H_{\mathrm{orb}} = 2\mu_{\mathrm{B}}\beta A_{\mathrm{n}}(r)\boldsymbol{\mu}_{\mathrm{n}} \cdot \boldsymbol{l}, \tag{5.35}$$

where $\boldsymbol{A}_{\mathrm{n}}(\boldsymbol{r})$ is the vector potential of the nucleus with magnetic moment μ_{n} and finite size r_{n}.

Figure 5.5 shows results for the hyperfine field B_{Ni} of Ni in the disordered alloy system FCC-Fe_xNi_{1-x} obtained using Equation (5.32).

While the concentration dependence of the experimental fields are reproduced rather well by the theoretical fields (a phase transition to the BCC structure occurs around 65% Fe), the later ones are obviously too small. This finding has been ascribed in the past to a shortcoming of plain spin density functional theory in dealing with the core polarization mechanism (Ebert *et al.* 1988a). Recent work done on the basis of the optimized potential method (OPM) gave results for the pure elements Fe, Co and Ni in very good agreement with experiment (Akai and Kotani 1999).

The core contribution to B_{Ni} nearly exclusively comes from the s electrons. For the valence band electrons, on the other hand, there are appreciable contributions from the non-s electrons as well. While p contributions are nonnegligible for 4d and 5d elements, they are extremely small for 3d elements. Accordingly, these will be ignored in the following. On the basis of the spin–orbit-induced orbital magnetic

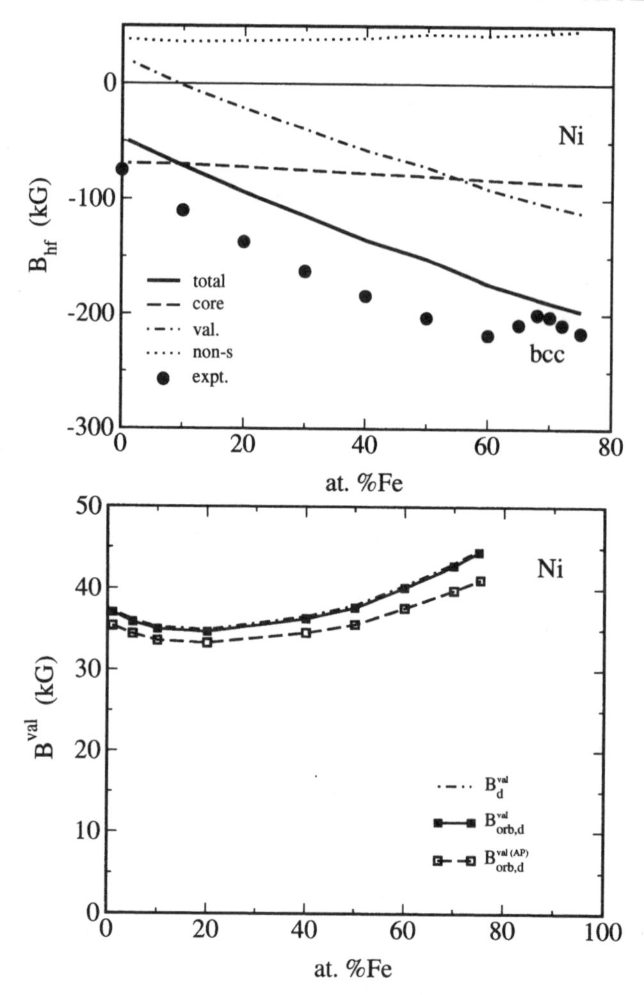

Figure 5.5 Top: hyperfine fields B_{hf} of Ni in the disordered alloy system Fe_xNi_{1-x} together with experimental data (Stearns 1987). Apart from the total field the contributions of the core, valence band as well as non-s electrons are given separately. Bottom: corresponding orbital and total valence hyperfine fields, $B^{val}_{orb,d}$ and B^{val}_{d}, respectively, for the d electrons. In addition, the orbital fields $B^{val(AP)}_{orb,d}$ according to the approximate expression due to Abragam and Pryce (Equation (5.36)) are given.

moments discussed above we may assume that the hyperfine fields of non-s electrons are also primarily of orbital origin. This can be confirmed in a simple way by using an approximate expression that connects the orbital magnetic moment and the hyperfine field (Abragam and Pryce 1951):

$$B^{val(AP)}_{orb,l} \approx 2\mu_B \langle r^{-3} \rangle_l \mu_{orb,l}. \tag{5.36}$$

Corresponding results for the d electrons of Ni are given in the right panel of Figure 5.5. Obviously, these fields are very close to the field $B_{\rm d}^{\rm val}$ corresponding to the full hyperfine interaction (Equation (5.32)). In Figure 5.5 results for the orbital hyperfine field calculated by use of the proper orbital hyperfine interaction operator $H_{\rm orb}$ (Equation (5.35)) are shown in addition. The corresponding field $B_{\rm orb,d}^{\rm val}$ nearly coincides with the total field $B_{\rm d}^{\rm val}$ indicating that the contributions of the Fermi-contact and spin–dipolar operators (see Equations (5.33) and (5.34)) are very small. Indeed it was found that the Fermi-contact field of the d electrons is completely negligible and the small difference between $B_{\rm d}^{\rm val}$ and $B_{\rm orb,d}^{\rm val}$ seen in Figure 5.5 can be ascribed to the spin–dipolar contribution.

5.3.2 Surfaces

In the past, the interest in surface magnetism was primarily caused by the enhancement of the spin moments at surfaces due to the reduced coordination. A typical example of this effect is iron, for which the bulk moment ($2.15\mu_{\rm B}$) is enhanced at the (001) surface to $2.9\mu_{\rm B}$ (Ohnishi *et al.* 1983). Similar enhancements are also found in ultrathin magnetic films (Blügel *et al.* 1989) and in particular in 3d monolayers on noble-metal surfaces (Blügel and Dederichs 1989), where the 3d moments approach the atomic values given by Hund's first rule. Also some 4d monolayers seem to be magnetic. For example, for Rh and Ru on an Ag or Au surface, moments of about 0.62 and $0.29\mu_{\rm B}$, respectively, have been calculated (Eriksson *et al.* 1991). Even larger moments have been obtained for single transition-metal adatoms on surfaces (Lang *et al.* 1994; Stepanyuk *et al.* 1996). Here the 4d and 5d atoms, being nonmagnetic as impurities in the bulk, show as adatoms very large local moments comparable with the free-atom values. Sizeable moments also survive when these atoms are incorporated into the first atomic layer at the surface.

In contrast to the spin magnetism, the orbital magnetism in transition-metal solids has in general its origin in the spin–orbit interaction. While the orbital magnetic moments in bulk transition-metal systems are nearly quenched due to the strong hybridization with neighbouring atoms, they are usually enhanced at surfaces because of the reduced coordination. For example, an increase by 50% for the surface layer of Co (0001) ($0.090\mu_{\rm B}$) has been found compared with the bulk value ($0.058\mu_{\rm B}$). Even larger orbital moments were obtained for 3d monolayers, e.g. for a Co monolayer on Cu(100) ($0.121\mu_{\rm B}$) (Hjortstam *et al.* 1996; Tischer *et al.* 1995). Thus, at surfaces the quenching of the orbital moments is less pronounced due to the reduced hybridization. Nevertheless, we should note that these enhanced orbital moments are still an order of magnitude smaller than the corresponding free-atom values, as given by Hund's second rule.

However, there are exceptions to this rule. Riegel and co-workers have already shown that Fe impurities, being injected into alkali metals, show hyperfine properties that indicate very large orbital moments close to the full atomic values (Riegel *et al.* 1986). This seems to be due to the weak hybridization with the host electronic states

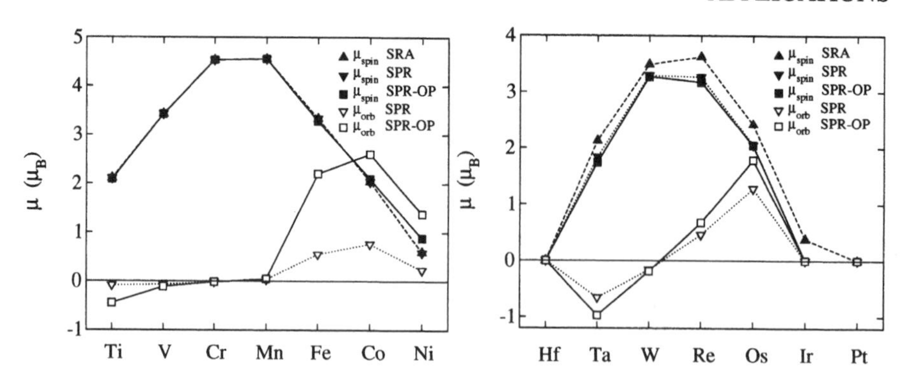

Figure 5.6 Calculated spin (μ_{spin}) and orbital (μ_{orb}) moments of 3d adatoms (left) and 5d adatoms (right) on the Ag (001) surface. The results have been obtained in a scalar (SRA) as well as fully relativistic (SPR) procedure. The label OP indicates that the orbital polarization term has been included in the Dirac equation.

caused by the low electron density of the alkali metals. In line with this experience, very pronounced enhancement effects for the spin as well as orbital magnetic moments have been obtained recently for adatoms on Ag and Au surfaces (Nonas *et al.* 2001).

Figure 5.6 shows the corresponding results for the local spin and orbital moments for the 3d (left) and 5d (right) atoms as adsorbate atoms on the Ag (001) surface.

For comparison, the spin moments have also been calculated in the scalar relativistic approximation. As we can see, the 3d spin moments are very large and agree very well with the fully relativistic data. Similar behaviour has also been found for 4d and 5d transition-metal adatoms. The spin moments of the 5d adatoms are considerably smaller than the 3d ones due to the larger extent of the 5d wave functions and the resulting stronger hybridization with the Ag substrate electrons. Nevertheless, these moments show the same parabolic variation as a function of the valence of the adatoms. In contrast to the 3d elements, the spin magnetic moments of the 5d elements are slightly reduced compared with a scalar relativistic calculation. For instance, for Os the scalar relativistic moment of $2.42\mu_{\text{B}}$ is reduced to $2.06\mu_{\text{B}}$ by inclusion of the spin–orbit coupling. Most significant is the reduction for Ir, for which the moment vanishes in a relativistic treatment. Qualitatively, the reduction of the spin moments can be understood from the broadening of the local density of states (LDOS) due to the spin–orbit splitting. Due to the larger spin–orbit coupling parameter these effects are larger for the 5d atoms.

In addition to the spin moments, Figure 5.6 also shows the orbital moments of the adatoms, calculated self-consistently in the fully relativistic scheme. For the 3d and 5d atoms we observe a change of the orbital moment from negative values in the first half of the series to positive values in the second half, with reduced values at the beginning and at the end of the series. In the 3d series by far the largest orbital moments are obtained for Fe and Co adatoms, with values of 0.55 and $0.76\mu_{\text{B}}$. Apart from these two cases, the 5d orbital moments are considerably larger than the 3d

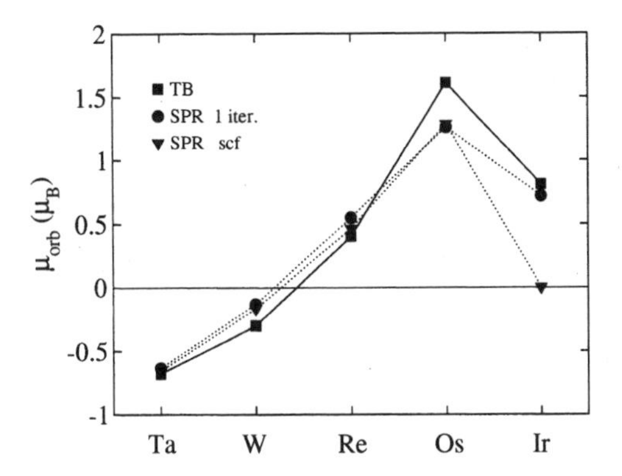

Figure 5.7 Orbital moments (μ_{orb}) of single 5d adatoms on the Ag (001) surface. The results labelled TB have been obtained on the basis of Equation (5.37), while the other data stem from a non- (SPR 1 iter.) and a fully converged (SPR scf) self-consistent relativistic calculation, respectively.

ones. This is a consequence of the much larger spin–orbit coupling parameter of the 5d transition-metal atoms.

The trends for the calculated orbital moments presented in Figure 5.6 can be understood qualitatively if we treat the spin–orbit term in first-order perturbation theory. In this order, the spin–orbit perturbation ΔV can be written as follows,

$$\Delta V = \tfrac{1}{2}\xi \boldsymbol{l} \cdot \boldsymbol{\sigma} \simeq \tfrac{1}{2}\xi l_z \sigma_z, \tag{5.37}$$

leaving the spin as a good quantum number. In a simple tight-binding model (Bruno 1993; Ebert *et al.* 1990) we obtain for the expectation value of the orbital angular momentum operator

$$\mu_{\text{orb}} \simeq -\xi(n^{\uparrow}(E_{\text{F}}) - n^{\downarrow}(E_{\text{F}}))\,\mu_{\text{B}},$$

where $n^{\uparrow}(E_{\text{F}})$ and $n^{\downarrow}(E_{\text{F}})$ are the spin-up and spin-down LDOS at the Fermi energy E_{F} for the considered adatoms. This expression explains the findings in Figure 5.6 in a natural way. Due to the reduced coordination, the adatoms have rather narrow virtual bound states with a high densities of states, and thus quite large orbital magnetic moments emerge. At the end of the series the widths of the virtual bound states are narrower, so that the orbital moments become larger, while the sign changes from negative to positive, since then the minority states are filled.

The validity of this model is illustrated in Figure 5.7 for the orbital moments of 5d adatoms on the Ag (001) surface.

This figure shows the orbital moments as calculated from Equation (5.37) with tabulated spin–orbit parameters (Mackintosh and Andersen 1980) and the LDOS obtained from a scalar relativistic treatment. These data are compared with the results of a nonself-consistent relativistic calculation, i.e. the Dirac equation has been solved using converged scalar relativistic potentials as input. Obviously, the trends are very well described by the above model. The self-consistent fully relativistic calculation

Table 5.1 Spin moment, orbital moment and anisotropy of the orbital moment $\Delta\mu_{\text{orb}}$ (in μ_{B}/atom) and magnetocrystalline anisotropy energy per cluster atom ΔE (in meV) of Ru clusters on Ag (001) surface. N is the number of atoms in the cluster and N_{c} is the coordination number for the various atom types.

N	N_{c}	position of atom	μ_{spin}	μ_{orb}	$\Delta\mu_{\text{orb}}^{x-z}$	ΔE^{x-z}	$\Delta\mu_{\text{orb}}^{y-z}$	ΔE^{y-z}
1	0		2.23	0.75	−0.19	+6.44		
2	1		2.19	0.36	−0.07	+2.54	+0.09	+3.37
3	2	centre	2.17	0.27	−0.05	—	+0.25	—
	1	border	1.97	0.42	−0.05	—	+0.06	—
	1.3	average	2.04	0.37	−0.05	+1.05	+0.12	+2.35
4	2		1.85	0.17	+0.07	−0.48		
5	4	centre	1.68	0.12	+0.06	—		
	1	border	2.05	0.43	+0.02	—		
	1.6	average	1.98	0.37	+0.03	+1.95		

gives practically the same results, except for Ir, for which the spin and orbital moments vanish in a self-consistent treatment.

Equation (5.37) also gives a straightforward explanation for the dependency of the orbital magnetic moment on the atomic coordination. In particular, it suggests that the formation of atomic clusters on surfaces should lead to a reduction in the orbital magnetic moments because of the broadening of the electronic bands. This behaviour was demonstrated recently by calculations for transition-metal clusters on an Ag (001) surface that contained up to five atoms (Cabria *et al.* 2002). The atoms of these clusters were assumed to occupy hollow surface sites with a linear arrangement for less than four atoms and a most compact one for the larger clusters. The resulting spin and orbital magnetic moments for Ru on the Ag (001) surface are given in Table 5.1.

As we can see, the tendency for the spontaneous formation of a spin magnetic moment reduces with increasing coordination number N_{c}. Nevertheless, there is still a finite moment in the limit of complete coverage of the Ag (001) surface. For Os, on the other hand, the spin magnetic moment found for small clusters depends strongly on the atomic configuration and collapses upon growing due to the increased hybridization.

Although there is no one-to-one relationship (see above), a similar behaviour is found for the orbital magnetic moment. As Table 5.1 shows, there is also a strong decrease with increasing coordination. However, quite different from the behaviour of the spin magnetic moment, the orbital moment shows a very pronounced anisotropy. This is represented in Table 5.1 by the difference of the orbital magnetic moment $\Delta\mu_{\text{orb}}^{x(y)-z}$ for an orientation of the moment along the x- or y-axis lying in the surface plane and the z-axis that coincides with the surface normal.

The anisotropy $\Delta\mu_{\text{orb}}^{x(y)-z}$ of the orbital magnetic moment has been connected by several authors (Bruno 1989; van der Laan 1998) to the corresponding magnetocrystalline anisotropy energy $\Delta E^{x(y)-z}$. Results obtained from band-structure calculations for layered systems were found in the past to be in rather satisfying agreement with the proposed interrelationship (Újfalussy *et al.* 1997). In particular, the anisotropy energy deduced from the anisotropy of the orbital magnetic moment using van der Laan's expression was found to be very close to the directly calculated ones (Cabria *et al.* 2001). For atomic clusters on an Ag (001) surface, however, the situation seems to be quite different. First of all, note that the magnetocrystalline anisotropy energies are very large and depend sensitively on the atomic number and specific configuration (Cabria *et al.* 2002). For Ru, for example, ΔE is positive apart from the cluster with four atoms, implying that we have an out-of-plane anisotropy. In the case of Fe, on the other hand, the easy axis was found to be in-plane for the monomer and dimer. In contrast to the previous work on layered systems, no clear interrelationship between the anisotropy of the orbital magnetic moment $\Delta\mu_{\rm orb}$ and the magnetocrystalline anisotropy energy ΔE could be found by the investigation of 3d, 4d and 5d transition-metal clusters on the (001) surface of Ag and Au (Cabria *et al.* 2002). This finding indicates that the assumptions made in the past to correlate $\Delta\mu_{\rm orb}$ and ΔE are no more justified for these relatively complex systems.

As shown in the last section, electron–electron interactions may have a rather pronounced influence on the spin–orbit-induced orbital magnetic moment of transition metals. The importance of these has also been studied for adatoms on the Ag (001) surface using Brooks's orbital polarization (OP) formalism. Figure 5.6 shows that the orbital magnetic moments of the 5d elements are enhanced by the OP term to an extent comparable with bulk 3d metals. For Fe, Co and Ni as adatoms, however, an enhancement by a factor of 4–5 has been found, leading to orbital magnetic moments that are larger than the spin magnetic moments for Co and Ni. This finding indicates that the large orbital magnetic moments shown in Figure 5.6 are primarily due to a Stoner-like instability of the state without orbital moment (Nordström 1991) and to a lesser extent to spin–orbit coupling. This is just the behaviour expected for the atomic limit and described by Hund's second rule. To investigate this point in more detail additional calculations have been made with the strength of the spin–orbit coupling reduced but with the OP term kept unchanged (Ebert *et al.* 1996a). Even for vanishing spin–orbit coupling, i.e. in the scalar relativistic limit, the orbital magnetic moments were only slightly reduced compared with a fully relativistic calculation: for example, from 2.57 to $2.40\mu_{\rm B}$ for Co and from 1.78 to $1.75\mu_{\rm B}$ for Os. This spontaneous formation of an orbital magnetic moment via a Stoner-like mechanism was not observed before for a transition-metal system. The reason that an orbital magnetic moment forms spontaneously for the investigated adatom systems is obviously a sharp virtual bound state like the peak at the Fermi level with a strong spin polarization. This is a direct consequence of the low coordination number and the weak hybridization with the substrate.

5.3.3 Noncollinear spin structures

For the previous examples it was assumed that the magnetization of the system is oriented either parallel or antiparallel to a common axis. There are many cases where this assumption is not well justified or does not apply at all. At finite temperatures or in disordered and amorphous alloys it is quite obvious that the orientation of the atomic magnetic moment may change from site to site. In fact it has been claimed recently that this behaviour plays an important role for the invar effect of $Fe_{0.65}Ni_{0.35}$ (van Schilfgaarde *et al.* 1999).

Apart from the disordered noncollinear spin structures there are also ordered ones. A prominent example of these is the transition-metal compound γ-FeMn, which has a magnetic unit cell that is not much larger than the chemical one. On the other hand, magnetic spiral structures, which are quite common in rare earth metals, can possess very large or even infinite magnetic unit cells.

Theoretical investigations of systems showing a noncollinear spin structure require an appropriate extension of SDFT that is normally formulated assuming a collinear spin structure (Sandratskii 1998). This important aspect has recently been reconsidered by Fritsche and co-workers in great detail (Reinert 2000). For a system with an ordered noncollinear spin structure the calculation of its electronic structure is dramatically simplified by making use of group theory (Sandratskii 1998). For this purpose we introduce the concept of spin-space groups, whose elements $\{\alpha_S|\alpha_R|\boldsymbol{t}\}$ are combinations of spatial translations and rotations, $\boldsymbol{t}$ and α_R, and rotations α_S, which act only on the spin part of the wave function. If the spin–orbit coupling is ignored, the allowed rotations α_R and α_S are completely independent from one another. As a consequence, a generalized Bloch theorem can be derived that describes the symmetry properties of Bloch states in systems with a spin spiral structure. This important step allows one, even for noncommensurate spin spirals, to restrict the electronic wave vector $\boldsymbol{k}$ to the first Brillouin zone, which is defined in the usual way, and to consider only the chemical unit cell.

Within a nonrelativistic or scalar relativistic approach the electronic structure of a magnetic solid depends only on the relative orientation of the magnetic moments on the various atomic sites. As a consequence, we expect the same total energy and magnitude of the moments for spin configurations that have the same relative orientation of the magnetic moments. This has been demonstrated for example in the case of the metallic perovskite-like compound Mn_3GaN for two noncollinear spin structures (Kübler *et al.* 1988). Because the relative orientation of the magnetic moments with respect to the lattice is different for these spin configurations, their energetic degeneracy is removed if the spin–orbit coupling is accounted for. In addition, inclusion of the spin–orbit coupling requires the spatial and spin rotations mentioned above to be identical ($\alpha_R = \alpha_S$). For that reason the generalized Bloch theorem for spin spiral structures does not hold any more. As a consequence, we have to deal with the proper magnetic unit cell in the case of commensurate spin spiral structures, while noncommensurate configurations cannot be handled any more by a band-structure scheme that relies on translational symmetry.

Most electronic structure calculations for noncollinear spin structures have so far been done assuming spherical symmetric potential terms within an atomic cell. During the last few years a number of computational schemes have been developed that take the noncollinearity of the intra-atomic magnetization in atoms (Eschrig and Servedio

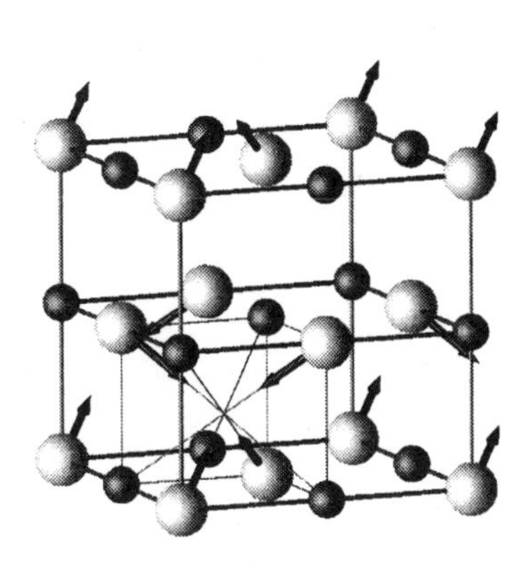
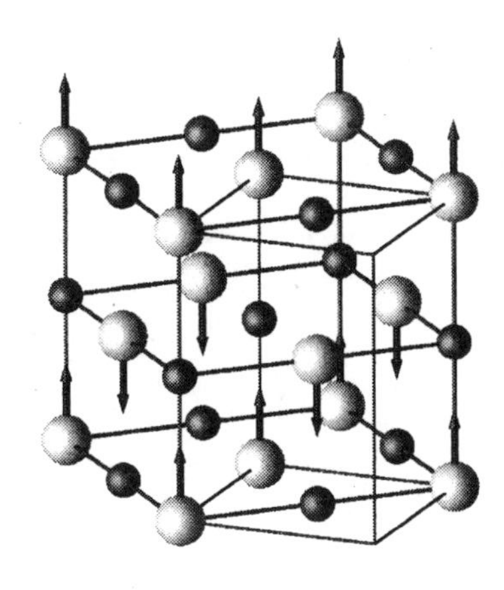

Figure 5.8 Triple-$\boldsymbol{k}$ structure (left) and single-$\boldsymbol{k}$ structure (right) in USb. Reproduced with permission from Knöpfle and Sandratskii (2000) © 2000 from the American Physical Society.

1999) and in solids (Knöpfle *et al.* 2000; Nordström and Singh 1996; Oda *et al.* 1998) into account. A corresponding version of the MASW (modified augmented spherical wave) method that uses two-component wave functions and includes the spin–orbit coupling in the variational step has been recently applied for an investigation of the compound USb (Knöpfle and Sandratskii 2000). In contrast to other U-monopnictides, USb possesses a complex triple-$\boldsymbol{k}$ magnetic structure that is shown in Figure 5.8 ($\boldsymbol{k}$ denotes here the wave vector of superimposed spin waves).

Under pressure, a phase transition to a collinear, anti-ferromagnetic single-$\boldsymbol{k}$ structure occurs (see Figure 5.8). While the single-$\boldsymbol{k}$ structure has a body-centred tetragonal magnetic unit cell with two formula units per cell, the triple-$\boldsymbol{k}$ structure is simple cubic with four formula units in the magnetic unit cell. The different symmetry of the two magnetic configurations is represented in detail in Table 5.2, where the allowed symmetry operations are listed.

Due to the inclusion of the spin–orbit coupling there is no need to distinguish between rotations in real and spin space. As can be seen in Table 5.2, the operations $\{\alpha_R|\boldsymbol{t}\}$ can be accompanied by the time reversal T. Because the operator T reverses the direction of the magnetic moments, time reversal T itself is of course no symmetry operation.

By including the OP term in the underlying Hamiltonian, the above-mentioned MASW calculations could improve the agreement of the calculated moments of the U atoms ($2.24\mu_B$) compared with previous work ($1.54\mu_B$) (Yamagami 2000). The strong influence of the OP term is understandable, because the U moment is primarily of orbital origin ($+4.46\mu_B$) with the spin moment ($-2.22\mu_B$) oriented antiparallel. In agreement with experiment, the MASW calculations found the triple-$\boldsymbol{k}$ structure to be the ground-state configuration. Also the transition to the single-$\boldsymbol{k}$ configuration under pressure could be reproduced.

The magnetic configurations of USb shown in Figure 5.8 are both spin compen-

Table 5.2 Symmetry operations for the single- and triple-$\boldsymbol{k}$ structures in of USb. Here $\boldsymbol{t}_x = (\frac{1}{2}00)$, $\boldsymbol{t}_y = (0\frac{1}{2}0)$, $\boldsymbol{t}_{xy} = (\frac{1}{2}\frac{1}{2}0)$, $\boldsymbol{t} = (\frac{1}{2}\frac{1}{2}\frac{1}{2})$ and so on.

	symmetry operations $\{\alpha_R\|\boldsymbol{t}\}$			
single-$\boldsymbol{k}$	$\{E\|0\}$	$\{C_{2x}\|0\}$	$\{C_{2y}\|0\}$	$\{C_{2z}\|0\}$
	$\{C_{2b}\|\boldsymbol{t}_x\}$	$\{C_{4z}^-\|\boldsymbol{t}_y\}$	$\{C_{4z}^+\|\boldsymbol{t}_x\}$	$\{C_{2a}\|\boldsymbol{t}_y\}$
	$\{I\|\boldsymbol{t}\}$	$\{\sigma_x\|\boldsymbol{t}_z\}$	$\{\sigma_y\|\boldsymbol{t}_z\}$	$\{\sigma_z\|\boldsymbol{t}_z\}$
	$\{\sigma_{db}\|\boldsymbol{t}_{yz}\}$	$\{S_{4z}^+\|\boldsymbol{t}_{xz}\}$	$\{S_{4z}^-\|\boldsymbol{t}_{yz}\}$	$\{\sigma_{da}\|\boldsymbol{t}_{xz}\}$
triple-$\boldsymbol{k}$	$\{E\|0\}$	$\{C_{2x}\|0\}$	$\{C_{2y}\|0\}$	$\{C_{2z}\|0\}$
	$\{C_{31}^-\|0\}$	$\{C_{32}^-\|0\}$	$\{C_{33}^-\|0\}$	$\{C_{34}^-\|0\}$
	$\{C_{31}^+\|0\}$	$\{C_{34}^+\|0\}$	$\{C_{32}^+\|0\}$	$\{C_{33}^+\|0\}$
	$T\{C_{2b}\|\boldsymbol{t}\}$	$T\{C_{4z}^-\|\boldsymbol{t}\}$	$T\{C_{4z}^+\|\boldsymbol{t}\}$	$T\{C_{2a}\|\boldsymbol{t}\}$
	$T\{C_{2e}\|\boldsymbol{t}\}$	$T\{C_{2c}\|\boldsymbol{t}\}$	$T\{C_{4y}^-\|\boldsymbol{t}\}$	$T\{C_{4y}^+\|\boldsymbol{t}\}$
	$T\{C_{2f}\|\boldsymbol{t}\}$	$T\{C_{4x}^+\|\boldsymbol{t}\}$	$T\{C_{2d}\|\boldsymbol{t}\}$	$T\{C_{4x}^-\|\boldsymbol{t}\}$
	$\{I\|\boldsymbol{t}\}$	$\{\sigma_x\|\boldsymbol{t}\}$	$\{\sigma_y\|\boldsymbol{t}\}$	$\{\sigma_z\|\boldsymbol{t}\}$
	$\{S_{61}^+\|\boldsymbol{t}\}$	$\{S_{62}^+\|\boldsymbol{t}\}$	$\{S_{63}^+\|\boldsymbol{t}\}$	$\{S_{64}^+\|\boldsymbol{t}\}$
	$\{S_{61}^-\|\boldsymbol{t}\}$	$\{S_{64}^-\|\boldsymbol{t}\}$	$\{S_{62}^-\|\boldsymbol{t}\}$	$\{S_{63}^-\|\boldsymbol{t}\}$
	$T\{\sigma_{db}\|0\}$	$T\{S_{4z}^+\|0\}$	$T\{S_{4z}^-\|0\}$	$T\{\sigma_{da}\|0\}$
	$T\{\sigma_{de}\|0\}$	$T\{\sigma_{dc}\|0\}$	$T\{S_{4y}^+\|0\}$	$T\{S_{4y}^-\|0\}$
	$T\{\sigma_{df}\|0\}$	$T\{S_{4x}^-\|0\}$	$T\{\sigma_{dd}\|0\}$	$T\{S_{4x}^+\|0\}$

sated, i.e. the net magnetic moment of the compound vanishes. However, this is not a necessary condition for the occurrence of a noncollinear spin structure as exemplified by the series of compounds U_3X_4 (X = P, As, Sb, Bi). These compounds crystallize in the Th_3P_4 structure, which has a BCC lattice with four formula units per unit cell. Experimentally, it is found that U_3P_4 and U_3As_4 have their net magnetic moment oriented along the [111]-axis, while for U_3Sb_4 and U_3Bi_4 an orientation along the [001]-axis is found. The corresponding configurations of the individual U moments are shown in Figure 5.9.

The change of the orientation of the easy axis along the series could be reproduced by band-structure calculations. Recent work done using the MASW method (Knöpfle *et al.* 2000) led to a continuous change of the anisotropy energy $\Delta E = E_{[111]} - E_{[001]}$, defined as the difference in total energy for an orientation along the [111]- and [001]-axes, from −0.15 mRy/U-atom for U_3P_4 to +0.15 mRy/U-atom for U_3Bi_4. This variation of ΔE has been ascribed to the different change of the U–U and U–X hybridizations along the series (X = P, As, Sb, Bi); the U–U hybridization decreases, while the U–X hybridization slightly increases. Also the experimental magnetic moment of the U atoms could be reproduced in a satisfying way by the calculations, as can be seen in Table 5.3.

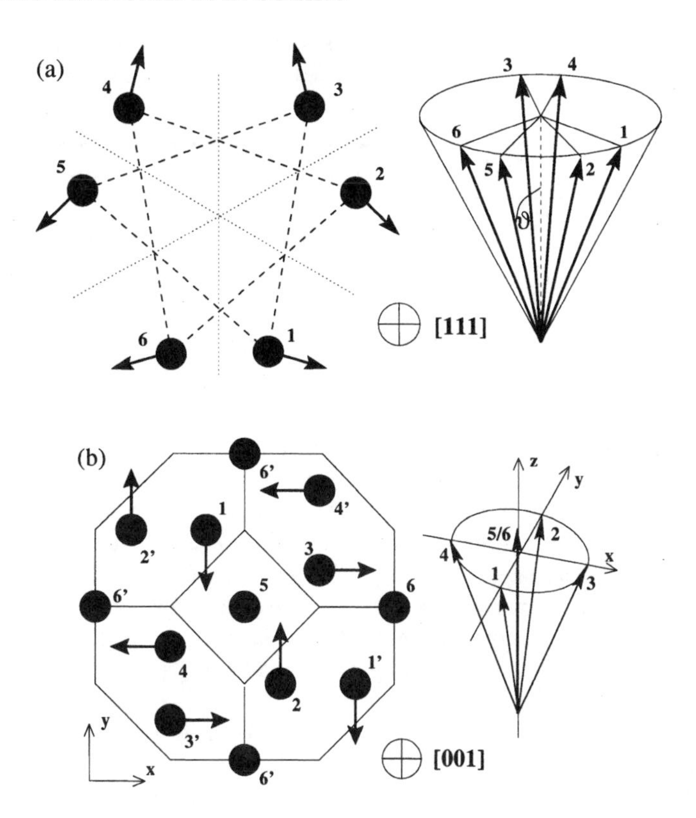

Figure 5.9 Magnetic structures with net moment along the [111]-axis (a) and [001]-axis (b). The projections of the moments onto the plane perpendicular to these axes and the corresponding conical structures are shown. Reproduced from Knöpfle *et al.* (2000). © (2000), with permission from Elsevier Science.

Table 5.3 Spin, orbital and total moments for the U_3X_4 (X = P, As, Sb, Bi) compounds calculated with the MASW together with the experimental values (Wiśniewski *et al.* 1999).

	Axis	Atom	μ_{spin}	μ_{orb}	μ_{tot}	Exp.
U_3P_4	[111]	U1-6	−1.62	3.01	1.40	1.34
U_3As_4	[111]	U1-6	−1.71	3.45	1.74	1.82
U_3Sb_4	[001]	U1-4	−1.78	3.91	2.14	1.72
		U5-6	−1.86	4.09	2.24	2.36
		average			2.17	1.93
U_3Bi_4	[001]	U1-4	−2.01	4.34	2.33	1.96
		U5-6	−2.05	4.41	2.36	2.31
		average			2.34	2.08

Table 5.4 Deviation angles of spin, orbital and total moment calculated with the MASW method. The last column gives experimental values (Gukasov *et al.* 1996; Wiśniewski *et al.* 1999).

	Axis	Spin (deg)	Orbital (deg)	Total (deg)	Exp. (deg)
U_3P_4	[111]	0.96	1.10	1.27	0.0 ± 2.3
U_3As_4	[111]	1.83	1.69	1.56	3.1 ± 0.5
U_3Sb_4	[001]	9.37	7.47	5.85	3.7 ± 3.3
U_3Bi_4	[001]	13.20	11.48	9.98	collinear

Again, inclusion of the OP-correction term is necessary to obtain the large orbital magnetic moments that by far exceed the spin moments in all cases.

The data given in Table 5.3 reflect the fact that there are two inequivalent U sites if the net magnetization points along the [001]-direction (see Figure 5.9). This is a direct consequence of the restriction imposed by symmetry for the two considered magnetic configurations. However, symmetry does not require the individual moments to be collinear with the net magnetic moment. For the U_3X_4 compounds considered, the deviations from the common axis are shown in Figure 5.9 by the projection of the U moments onto the plane perpendicular to the net magnetic axis and the corresponding conical structures. Because the canting of the total magnetic moment of a U atom does not affect the symmetry, there is no need to have the same canting angle for the spin and the orbital moments. In fact, for the series U_3X_4 different canting angles are found. As can be seen in Table 5.4, these are in reasonable quantitative agreement with corresponding experimental data, which have a rather high uncertainty.

Most importantly, however, the results summarized in Table 5.4 are in full agreement with a criterion formulated by Sandratskii and Kübler for the instability of a collinear spin structure:

> if the collinear magnetic structure under consideration is not distinguished by symmetry compared with the noncollinear structures obtained with infinitesimal deviations of the magnetic moments from collinear directions, this structure is unstable.
>
> (Sandratskii and Kübler 1995)

5.3.4 Linear response

Using the Green function formalism for a description of the underlying electronic structure gives several important advantages when dealing with response quantities. In particular, it allows us to investigate impurity systems (Terakura *et al.* 1982), surfaces (Freyer *et al.* 1999) and disordered alloys (Butler 1985). In the later case we have to deal with the configurational average of the product of two Green's functions. This problem has been studied in a rather detailed way within the framework of the KKR-CPA (Staunton 1982) leading to corresponding expressions for the so-called vertex corrections (Butler 1985). Using the relativistic version of the KKR-CPA in particular accounts for the spin–orbit coupling that can play a very important role for response functions. In the case of the susceptibility, Knight shift or induced magnetic form factor, which describes the response of a paramagnetic solid to an external magnetic field, the spin–orbit coupling leads to spin–orbit cross terms that do not occur within a nonrelativistic approach (Yasui and Shimizu 1985). In the case of transport properties of magnetic solids it gives rise to the galvanomagnetic effects, which are of great technological importance (McGuire and Potter 1975).

Static magnetic susceptibility and Knight shift

The response of a solid to a perturbation ΔH can be described straightforwardly by means of the Dyson equation. Restricting to a linear response, the Green function G^{B} of the distorted system is given by

$$G^{\mathrm{B}} = G + G\,\Delta H\,G. \tag{5.38}$$

Assuming that the perturbation ΔH stems from a coupling of an external magnetic field $\boldsymbol{B}_{\mathrm{ext}} = B_{\mathrm{ext}}\hat{\boldsymbol{e}}_z$ to the spin of the electrons we may write for the induced spin magnetization,

$$m^n_{\mathrm{spin}}(\boldsymbol{r}) = -\frac{\mu_{\mathrm{B}}}{\pi}\,\mathrm{Im}\int^{E_{\mathrm{F}}} \mathrm{d}E \sum_{n'} \int_{\Omega_{\mathrm{WS}_{n'}}} \mathrm{d}^3r'\,\beta\sigma_z\,G(\boldsymbol{r},\boldsymbol{r}',E) \\ \times (\beta\sigma_z + \beta\sigma_z\,K^{\mathrm{xc},n}_{\mathrm{spin}}(\boldsymbol{r}')\,\gamma^{n'}(\boldsymbol{r}')\,\chi^{n'}_{\mathrm{spin}})B_{\mathrm{ext}}\,G(\boldsymbol{r}',\boldsymbol{r},E), \tag{5.39}$$

where $\boldsymbol{r}'$ is restricted to the atomic cell n'. The first term in Equation (5.39) stands for the conventional Zeeman term, while the second accounts for the change $\Delta V_{\mathrm{xc}}(\boldsymbol{r})$ in the exchange correlation potential due to the induced spin magnetization $m_{\mathrm{spin}}(\boldsymbol{r})$. It is well justified to assume that $\Delta V_{\mathrm{xc}}(\boldsymbol{r})$ depends linearly on $m_{\mathrm{spin}}(\boldsymbol{r})$ with a corresponding interaction kernel $K^{\mathrm{xc},n}_{\mathrm{spin}}(\boldsymbol{r})$ (Gunnarsson 1976). In Equation (5.39) $m_{\mathrm{spin}}(\boldsymbol{r})$ has been replaced by the product $\gamma^n(\boldsymbol{r})\,\chi^n_{\mathrm{spin}}B_{\mathrm{ext}}$ with $\gamma^n(\boldsymbol{r})$ the normalized spin density and χ^n_{spin} the local Pauli spin susceptibility for site n. For a pure system χ^n_{spin} does not depend on the site n and we get the conventional expression for the Stoner enhanced Pauli spin susceptibility χ^n_{spin}:

$$\chi_{\mathrm{spin}} = S\chi^0_{\mathrm{spin}}.$$

Here S is the so-called Stoner enhancement factor, usually written as

$$(1 - I\chi^0_{\mathrm{spin}})^{-1},$$

with I the Stoner exchange correlation integral (Gunnarsson 1976) and χ^0_{spin} the unenhanced Pauli spin susceptibility obtained if the second term in Equation (5.39) is ignored. For more complex systems with more than one atom type, Equation (5.39) leads to a system of linear equations for the local susceptibilities χ^n_{spin}. As a consequence, the Stoner enhancement factor will also depend on the lattice site or atom type, respectively (Staunton 1982).

Within a nonrelativistic theory, the coupling to the orbital degree of freedom leads, in addition to the spin susceptibility, to the Langevin and Landau diamagnetic as well as to the Van Vleck paramagnetic susceptibility (Benkowitsch and Winter 1983). The Langevin diamagnetic susceptibility χ_{dia} can be calculated straightforwardly in a relativistic manner (Mendelsohn *et al.* 1970). This also applies for the Van Vleck susceptibility χ_{VV} derived from the expectation value of the z-component of the orbital angular momentum operator l_z and using the perturbation Hamiltonian (Deng *et al.* 2000):

$$\Delta H_{\text{orb}} = \beta l_z B_{\text{ext}}.$$

Calculating χ_{VV} within the framework of plain spin density functional theory (SDFT), there is no modification of the electronic potential due to the induced orbital magnetization. Working instead within the more appropriate current density functional theory, however, there would be a correction to the exchange correlation potential just as in the case of the spin susceptibility giving rise to a Stoner-like enhancement. Alternatively, this effect can be accounted for by adopting Brooks's orbital polarization formalism (Brooks 1985).

Within a nonrelativistic formalism the spin and orbital degrees of freedom are completely decoupled; this means that the cross terms $\langle \sigma_z G l_z G \rangle$ and $\langle l_z G \sigma_z G \rangle$ vanish. However, Yasui and Shimizu (1985) pointed out that these cross terms lead to corresponding contributions χ_{SO} and χ_{OS}, respectively, to the susceptibility if the spin–orbit coupling is accounted for.

In analogy to the treatment of the susceptibility, any other magnetic response function can be derived from Equation (5.38). For the Knight shift K this leads to the expression,

$$K = -\frac{e}{\pi B_{\text{ext}}} \operatorname{Im} \int^{E_{\text{F}}} \mathrm{d}E \, \frac{(\boldsymbol{r} \times \boldsymbol{\alpha})_z}{r^3} \, G^{\text{B}}(\boldsymbol{r}, \boldsymbol{r}, E),$$

where the observable in the integral corresponds to the vector potential stemming from the nuclear magnetic dipole moment (Rose 1961). As for the susceptibility, there are contributions to K because of the coupling of an external magnetic field to the spin (ΔH_{spin}) as well as to the orbital degree of freedom (ΔH_{orb}) of the electrons. Furthermore, note that the Stoner mechanism also leads to an enhancement of the corresponding Knight shift contributions.

The formalism outlined above has been recently applied to supply a firm basis for the interpretation of the results of susceptibility and NMR measurements on a number of disordered alloy systems (Deng 2001). In the case of $\text{Ag}_x\text{Pt}_{1-x}$ the most prominent feature of its electronic band structure is the rather narrow d band of Ag (3.1 eV) lying about 3.2 eV below the Fermi level. Pt, on the other hand, has a d bandwidth of about 7.4 eV with the Fermi level cutting the d band complex. As a consequence, the density of states (DOS) at the Fermi energy is dominated by the Pt d states for most concentrations. Starting from pure Pt, an increase in Ag content leads to a shift of the Pt bands towards lower energies. This gives rise to a monotonous and rapid decrease of the DOS at E_{F} with the filling of the Pt d bands. In the past it has been assumed that this behaviour should be reflected directly by the magnetic susceptibility

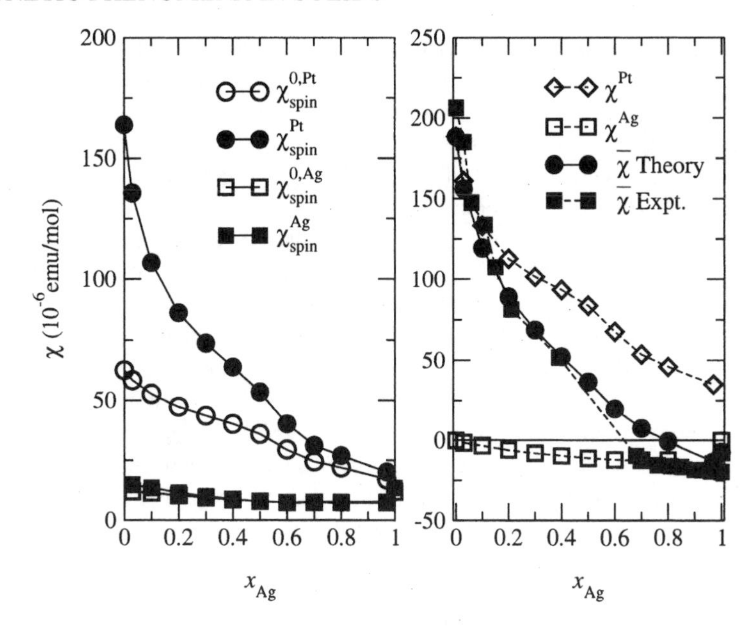

Figure 5.10 Left: calculated component resolved Pauli spin contributions to the magnetic susceptibilities of Ag_xPt_{1-x} as a function of the Ag concentration. The Pauli spin susceptibilities are given without Stoner enhancement $\chi_{spin}^{0,\alpha}$ and with Stoner enhancement χ_{spin}^{α}. Right: calculated partial magnetic susceptibility for Ag and Pt in Ag_xPt_{1-x} as well as the total susceptibility as a function of the Ag concentration together with corresponding experimental data (Ebert *et al.* 1984).

χ (Hoare *et al.* 1953). However, this point of view, based on the rigid band model, is oversimplified for several reasons, as can be demonstrated by calculations for the susceptibility.

Figure 5.10 shows theoretical results for the Pauli contributions to the magnetic susceptibility of the various alloy partners in Ag_xPt_{1-x}.

As expected from the behaviour of the partial DOS, the relatively high unenhanced Pauli susceptibility $\chi_{spin}^{0,Pt}$ of pure Pt decreases rapidly with increasing Ag content. The partial susceptibility $\chi_{spin}^{0,Ag}$ of Ag, on the other hand, is rather small and hardly varies with concentration. As we can see from Figure 5.10, the Stoner enhancement of the Pauli spin susceptibility of pure Pt is rather large. Accordingly, Pt dominates the total magnetic susceptibility on the Ag-poor side of Ag_xPt_{1-x}. Because the Stoner enhancement also gets weaker when the spin susceptibility χ_{spin}^{0} decreases, the partial susceptibility of Pt decreases more rapidly with increasing x than the bare density of states at the Fermi level suggests. For the alloy partner Ag, on the other hand, the situation is quite different. Here the Stoner enhancement amounts to only a few per cent. Only on the Ag-poor side is it somewhat more pronounced. On the Ag-rich side, however, it can more or less be ignored.

As for the spin susceptibility, the partial Van Vleck susceptibility χ_{VV}^{α} of Pt is found to be much larger than for Ag. Because $\chi_{\mathrm{VV}}^{\alpha}$ is primarily determined by the filling and width of the d band, it shows only a rather weak concentration dependence. As a consequence, $\chi_{\mathrm{VV}}^{\mathrm{Pt}}$ exceeds $\chi_{\mathrm{spin}}^{\mathrm{Pt}}$ on the Ag-rich side of the system although it is only about one-quarter of $\chi_{\mathrm{spin}}^{\mathrm{Pt}}$ for pure Pt. Not only their concentration dependency, but also the Stoner enhancement of $\chi_{\mathrm{spin}}^{\alpha}$ and $\chi_{\mathrm{VV}}^{\alpha}$ is quite different. Using the OP formalism we find for Pt only about 10% and negligible corrections in the case of Ag.

The negative Langevin susceptibility $\chi_{\mathrm{dia}}^{\alpha}$ hardly varies with concentration x for both components in $\mathrm{Ag}_x\mathrm{Pt}_{1-x}$. It more or less compensates for the positive orbital contribution $\chi_{\mathrm{VV}}^{\alpha}$ of the corresponding atom. A variation in $\chi_{\mathrm{dia}}^{\alpha}$ with concentration should primarily arise from the change of the lattice parameter with concentration (Banhart *et al.* 1986). It seems that this effect is not very pronounced for $\mathrm{Ag}_x\mathrm{Pt}_{1-x}$.

In accordance with the strong relativistic effects that can be expected for Pt in $\mathrm{Ag}_x\mathrm{Pt}_{1-x}$, the spin–orbit cross-term contributions $\chi_{\mathrm{SO}}^{\mathrm{Pt}}$ and $\chi_{\mathrm{OS}}^{\mathrm{Pt}}$, which are almost of the same magnitude, should not be ignored over the whole concentration range. For Ag, on the other hand, the cross-term contributions can be ignored.

In Figure 5.10, the resulting partial susceptibility

$$\chi^{\alpha} = \chi_{\mathrm{spin}}^{\alpha} + \chi_{\mathrm{VV}}^{\alpha} + \chi_{\mathrm{SO}}^{\alpha} + \chi_{\mathrm{OS}}^{\alpha} + \chi_{\mathrm{dia}}^{\alpha}$$

as well as the total susceptibility

$$\chi = \sum_{\alpha} x_{\alpha} \chi^{\alpha}$$

of $\mathrm{Ag}_x\mathrm{Pt}_{1-x}$ are shown together with the corresponding experimental data measured at 4.2 K (Ebert *et al.* 1984). These results demonstrate that the SPR-KKR-CPA method allows us to reproduce the experimental susceptibilities of disordered alloys in a very reliable way.

It was found in many experimental investigations that the Knight shifts of the noble metals Au, Ag and Cu are positive but become negative if they are dissolved as impurities in the transition metals Pd and Pt (Ebert *et al.* 1984; Kobayashi *et al.* 1963; Narath 1968). It was assumed quite early on (Kobayashi *et al.* 1963) that these negative shifts are not due to the polarization of the core electrons, as in the case of pure Pd or Pt (Krieger and Voitländer 1980), but due to the influence of the host metals Pd and Pt coming from their strongly enhanced spin susceptibility. The underlying mechanism giving rise to the observed negative Knight shift was always assumed to be essentially the same as that leading to the negative hyperfine field of the noble metals dissolved in a ferromagnetic host metal (Narath 1968). This assumption could be verified by the calculation of the spin contributions to the Knight shifts in $\mathrm{Ag}_x\mathrm{Pt}_{1-x}$.

Due to the large spin magnetization of Pt on the Ag-poor side of $\mathrm{Ag}_x\mathrm{Pt}_{1-x}$, the core polarization contribution K_{cp} of Pt is quite large. As can be seen in Figure 5.11, it rapidly decreases with increasing Ag concentration x because of the decreasing Pt spin magnetization. For Ag, on the other hand, the core polarization contribution is nearly zero. In contrast to this, the Stoner enhanced spin contribution to the Knight shift of Ag in $\mathrm{Ag}_x\mathrm{Pt}_{1-x}$ has a relatively large absolute value with a negative sign on the

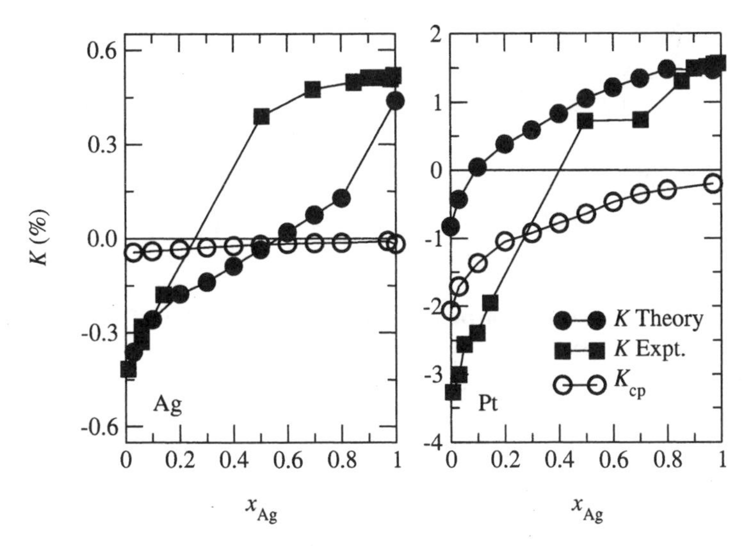

Figure 5.11 Calculated total Knight shifts K^{α} of Ag (left) and Pt (right) in Ag_xPt_{1-x} in comparison with the experimental data (Ebert *et al.* 1984). In addition, the theoretical core polarization contribution K^{α}_{cp} is shown.

Ag-poor side, which is attributed to interatomic effects arising from the strong induced spin magnetization of Pt. This finding is completely in line with the results of previous calculations (Gräf *et al.* 1993) that included the effect of an external magnetic field in the SCF cycle. The analysis of the induced spin density of the various components for the Pd-rich side of Ag_xPd_{1-x} and Cu_xPd_{1-x} demonstrated that for Pd the core polarization dominates the magnetization at the nuclear site. For Ag and Cu, on the other hand, the Stoner enhanced spin density of the valence electrons has negative sign and is responsible for the main part of the magnetization at the nuclear site. Since the spin density at a nuclear site leads to the Fermi contact contributions to the Knight shift, it was concluded that the negative Knight shift of the noble metal arises from the large induced spin moment of neighbouring Pd atoms in the alloy, instead of from the much smaller magnetization within the atomic cell of Ag or Cu.

The calculated total Knight shifts $K_{\text{theory}} = K_{\text{spin}} + K_{\text{VV}} + K_{\text{cp}} + K_{\text{dia}}$ for the various components in Ag_xPt_{1-x}, compared with the corresponding experimental data, are shown in Figure 5.11. For the Ag-poor side of the alloy systems note a rather pronounced deviation of the calculated Knight shifts of the alloy partner Pt from the corresponding experimental data. This has to be attributed to a large extent to the problems in dealing with the core polarization mechanism within the framework of plain SDFT. The same problem is present when dealing with the corresponding core polarization hyperfine field in spontaneously magnetized solids that in general is also found to be too small compared with experiment. Recently, it could be demonstrated that using the so-called optimized potential method (OPM) the core polarization hyperfine field in ferromagnets is strongly increased compared with plain SDFT-type calculations leading to a satisfying agreement with experiment (Akai and Kotani 1999). From this we can conclude that an improved treatment of the core polarization Knight shift $K_{\rm cp}$ should also improve the agreement with experiment in Figure 5.11.

Transport properties

The resistivity tensor $\boldsymbol{\rho}$ of a paramagnetic cubic solid is diagonal with all elements identical, i.e. its resistivity is isotropic. For a ferromagnetic cubic solid, however, this does not apply. In particular, we find that the form of the resistivity tensor depends on the direction of the magnetization reflecting the lowering of the symmetry of the system upon magnetic ordering. On the basis of group theoretical considerations (Huhne 2000; Kleiner 1966), we find the following form for $\boldsymbol{\rho}$ of a cubic solid with the magnetization along the z-axis:

$$\boldsymbol{\rho} = \begin{pmatrix} \rho_\perp & -\rho_{\rm H} & 0 \\ \rho_{\rm H} & \rho_\perp & 0 \\ 0 & 0 & \rho_\parallel \end{pmatrix} = \boldsymbol{\sigma}^{-1}.$$

Here $\rho_\perp$, $\rho_\parallel$ and $\rho_{\rm H}$ are the transverse, the longitudinal and the spontaneous or anomalous Hall resistivities, with $\boldsymbol{\sigma}$ the corresponding conductivity tensor. In addition, we define the spontaneous magnetoresistance anisotropy (SMA) or anomalous magnetoresistance (AMR) ratio $\Delta\rho/\bar{\rho}$ (McGuire and Potter 1975), with $\Delta\rho = \rho_\parallel - \rho_\perp$ and the isotropic resistivity $\bar{\rho} = \frac{1}{3}(2\rho_\perp + \rho_\parallel)$. Because these quantities depend on the magnetic domain structure, they are determined experimentally from measurements at high magnetic fields $B_{\rm ext}$ with a subsequent extrapolation to $B_{\rm ext} = 0$.

A sound theoretical framework for dealing with the galvanomagnetic properties of disordered alloys, i.e. their anomalous magnetoresistance and Hall resistivity, is supplied by the Kubo formalism. Representing the electronic structure in terms of the Green function leads to the following expressions for the elements of the conductivity tensor $\boldsymbol{\sigma}$ (Greenwood 1958; Střda and Smrčka 1975),

$$\sigma_{\mu\mu} = \frac{\hbar}{\pi V_{\rm cryst}} \operatorname{Tr}\langle j_\mu \operatorname{Im} G^+(E_{\rm F})\, j_\mu \operatorname{Im} G^+(E_{\rm F})\rangle_{\rm conf}, \tag{5.40}$$

$$\sigma_{\mu\nu} = \frac{-i\hbar}{\pi V_{\rm cryst}} \int_\infty^{E_{\rm F}} {\rm d}E \operatorname{Tr}\left\langle j_\mu \frac{{\rm d}G^+(E)}{{\rm d}E}\, j_\nu \operatorname{Im} G^-(E) - j_\mu \operatorname{Im} G^+(E)\, j_\nu \frac{{\rm d}G^-(E)}{{\rm d}E}\right\rangle_{\rm conf}, \tag{5.41}$$

where j_μ is the μth spatial component of the electronic current density operator $\boldsymbol{j} = ec\boldsymbol{\alpha}$. In the following it is assumed that a finite conductivity or resistivity of the investigated system stems exclusively from chemical disorder, i.e. contributions caused by lattice imperfections, grain boundaries, phonons, magnons, and so on are ignored. This implies in particular that we are dealing with the residual resistivity for $T = 0$ K and that $\langle\cdots\rangle_{\rm conf}$ in Equations (5.40) and (5.41) denotes the atomic

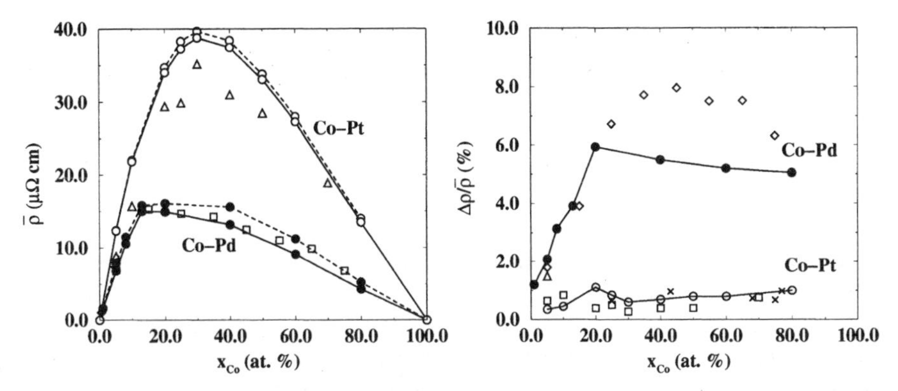

Figure 5.12 Left: residual isotropic resistivity $\bar{\rho}$ of disordered Co_xPd_{1-x} (•) and Co_xPt_{1-x} (○) alloys. Full lines, calculated including vertex corrections; broken lines, calculated omitting vertex corrections. Right: calculated anomalous magnetoresistance (AMR) ratio $\Delta\rho/\bar{\rho}$ of Co_xPd_{1-x} (•) and Co_xPt_{1-x} (○) alloys. Experimental data denoted by open squares, white diamonds, triangles and crosses arise from various sources (Ebert *et al.* 1996b).

configuration average for a disordered alloy. This can be achieved in a very reliable way within the framework of the KKR-CPA formalism (Butler 1985).

The magnetization of transition-metal systems is primarily due to the exchange-splitting of their spin subsystems. Accordingly, the aforementioned symmetry reduction due to magnetic ordering affects the transport properties only due to the presence of the spin–orbit coupling. This implies that the galvanomagnetic properties discussed here are of pure relativistic origin. As a consequence, to deal with these, Butler's non-relativistic approach had to be generalized by adopting the SPR-KKR-CPA formalism (Banhart and Ebert 1995).

As an example of an application of the formalism sketched above, the calculated isotropic resistivities $\bar{\rho}$ for the alloy systems Co_xPd_{1-x} and Co_xPt_{1-x} are shown in the left part of Figure 5.12 (Ebert *et al.* 1996b) together with corresponding experimental data measured at low temperature. As we can see, the agreement between calculated and measured resistivities is very good for Co_xPd_{1-x}. The maximum value of the resistivity in this system (16 μΩ cm) as well as the composition for which the maximum occurs (about 20% Co) are well reproduced by the calculations.

Using Butler's approach in dealing with Equations (5.40) and (5.41) we account for the so-called vertex corrections within the framework of the CPA. For Co_xPd_{1-x} it was found that their contribution increases from about 2% for 5 at.% Co to about 25% for 80 at.% Co.

For the system Co_xPt_{1-x} the calculated resistivities are much higher than for Co_xPd_{1-x}, reaching almost 40 μΩ cm for 30 at.% Co. This agrees in a satisfying way with the experimental maximum of about 35 μΩ cm at that composition. In contrast to Co_xPd_{1-x}, the vertex corrections are quite small for Co_xPt_{1-x}, contributing less than 3% to the total conductivity over the entire composition range. Previous investigations of paramagnetic alloy systems (Banhart *et al.* 1994) led to the conclusion that the vertex corrections are more important the lower the d-like DOS at the

Fermi level is. For Cu_xPt_{1-x} (Banhart *et al.* 1994), for example, this applies to the noble-metal-rich side of this system. For ferromagnetic systems, on the other hand, the vertex corrections seem to be more important if the d-like DOS at the Fermi level is low at least for one spin subsystem. For this reason, they are more pronounced for Co_xPd_{1-x} compared with Co_xPt_{1-x} and more important on the Co-rich side of both systems. The anomalous magnetoresistance (AMR) ratios for the two alloy systems Co_xPd_{1-x} and Co_xPt_{1-x} are shown in the right part of Figure 5.12. Experimental values for both systems are included for comparison. Co_xPd_{1-x} shows remarkably high AMR values of more than 6% for concentrations higher than 20 at.% Co (Jen 1992; Senoussi *et al.* 1977). The calculations reproduce the increase in the experimental data at low Co concentrations very well. For higher Co concentrations the calculated values are slightly too low. Here it should be noted that the AMR in Co_xPd_{1-x} is still as large as 1.5% even for very low Co contents (Hamzić *et al.* 1978; Senoussi *et al.* 1977), which was attributed to local orbital moments on the magnetic sites (Senoussi *et al.* 1977). In contrast to Co_xPd_{1-x}, the AMR for Co_xPt_{1-x} was found to be below 1% throughout the whole concentration range (Jen *et al.* 1993; McGuire *et al.* 1984). These findings are perfectly reproduced by the relativistic calculation, which reflects the weak variation of the AMR in Co_xPt_{1-x} with concentration.

For the discussion of experimental galvanomagnetic properties, a number of phenomenological descriptions have been developed and used in the past. These approaches were based on Mott's two-current model that ascribe to each spin subsystem an independent current contribution and introduced a number of model parameters. The SPR-KKR-CPA formalism, on the other hand, does not rely on Mott's two-current model and allows for a parameter-free and quantitative investigation of galvanomagnetic properties. By manipulating the strength of the spin–orbit coupling it was possible in particular to demonstrate numerically the dependency of the AMR and the AHR on the spin–orbit coupling (Banhart *et al.* 1996). In addition, it could be shown that even the isotropic resistivity $\bar{\rho}$ can be strongly influenced by the spin–orbit coupling, as previously predicted (Mertig *et al.* 1993).

Further insight into the mechanisms giving rise to galvanomagnetic effects could be obtained by a decomposition of the spin–orbit coupling. To demonstrate this, corresponding results for $\bar{\rho}$ and AMR ratio $\Delta\rho/\bar{\rho}$ are given in Figure 5.13.

The left part of this figure shows the isotropic residual resistivity $\bar{\rho}$ of BCC-Fe_xCo_{1-x} obtained from calculations using the full spin–orbit coupling (ξ). Note that the variation of $\bar{\rho}$ with composition is strongly asymmetric. This agrees well with the experimental findings (Beitel and Pugh 1958; Freitas and Berger 1988). The deviation from a parabolic shape can be qualitatively explained by the change of the DOS at the Fermi energy $n(E_F)$ that decreases monotonously with increasing Fe content. Keeping only the spin mixing part ξ_{xy} of the spin–orbit coupling (see Section 5.2.3), we find that $\bar{\rho}$ hardly changes. This already indicates that the coupling of the two spin subsystems is the primary source for the relativistic enhancement of $\bar{\rho}$. This spin mixing or hybridization can be directly demonstrated by means of the spin-projected Bloch spectral function $A_B(\boldsymbol{k}, E)$. For FCC-$Fe_{0.2}Ni_{0.8}$, for example, we find in this way that there is an appreciable minority spin character admixed to

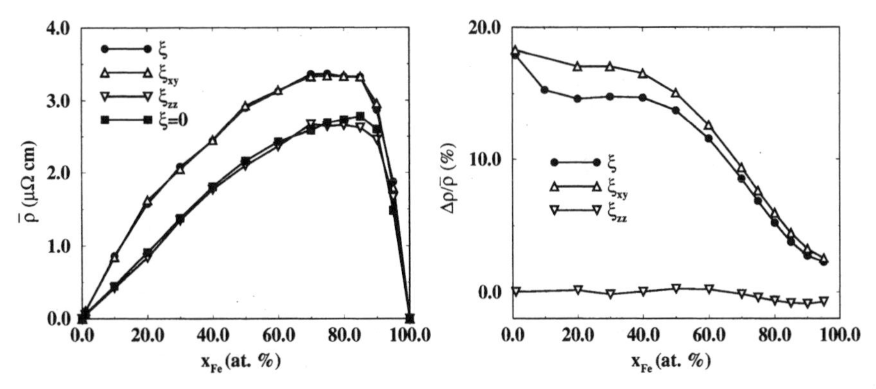

Figure 5.13 Isotropic residual resistivity $\bar{\rho}$ (left), and spontaneous magnetoresistance anisotropy ratio $\Delta\rho/\bar{\rho}$ (right) of disordered BCC-Fe_xCo_{1-x} alloys calculated in four different ways. The results obtained using the full spin–orbit coupling, indicated by ξ, are represented by full circles. Open triangles give the results obtained keeping the xy-part ξ_{xy} and zz-part ξ_{zz}, respectively, of the spin–orbit coupling. Full squares give the result with the spin–orbit coupling completely suppressed ($\xi = 0$).

the majority spin states which form a Γ-centred sheet of the Fermi surface and which primarily carry the electronic current (Ebert *et al.* 1997c; Mertig *et al.* 1993). Admixture of minority spin character opens a new scattering channel for these states that is very effective because of the high DOS $n^{\downarrow}(E_F)$ at the Fermi energy with minority spin character. As a consequence, the total resistivity has to go up in a pronounced way compared with a calculation based on the two-current model (Banhart *et al.* 1997). This interpretation is confirmed by the results obtained by keeping just the spin-diagonal part ξ_{zz} of the spin–orbit coupling, i.e. suppressing the spin mixing effect. In Figure 5.13 we can see that this manipulation leads to a strong reduction of the total resistivity throughout the whole range of concentration. To demonstrate that the remaining part ξ_{zz} of the spin–orbit coupling has practically no influence on $\bar{\rho}$, an additional calculation has been carried out with the spin–orbit coupling completely suppressed ($\xi = 0$). The corresponding results nearly completely coincide with the ξ_{zz}-data, confirming this expectation. Here we should note that the latter calculational mode ($\xi = 0$), although technically somewhat different, corresponds essentially to a calculation on the basis of the two-current model, where the electronic structure is calculated in a scalar relativistic way, i.e. with the relativistic corrections Darwin- and mass–velocity-terms taken into account (Banhart *et al.* 1996).

For the spin–orbit-induced AMR ratio the results obtained by the various calculations are given in the right panel of Figure 5.13. Here we find that keeping only ξ_{xy} slightly reduces $\Delta\rho/\bar{\rho}$. This means that in contrast to $\bar{\rho}$, ξ_{zz} has some small effect on this quantity. Nevertheless, we find that keeping ξ_{zz} alone brings $\Delta\rho/\bar{\rho}$ essentially to zero. From this result it can be concluded that the part ξ_{zz} of the spin–orbit coupling can in general be neglected as a source for the AMR compared with ξ_{xy}. Finally, setting $\xi = 0$ of course reduces $\Delta\rho/\bar{\rho}$ exactly to zero (not shown here) in agreement with the above statement that galvanomagnetic properties are caused by spin–orbit coupling (Banhart *et al.* 1996).

The model calculations performed for the residual resistivity tensor elements of Fe_xCo_{1-x} allow us to check the above-mentioned phenomenological models for the galvanomagnetic effects. For example, Smit (1951) ascribed the occurrence of the AMR to the spin hybridization caused by the spin–orbit coupling. From an analysis of experimental data, on the basis of corresponding expressions for $\Delta\rho/\bar{\rho}$, Jaoul *et al.* (1977) concluded that there should be an additional contribution due to the spin-diagonal part of the spin–orbit coupling. The results presented in Figure 5.13 clearly demonstrate that the mechanism discussed by Jaoul *et al.* can be neglected for the isotropic resistivity $\bar{\rho}$ and has only a minor contribution to the AMR in the case of the alloy system BCC-Fe_xCo_{1-x}.

Finally, it should be mentioned that the expression for the diagonal conductivity tensor elements $\sigma_{\mu\mu}$ given in Equation (5.40) has been generalized to deal with the giant magnetoresistance (GMR) of multilayer systems (Butler *et al.* 1995). A corresponding spin-polarized relativistic formulation has been given by Weinberger *et al.* that accounts in particular for the influence of the spin–orbit coupling (Weinberger *et al.* 1996). A straightforward extension of Equation (5.40) to finite frequencies ω giving the absorptive part of the optical conductivity tensor element $\sigma_{\mu\mu}(\omega)$ has been presented by Banhart for the visible regime of the light and for paramagnetic alloy systems (J. Banhart 1998, personal communication). With the full optical conductivity tensor for a ferromagnetic available, we get access to a treatment of the spin–orbit-induced magneto-optical Kerr effect (MOKE) (Ebert 1996). A corresponding fully relativistic expression has been recently derived within the framework of the SPR-KKR formalism (Huhne and Ebert 1999; Szunyogh and Weinberger 1999). This new approach allows us to deal with, in particular, the Kerr effect for surface-layer systems (Huhne 2000).

5.3.5 Spectroscopy

Spectroscopy plays a central role in the investigation and understanding of the various properties of solids. One of the early hints for the importance of spin–orbit coupling for the electronic structure of solids stems from X-ray absorption experiments on Pt (Cauchois and Manescu 1940). While for the L_3-edge a pronounced *white line* was observed, none was found for the L_2-edge. Mott (Mott 1949) ascribed this finding to the spin–orbit coupling, which should cause the d states of Pt above the Fermi energy to have predominantly $d_{5/2}$ character. As a consequence of this and because of the dipole selection rules $\Delta j = 0, \pm 1$, we expect strong absorption for the L_3-edge but not for the L_2-edge.

Another consequence of spin–orbit coupling is the occurrence of the Fano effect in photoemission (Fano 1969a,b). This term denotes the phenomenon where we obtain a spin-polarized photoelectron current even for a paramagnetic solid if circularly polarized light is used for excitation (Heinzmann *et al.* 1972). Due to time-reversal symmetry the spin polarization gets just reversed if the helicity of the radiation is reversed. For a magnetic solid, however, time-reversal symmetry is broken, leading to a large number of interesting magneto-optical effects.

As an example of spin–orbit-induced magneto-optical properties of magnetic solids, magnetic dichroic effects in X-ray absorption and photo-emission will be discussed in the following. In supplying an adequate theoretical description for these we have in principle to account for the fact that the corresponding experiment in general transfers the investigated system from its ground state to an excited state. This can be done, for example, on the basis of Hedin's GW-approximation (Hedin and Lundqvist 1969), which proved extremely successful when dealing with the band-gap problem of semiconductors (Godby *et al.* 1988). Although many-body effects may also have a strong impact on the spectroscopic properties of transition-metal systems (Liebsch 1979), it is often well justified to discuss these on the basis of the electronic structure of their ground state alone. This point of view will be adopted in the following.

Magnetic circular dichroism in X-ray absorption

Guided by their experience with the magneto-optical Kerr effect (MOKE), Erskine and Stern (1975) suggested that there should be a corresponding magnetic dichroism in X-ray absorption when circularly polarized radiation is used. This magnetic circular X-ray dichroism (MCXD) could be demonstrated for the first time for transition metals by Schütz *et al.* (1987) by measurements at the K-edge of Fe in BCC-Fe in the XANES (X-ray absorption near edge structure) region.

This pioneering work was followed by many investigations into a great variety of systems (Ebert 1996). In particular, extensive work has been done in the soft X-ray regime at the $L_{2,3}$-edges of 3d transition metals to probe their magnetic properties (see below). Schütz *et al.* (1989) also observed the magnetic dichroism in the EXAFS (extended X-ray absorption fine structure) region by investigating the $L_{2,3}$-edges of Gd in HCP-Gd. This opened the way to magnetic EXAFS as a new type of spectroscopy (Knülle *et al.* 1995; Schütz and Ahlers 1997).

Although there are also various interesting forms of linear magnetic dichroism (Kortright and Kim 2000), most experimental investigations of the magnetic dichroism in X-ray absorption spectroscopy use circularly polarized radiation because the circular dichroism is most pronounced. To allow a sound interpretation of the corresponding dichroic signal $\Delta\mu = \mu^+ - \mu^-$, given by the difference in absorption of left- and right-circularly-polarized radiation, a set of sum rules have been derived by several authors (Carra *et al.* 1993; Schütz *et al.* 1993; Thole *et al.* 1992; Wienke *et al.* 1991). The main virtue of these rules is that they should allow us to obtain a reasonable estimate for the expectation values $\langle\sigma_z\rangle$ and $\langle l_z\rangle$ of an absorber atom from its energy integrated dichroic signals $\int \Delta\mu(E)\,\mathrm{d}E$. Of course, this is a very appealing property because these quantities are directly proportional to the spin and orbital magnetic moments, μ_{spin} and μ_{orb}, respectively. However, in applying the sum rules one of the main problems is to fix the upper energy integration limit. For that reason it has been suggested the sum rules be applied in their differential form and the dichroic spectra $\Delta\mu(E)$ discussed directly. For the $L_{2,3}$-edges these differential sum rules are given by (Ebert *et al.* 1999)

$$3[\Delta\mu_{L_3} - 2\Delta\mu_{L_2}] = C_d\left(\frac{d}{dE}\langle\sigma_z\rangle_d + 7\frac{d}{dE}\langle T_z\rangle_d\right), \tag{5.42}$$

$$2[\Delta\mu_{L_3} + \Delta\mu_{L_2}] = C_d\frac{d}{dE}\langle l_z\rangle_d. \tag{5.43}$$

Here C_d is a normalization constant and T_z is the magnetic dipole operator, which often can be ignored. Thus, the basic information to be deduced from the dichroic signal at the $L_{2,3}$-edges is the spin and orbital polarization,

$$\frac{d}{dE}\langle\sigma_z\rangle_d \quad \text{and} \quad \frac{d}{dE}\langle l_z\rangle_d,$$

respectively, of final states with d character.

Motivated by the MCXD measurements on BCC-Fe, Ebert *et al.* (1988b,c) developed a corresponding fully relativistic description based on the SPR-KKR formalism and which has since been applied to a great variety of different systems (Ebert 1996). This approach was later extended to deal with magnetic EXAFS (MEXAFS) by making use of the cluster approximation for the multiple scattering representation of the final states. Using the SPR-KKR formalism, the X-ray absorption coefficient $\mu^{q\lambda}(\omega)$ is given by (Ebert 1996)

$$\mu^{q\lambda}(\omega) \propto \operatorname{Im}\sum_{i\,\mathrm{occ}}\left[\sum_{\Lambda\Lambda'} M^{q\lambda *}_{\Lambda i}(E_i + \hbar\omega)\tau^{nn}_{\Lambda\Lambda'}(E_i + \hbar\omega)M^{q\lambda}_{\Lambda' i}(E_i + \hbar\omega) + \sum_{\Lambda} I^{q\lambda}_{\Lambda i}(E_i + \hbar\omega)\right]. \tag{5.44}$$

Here the sum i runs over all involved core states with energy E_i and wave function Φ_i. The electron–photon interaction operator $X_{q\lambda}$, occurring in the matrix elements $M^{q\lambda}_{\Lambda i}$, carries, in particular, information on the wave vector $\boldsymbol{q}$ of the radiation and on its polarization λ. The last term $I^{q\lambda}_{\Lambda i}$ in Equation (5.44) is an atomic-like matrix element (Ebert 1996) and is connected to the term in the Green function involving the irregular solution to the Dirac equation (see Equation (5.31)). Accordingly, it contributes only when working with complex energies to account for finite lifetime effects.

The magnetic dichroism of the $L_{2,3}$-edges spectra of Pt in the disordered alloy system Fe_xPt_{1-x} has been studied experimentally as well as theoretically in great detail in the past (Baudelet *et al.* 1997; Ebert and Akai 1993; Ebert *et al.* 1993; Maruyama *et al.* 1992; Stähler *et al.* 1993). Typically, for Pt $L_{2,3}$-spectra it was found that the white lines at the L_2- and L_3-edges are quite different because of the influence of the spin–orbit coupling acting on the final states. This finding makes clear that a fully relativistic approach is indispensable to achieve a quantitative description of the $L_{2,3}$-absorption spectra of Pt. This applies in particular if we are dealing with magnetic EXAFS (MEXAFS). The top panel of Figure 5.14 shows the results of calculations

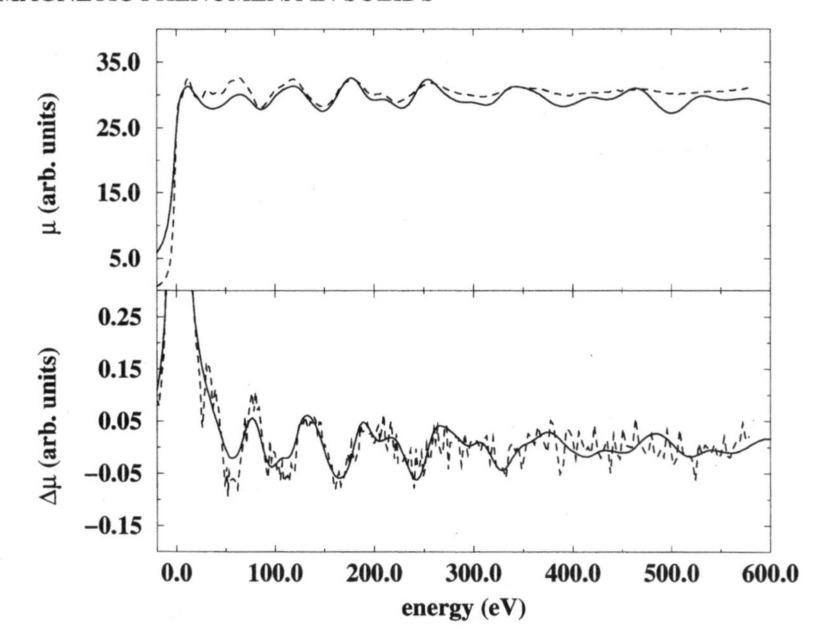

Figure 5.14 EXAFS (top) and MEXAFS spectra (bottom) at the L_2-edge of Pt in Fe_3Pt. Calculations for the ordered compound (full line), compared with the experimental data for the $Fe_{0.72}Pt_{0.28}$ (dashed line) (Ahlers 1998). The corresponding calculations for the scattering path operator $\tau^{nn}_{\Lambda\Lambda'}$ have been done using the matrix inversion technique for a cluster of 135 atoms in the XANES and 55 atoms in the EXAFS region, including the central absorber site. The effects of self-energy corrections (Fujikawa *et al.* 1997; Mustre de Leon *et al.* 1991) have been accounted for after calculating the spectra.

for the L_2-edge EXAFS-spectra of Pt in ordered Fe_3Pt. Corresponding experimental data, obtained by Ahlers and co-workers (Ahlers 1998) for an ordered but slightly off-stoichiometric sample, are added. As we can see, the agreement of the theoretical and experimental spectra is quite satisfying. Additional calculations for disordered $Fe_{0.72}Pt_{0.28}$ led to spectra in rather poor agreement with experiment (Popescu *et al.* 1999). Obviously, these findings confirmed in particular that the experimental sample is indeed ordered.

The circular dichroic spectrum $\Delta\mu_{L_2}$ for the L_2-edge is shown in the bottom panel of Figure 5.14. Again a very satisfying agreement with the corresponding experimental results could be achieved. The results for $\Delta\mu_{L_2}$ clearly demonstrate that the occurrence of magnetic dichroism is by no means restricted to the white line region. Although the amplitude for $\Delta\mu_{L_2}$ is quite small compared with the white line region, it is present throughout the whole EXAFS range.

As mentioned above, the applicability of the sum rules in their conventional form seems to be somewhat doubtful because of these findings. Nevertheless, a clear-cut interpretation of the MEXAFS spectra can be given making use of the sum rules in their differential form.

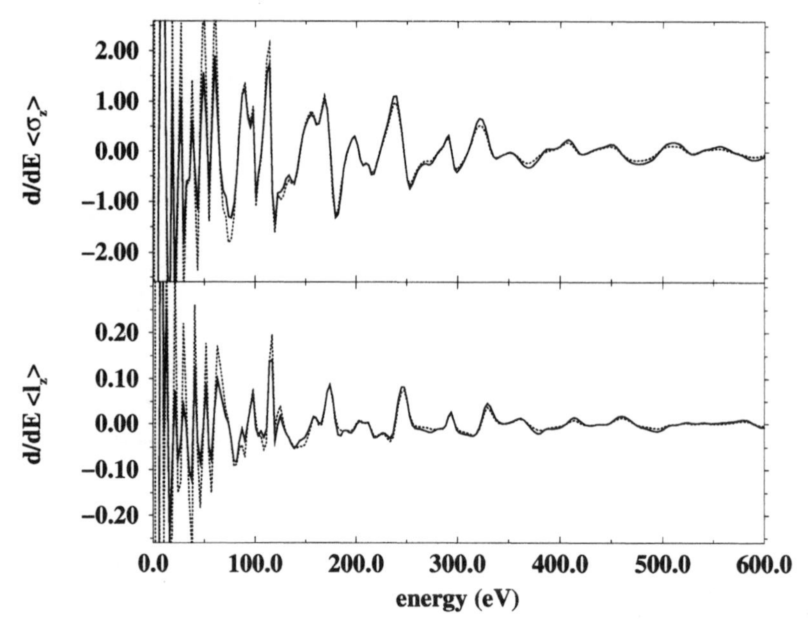

Figure 5.15 Top: spin polarization $(\mathrm{d}/\mathrm{d}E)\langle\sigma_z\rangle_\mathrm{d}$ for the d states of Pt (full line) compared with that derived from the MCXD spectra using Equation (5.42) (dotted line). Bottom: orbital polarization $(\mathrm{d}/\mathrm{d}E)\langle l_z\rangle_\mathrm{d}$ for the d states of Pt (full line) compared with that derived from the MCXD spectra using Equation (5.43) (dotted line). To compare the spectroscopic data with the band-structure results, the normalization factor C_d in Equations (5.42) and (5.43) has been used as a free scaling parameter using the same value for the upper and lower parts of the figure.

In the upper part of Figure 5.15 a superposition of the theoretical magnetic dichroism spectra $\Delta\mu_{\mathrm{L}_2}$ and $\Delta\mu_{\mathrm{L}_3}$ according to Equation (5.42) is given (here the very small contribution $(\mathrm{d}/\mathrm{d}E)\langle T_z\rangle_\mathrm{d}$ has been neglected). This is compared with the spin polarization $(\mathrm{d}/\mathrm{d}E)\langle\sigma_z\rangle_\mathrm{d}$ of the d states that have been obtained directly from the band-structure calculations. In the lower part of Figure 5.15 the superposition according to Equation (5.43) of the dichroic spectra is compared with the directly calculated orbital polarization $(\mathrm{d}/\mathrm{d}E)\langle l_z\rangle_\mathrm{d}$ of the d states.

The nearly perfect coincidence of the various curves in the upper and lower parts of Figure 5.15 convincingly demonstrates that the primary information that can be deduced from circular $\mathrm{L}_{2,3}$-MEXAFS spectra is the spin and orbital polarization for the final d-like states of the absorber atom. Of course, these are no pure atomic properties because the absorber atom is not decoupled from its surroundings. In particular, the variation of its spin and orbital polarization with energy strongly depends on the bonding to the neighbouring atoms. For this reason we may expect from the Fourier transform of a MEXAFS spectrum information on the magnetization distribution around the absorber atom (Ahlers 1998). Motivated by this consideration Schütz and co-workers extended the standard theoretical description of conventional EXAFS to deal with magnetic EXAFS (Ahlers and Schütz 1998; Schütz and Ahlers 1997). In this way it could be shown that the spin dependency of the scattering amplitude of an atom with a small magnetic moment should also be small. Accordingly, the atom should not show up as a backscatterer in the pair distribution function obtained by Fourier transformation of a MEXAFS spectrum. For a number of rare earth compounds containing oxygen, it was indeed found that oxygen, whose magnetic moment is always rather small, hardly shows up as a backscatterer in the MEXAFS spectra of the rare earth metals (Ahlers *et al.* 1998). This sensitivity of the MEXAFS spectra on the magnetic moment of the backscattering atoms could be exploited within a study on the interface roughness of Co/Cu multilayers (Schütz and Ahlers 1997). In this case, conventional EXAFS does not supply much information because the spin-averaged scattering amplitudes of Co and Cu are very similar. This does not hold for MEXAFS, because the magnetic moment of Co in Co/Cu is about two orders of magnitude larger than for Cu.

Magnetic dichroism in valence-band photoemission

The electronic structure of magnetic solids can be investigated in a very detailed way using spin- and angle-resolved photoemission. Adopting the so-called one-step model, the photocurrent at the detector is described by a 2×2 spin-density matrix (Braun 1996; Feder 1985),

$$\rho_{ss'} = \frac{1}{2\mathrm{i}}(\tilde{\rho}_{ss'} - \tilde{\rho}^{*}_{s's}),$$

where

$$\tilde{\rho}_{ss'}(E, \boldsymbol{k}_{\parallel}) = -\frac{1}{\pi}\langle E, \boldsymbol{k}_{\parallel}, s|HG(E - \hbar\omega, \boldsymbol{k}_{\parallel})H^{\dagger}|E, \boldsymbol{k}_{\parallel}, s'\rangle, \tag{5.45}$$

where $H = \boldsymbol{\alpha} \cdot \boldsymbol{A}$ is the interaction of the electrons and the photon field. In Equation (5.45) the initial band states are represented by the Green function $G(E - \hbar\omega, \boldsymbol{k}_{\parallel})$, while the final states $|E, \boldsymbol{k}_{\parallel}, s\rangle$ are time-reversed LEED states with $\boldsymbol{k}_{\parallel}$ being the surface parallel wave vector common to the initial and final states.

The above equations have been proved to supply an excellent framework for dealing with magnetic aspects in photoemission that are connected with spin–orbit coupling. A very prominent example of this is the Fano effect in paramagnetic, i.e. nonmagnetic, solids (Fano 1969a). Usually, the Fano effect denotes the observation that a spin-polarized photocurrent is created if circularly polarized radiation is used for excitation. However, a few years ago it was predicted by theory (Tamura *et al.* 1987) and later confirmed by experiment (Schmiedeskamp *et al.* 1988) that we can also have spin-polarization for s-polarized radiation impinging on (111) surfaces of cubic solids. Corresponding calculations have been done in the past making use of the muffin-tin geometry for the potential. This simplifying step has been removed recently by using the full potential version of relativistic multiple scattering theory (Fluchtmann *et al.* 1995). This step leads, even for a dense-packed metal like Cu, to noticeable changes in the photoemission spectra (Fluchtmann *et al.* 1998). This is demonstrated

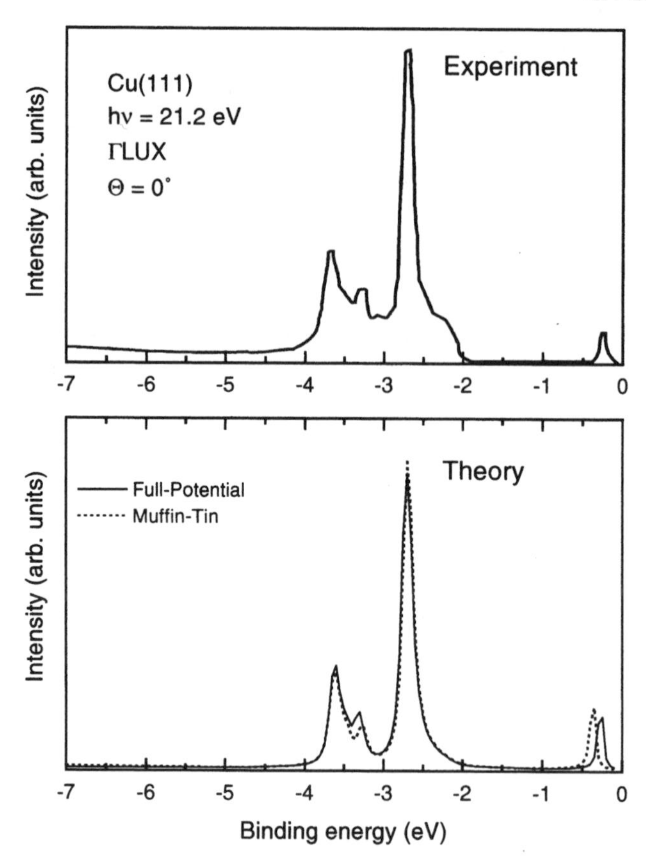

Figure 5.16 Top: experimental normal emission (Γ–L) ARUPS spectrum for Cu (111), $h\nu = 21.2$ eV (Gerlach *et al.* 1998). Bottom: corresponding calculated ARUPS spectrum using the muffin-tin approximation and the full potential version of relativistic multiple scattering theory. Reproduced with permission from Fluchtmann *et al.* (1998). © Elsevier Science.

in Figure 5.16, which shows theoretical results based on the muffin-tin as well as full-potential geometry in comparison with experimental data (Gerlach *et al.* 1998).

Using the full-potential geometry influences to some extent the peak positions and also the relative peak heights. Both modifications bring the theoretical spectrum in closer agreement with experiment. The same experiment has also been carried out for other systems, for example, GaAs (Fluchtmann *et al.* 1999), which is the standard source for spin-polarized electron beams by making use of the Fano effect (Meier 1985).

Due to time-reversal symmetry the spin-polarization from a paramagnetic solid using circularly polarized radiation is just reversed if the helicity of the radiation is reversed. This feature is of course removed if the solid is magnetically ordered, giving rise to magnetic circular dichroism in valence-band photoemission (Schneider *et al.* 1991). In a corresponding experiment we have in general the emission direction,

Table 5.5 Magnetic dichroic effects and photoelectron spin-polarization components P_i for perpendicular ($\boldsymbol{M} \| \hat{\boldsymbol{e}}_z$) and in-plane ($\boldsymbol{M} \| \hat{\boldsymbol{e}}_y$) magnetization $\boldsymbol{M}$ of surfaces with twofold, threefold or fourfold rotational axes. s, p and c stand for s, p and normally incident circularly polarized light. The sequence of signs indicates whether the respective component P_i occurs (+ sign) or not (− sign) if only spin–orbit coupling (first sign), only exchange splitting (second sign), or both (third sign) are present. MLD (L) and MCD (C) occur if a spin-polarization component parallel to $\boldsymbol{M}$ is produced by spin–orbit coupling in the nonmagnetic case, i.e. if there is a combination (+, +, +) (Feder and Henk 1996).

		***M* perpendicular**				***M* in-plane**			
Pol.	Sym.	P_x	P_y	P_z	I	P_x	P_y	P_z	I
s	2mm	−,−,−	−,−,−	+,+,+	L	−,−,+	−,+,+	+,−,+	
	4mm	−,−,−	−,−,−	−,+,+		−,−,+	−,+,+	−,−,+	
	3m	+,−,+	+,−,+	−,+,+		+,−,+	+,+,+	−,−,+	L
p	2mm	+,−,+	+,−,+	+,+,+	L	+,−,+	+,+,+	+,−,+	L
	4mm	+,−,+	+,−,+	−,+,+		+,−,+	+,+,+	−,−,+	L
	3m	+,−,+	+,−,+	−,+,+		+,−,+	+,+,+	−,−,+	L
c	2mm	−,−,−	−,−,−	+,+,+	C	−,−,+	−,+,+	+,−,+	
	4mm	−,−,−	−,−,−	+,+,+	C	−,−,+	−,+,+	+,−,+	
	3m	−,−,−	−,−,−	+,+,+	C	−,−,+	−,+,+	+,−,+	

the incoming photon beam as well as the magnetization perpendicular to the surface of the sample. The reversal of the helicity of the radiation or magnetization gives rise to a change in the intensity of the photocurrent, which is often also analysed with respect to the spin polarization. Apart from magnetic circular dichroism (MCD) we may also observe magnetic linear dichroism (MLD) (Rampe *et al.* 1996). In a typical MLD experiment with normal electron emission the magnetization lies in the plane. The photon beam is p-polarized, at an angle of 90° with the magnetization vector. Reversing, for example, the orientation of the magnetization gives rise to a magnetic linear dichroism. The various configurations that can lead to MLD and MCD in photoemission have been carefully analysed using symmetry arguments by Feder and co-workers (Henk *et al.* 1996). This analytical study, which accounted for the appropriate double-group symmetry of the half-space initial and final states, led to explicit expressions for the spin polarization vector of the photoelectrons, and the spin-averaged intensity and its change upon reversal of the magnetization direction. The most important results of this study, which elucidated the origin of spin polarization and dichroism in terms of an interplay between spin–orbit coupling and exchange splitting, are summarized in Table 5.5.

These analytical considerations support the analysis and interpretation of corresponding experimental and theoretical dichroic spectra in an appreciable way. As an example of such a combined investigation, valence-band photoemission spectra for a

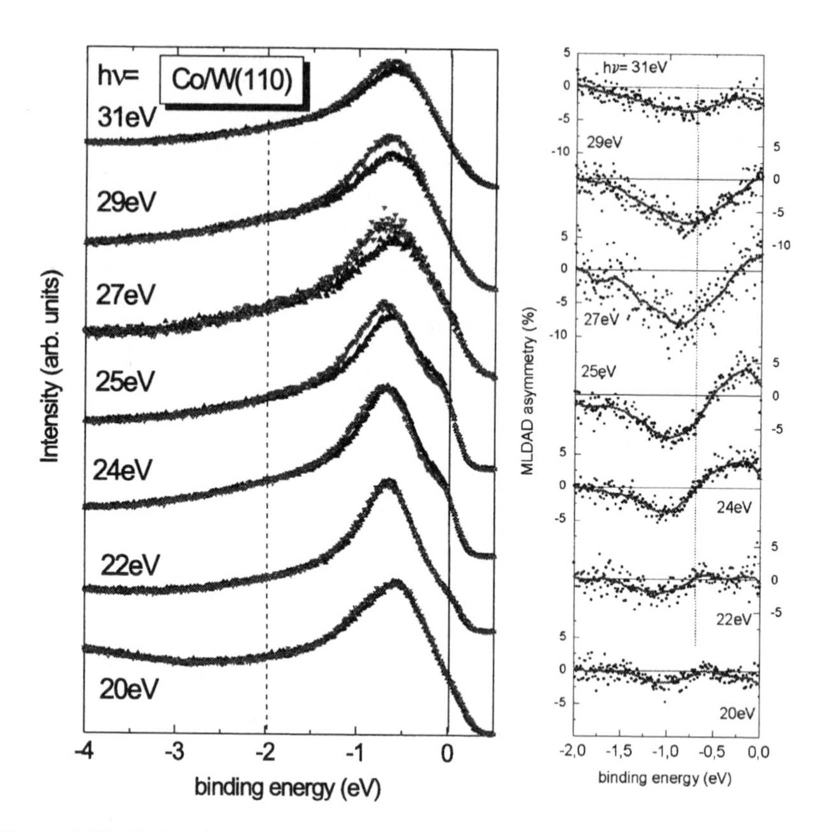

Figure 5.17 Left: valence-band photoemission spectra from a seven-monolayer-thick HCP (0001) Co film on W taken with linearly polarized radiation for opposite magnetization directions. Right: corresponding asymmetry values for binding energies between -2 eV and E_F. Experiment (Bansmann *et al.* 2000).

seven-monolayer-thick HCP(0001) Co film on top of a W(110) substrate are shown in Figure 5.17.

The experiment has been carried out in normal emission mode with the plane of incidence perpendicular to the magnetization that was aligned along the easy axis, i.e. in-plane along the W[1$\bar{1}$0]-direction. To investigate the MLD, the photoemission spectra shown in Figure 5.17 have been recorded with p-polarized light for two opposite orientations of the magnetization.

For a direct comparison of calculated spectra with experiment, these had to be shifted by about 2 eV, as indicated in the second panel of Figure 5.17. This finding is quite common and has to be ascribed to self-energy corrections. With this shift applied, the experimental spectra could be reproduced by the calculations in a rather satisfying way allowing for a detailed discussion of these. The observed main peak of the experimental spectra is found to be almost constant for the various photon energies

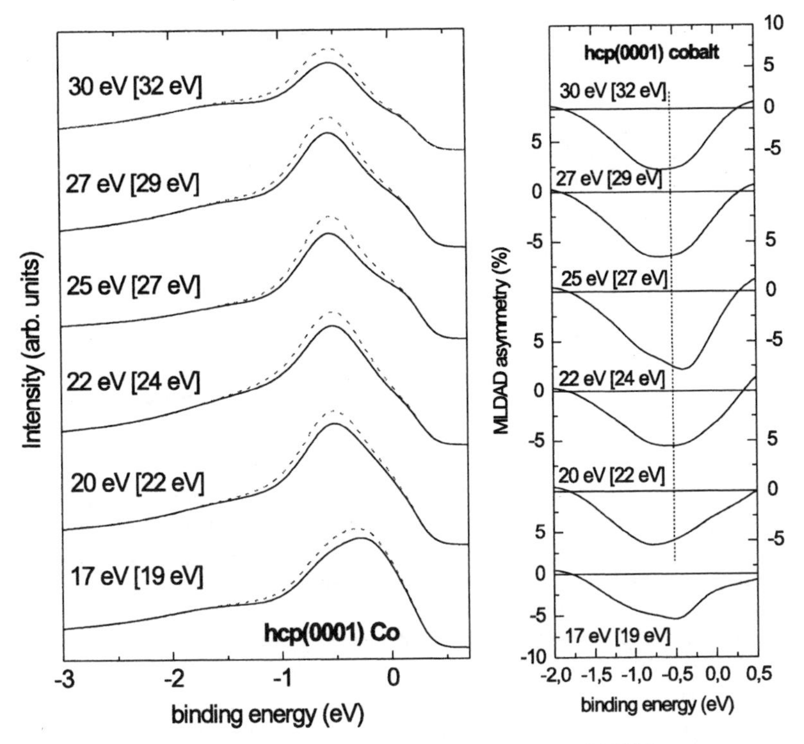

Figure 5.17 (*Cont.*) Theory (Bansmann *et al.* 2000).

at a binding energy of around 0.7 eV. This is in full agreement with the theoretical dispersion relation for HCP-Co, which has a rather flat band in that energy regime. For a photon energy of around 24 eV we find the onset of a peak very close to the Fermi level. This could be ascribed to an initial state close to E_{F} directly at the Γ-point.

Figure 5.17 shows that MLD is rather pronounced and energy dependent. As a possible explanation for this it was suggested that the involved transitions connect initial states with a strong mixing of the spin subsystems due to the spin–orbit coupling to final states with a nonvanishing exchange splitting (see also Table 5.5). As can be seen in Figure 5.17, the MLD is reproduced quite well by the calculations. A more detailed analysis of the theoretical spectra reveals in particular that the MLD is nearly exclusively due to transitions from initial states with d character to final states having f character.

Dichroism in superconductors

We now turn to relativistic effects in the spectroscopy of superconductors. Several recent experiments have reported the observation of dichroic phenomena in superconductors (Lawrence *et al.* 1992; Lihn *et al.* 1996; Lyons *et al.* 1990, 1991; Weber *et al.* 1990; Wu *et al.* 1996). On the theory side, the difference ΔP between the power absorption of left- and right-circularly-polarized light has been investigated on the

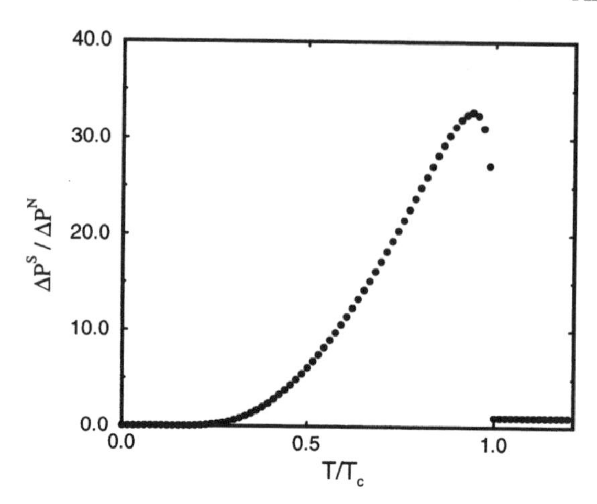

Figure 5.18 Difference in power absorption ΔP^{S} between left- and right-circularly-polarized light in the superconducting state, divided by the corresponding normal-state difference ΔP^{N}, as a function of temperature divided by the transition temperature T_{c}. The difference ΔP^{S} is proportional to $\mathrm{Im}[\hat{\sigma}^{\mathrm{S}}_{xy}(\omega, T, H)]$, the imaginary part of the off-diagonal elements of the conductivity tensor in the superconductor, as a function of frequency, temperature and magnetic field ($\omega = 4.5$ meV and $H = 0.05$ T for the data in the figure).

basis of the relativistic Bogoliubov–de Gennes equations described in Section 5.2.2. It has been demonstrated (Capelle *et al.* 1997, 1998) that in the presence of a static magnetic field, both ordinary spin–orbit coupling and the anomalous spin–orbit term, which involves the pair potential, can lead to circular dichroism (i.e. nonvanishing ΔP) in the electromagnetic response of superconductors. It is expected that the effect of ordinary spin–orbit coupling will be large for materials with heavy atoms in the lattice. Anomalous spin–orbit coupling, on the other hand, is expected to be important for strongly inhomogeneous superconductors (e.g. SNS multilayers or the vortex lattice) because then the gradients of the pairing potential will be large. Mechanisms, other than ordinary or anomalous spin–orbit coupling, have also been identified (Capelle *et al.* 1997, 1998). Those include orbital currents in the presence of a static magnetic field, as well as 'exotic' order parameters which break either inversion symmetry or time-reversal symmetry. The latter mechanism is unique in that it does not depend on the presence of an (external or internal) magnetic field. The observation of dichroism below T_{c} in the absence of such fields will be a strong hint for unconventional order parameters breaking either time-reversal or inversion symmetry.

Numerical calculations have been performed, so far, only for ordinary spin–orbit coupling in superconductors (Capelle *et al.* 1997, 1998). Figure 5.18 shows the difference ΔP^{S} in the superconducting phase divided by the corresponding quantity ΔP^{N} in the normal state as a function of (T/T_{c}) for a simple model superconductor. The figure shows that, below the critical temperature T_{c}, the dichroic response is dramatically modified compared with the normal state. Without spin–orbit coupling both numerator and denominator would be zero, while without superconducting coherence

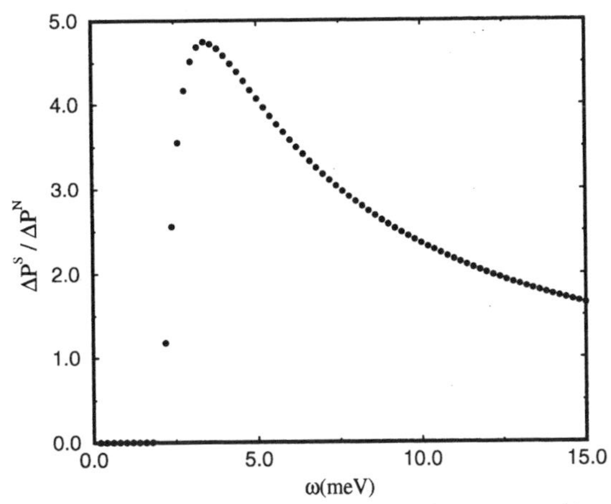

Figure 5.19 Dichroism ratio versus frequency ω. The plot corresponds to a reduced temperature T/T_C of approximately 0.45 (well below the strong peak in Figure 5.18) and a magnetic field of 0.1 T.

the curve would be flat ($= 1$) throughout. The strong peak seen immediately below the transition temperature thus arises only from the simultaneous presence of superconducting coherence and spin–orbit coupling. Figure 5.19 shows the same ratio as a function of the frequency ω of the incoming light. Here an absorption edge at $\omega = 2\Delta$ produced by pair-breaking processes becomes clearly visible.

Acknowledgment

The authors thank their colleagues and students D. Ahlers, J. Banhart, M. Battocletti, J. Braun, I. Cabria, K. Capelle, P. H. Dederichs, M. Deng, P. Fischer, H. Freyer, L. Fritsche, B. L. Gyorffy, J. Kübler, M. Marques, L. Sandratskii, G. Schütz, A. Vernes, A. Vetter, and R. Zeller for their fruitful collaboration.

6 Experimental and Theoretical Study of the Chemistry of the Heaviest Elements

J. V. Kratz
Institut für Kernchemie, Universität Mainz, Mainz, Germany

V. Pershina
Gesellschaft für Schwerionenforschung, Planckstr. 1, Darmstadt, Germany

6.1 Introduction

The study of chemical properties of the very heavy elements is an important and exciting task, since in this area of the periodic system relativistic effects are so strong that some deviations from well-established periodicities could be observed. Under theoretical and experimental investigations are presently elements with atomic number 104 and higher called transactinides. In these elements the filling of the 6d-shell takes place. Thus, they are expected to exhibit chemical behaviour analogous to that of the 4d and 5d transition elements. The task we are confronted with is therefore to prove experimentally and theoretically whether the chemical properties of these very heavy elements are typical for elements in the corresponding chemical groups, or whether they deviate from the trends found within the lighter homologues due to strong relativistic effects on their electronic shells.

Due to short half-lives and low production rates of these elements, there are many difficulties involved in the experimental investigation of their properties. To overcome these problems, special 'fast-chemistry' techniques of the one-atom-at-a-time chemistry have been developed (Kratz 1999). These are based on the principle of chromatographic separation, which has an advantage over other methods, since each of the few available atoms experiences many adsorption–desorption cycles thus ensuring a statistical chemical behaviour. The experiments are subdivided into two large groups: gas-phase chromatography experiments (Gäggeler 1994) studying volatility of the heavy elements and their compounds, and liquid chromatography experiments (Kratz 1999) studying the complex formation in aqueous solutions.

Results of these experiments provided strong evidence that elements 104, rutherfordium (Rf), 105, dubnium (Db), 106, seaborgium (Sg) and, very recently, element 107, bohrium (Bh) (Eichler *et al.* 2000) behave similarly to their lighter homologues

Relativistic Effects in Heavy-Element Chemistry and Physics. Edited by B. A. Hess

in the chemical groups and their place in the periodic system was confirmed. (Preparations are presently under way to conduct gas-phase experiments with element 108, hassium (Hs), which was found to have an isotope with a half-life of the order of 10 s (Hofmann 1996), suitable for chemical studies.) These experiments have, however, revealed that measured properties were often different from those expected from straightforward extrapolations within the groups, and trends were even reversed. Some further, finer experiments were therefore to be carefully planned and conducted in order to study the properties of these elements in more detail, as well as theoretical investigations to help explain these phenomena and/or predict the outcome of new experiments. Thus, combined experimental and theoretical research has been going on in our groups leading to a better understanding of the chemistry of these exotic elements.

In this chapter, results of the recent theoretical and experimental studies of the chemical properties of elements 105 and 106 performed within the REHE project are reviewed (for earlier reviews see Kratz 1999; Pershina 1996; Pershina and Fricke 1999).

6.2 Theory

A suitable computational approach for the investigation of electronic and geometric structures of transactinide compounds is the fully relativistic Dirac–Slater discrete-variational method (DS-DVM), in a modern version called the density functional theory (DFT) method, which was originally developed in the 1970s (Rosén and Ellis 1975). It offers a good compromise between accuracy and computational effort. A detailed description can be found in Chapter 4 of this book.

The total energy functional (in atomic units) is given by

$$E[n] = T_{\mathrm{s}} + E_{\mathrm{ext}}[n] + E_{\mathrm{H}}[n] + E_{\mathrm{xc}}[n] \tag{6.1}$$

with the relativistic form of kinetic energy, T_{s},

$$T_{\mathrm{s}} = \sum_i \int \mathrm{d}^3 r \, \psi_i^{\dagger}(\boldsymbol{r})([-\mathrm{i}c\boldsymbol{\alpha}\cdot\nabla + (\beta - 1)mc^2])\psi_i(\boldsymbol{r}).$$

The next term is the external potential energy with the nuclear charges Z_α at the positions $\boldsymbol{R}_\alpha$

$$E_{\mathrm{ext}}[n] = \int \mathrm{d}^3 r \, v_{\mathrm{ext}}(\boldsymbol{r}) + \frac{1}{2}\sum_{\alpha}\sum_{\beta\neq\alpha} \frac{Z_\alpha Z_\beta}{|\boldsymbol{R}_\alpha - \boldsymbol{R}_\beta|},$$

with the external potential

$$v_{\mathrm{ext}}(\boldsymbol{r}) = -\sum_{\alpha} \frac{Z_\alpha}{|\boldsymbol{R}_\alpha - \boldsymbol{r}|}.$$

The Hartree functional

$$E_{\mathrm{H}}[n] = \frac{1}{2} \int \mathrm{d}^3 r\, n(\boldsymbol{r}) v_{\mathrm{H}}(\boldsymbol{r})$$

is described by the Hartree potential

$$v_{\mathrm{H}}(\boldsymbol{r}) = \int \mathrm{d}^3 r' \frac{n(\boldsymbol{r}')}{|\boldsymbol{r} - \boldsymbol{r}'|}. \tag{6.2}$$

The last term in Equation (6.1) represents the exchange-correlation energy E_{xc}. The total energy can then be obtained by the solution of the relativistic Kohn–Sham equations (RKS)

$$([-\mathrm{i}c\boldsymbol{\alpha} \cdot \nabla + (\beta - 1)mc^2] + v_{\mathrm{ext}}(\boldsymbol{r}) + v_{\mathrm{H}}(\boldsymbol{r}) + v_{\mathrm{xc}}(\boldsymbol{r}))\psi_k(\boldsymbol{r}) = \epsilon_k \psi_k(\boldsymbol{r}) \tag{6.3}$$

on a three-dimensional point mesh using the highly accurate multicentre integration scheme developed by Boerrigter and co-workers (Boerrigter *et al.* 1988; te Velde and Baerends 1992). The exchange-correlation potential can be derived from E_{xc} according to

$$v_{\mathrm{xc}}(\boldsymbol{r}) = \frac{\delta E_{\mathrm{xc}}}{\delta n(\boldsymbol{r})}. \tag{6.4}$$

Both the local density approximation (LDA) and different types of semilocal functionals via the generalized gradient approximation (GGA) can be applied. The Becke (1988a) or the Perdew (1991) forms of the gradient corrected exchange functional are used. The correlation part is treated within the LDA by the parametrization of Vosko–Wilk–Nusair (Vosko *et al.* 1980) or by the semilocal functional of Perdew (1986a,b). All functionals are used in their relativistic extensions, denoted as RLDA and RGGA, respectively (Engel *et al.* 1995a, 1996).

The RKS Equation (6.3) can be recast as an algebraic eigenvalue problem

$$\boldsymbol{Hc} = \epsilon \boldsymbol{Sc}$$

with $\boldsymbol{H}$ being the Hamiltonian and $\boldsymbol{S}$ the overlap matrices, respectively. Since E_{xc} depends only on the density and its first derivative, Equation (6.3) are solved self-consistently on the RLDA level. After convergence, the GGA energies are evaluated by the density gradient. The difference between this efficient 'post-LDA' approach and fully self-consistent calculations is marginal (Becke 1992).

A further important methodological development of the DS-DVM to obtain accurate total energies was the implementation of a so-called variationally consistent procedure, which is based on a simplified but efficient treatment of the Hartree potential v_{H} (Bastug *et al.* 1995; Varga *et al.* 2000b).

This improved computational scheme was used to study the electronic structure and stability of various gas-phase compounds of the transactinides, such as, for example, $RfCl_4$ (Varga *et al.* 2000a), $BhOCl_3$ (Pershina and Bastug 1999) and HsO_4 (Pershina *et al.* 2001), and their lighter homologues in the corresponding chemical group.

Table 6.1 Binding energies D_e (in eV) of the MCl_4 (M = Ti, Zr, Hf, Rf) complexes (Varga *et al.* 2000a).

Molecule	RLDA	RGGA	AC	Theor.	Exp.[a]
$TiCl_4$	22.75	20.81	−0.58	20.23	17.81
$ZrCl_4$	23.80	21.98	−0.30	21.68	20.35
$HfCl_4$	23.30	20.46	−0.32	21.14	20.61
$RfCl_4$	21.65	19.75	−0.25	19.50	

[a]Calculated via a Born–Haber cycle using the data of Wagmann (1982).

Table 6.2 Optimized bond distances R_e (in a.u.) for MCl_4 (M = Ti, Zr, Hf, Rf).

Molecule	RLDA	RGGA	Exp.
$TiCl_4$	4.08	4.16	4.10[a]
$ZrCl_4$	4.39	4.43	4.38[b]
$HfCl_4$	4.36	4.43	4.376[b]
$RfCl_4$	4.46	4.54	

[a]Wagmann (1982); [b]Girichev *et al.* (1981).

In the following, we will present a methodology for choosing optimized basis sets for such calculations on the example of $RfCl_4$ (Varga *et al.* 2000a). As a first step, calculations of the total energy for a metal–ligand distance around the equilibrium geometry were performed with a minimal basis. The latter included all atomic occupied 1s to $(n-1)$d wave functions and ns ($n = 4, 5, 6, 7$ for Ti, Zr, Hf, Rf, respectively) of the (neutral) central atom and 1s to 2p wave functions of the (neutral) chlorine atoms. As a result of these calculations, the total energy as a function of the internuclear metal–ligand distance was obtained. Due to a large electronegativity of Cl, there is a polarization of the electronic charge distribution (a slightly positively charged central atom and negatively charged chlorine atoms) leading to fractional occupation numbers of the valence orbitals. The atomic basis functions were then recalculated with these new atomic occupations in order to include polarization effects in a consistent way. To be more complete, at a second step, further unoccupied np and nd atomic wave functions were added to the basis sets of the central atoms Ti, Zr, Hf and Rf. Additionally, wave functions of a 3d and 4s character shifted towards the binding region for slightly (approximately 2.0) ionized atoms were included for the basis sets of Cl. The inclusion of the additional basis sets resulted in an increase in the total energy by approximately 3 eV (compared with the minimal basis sets) reflecting the importance of adjusting the basis sets. The addition of further basis functions did not change the energies further. Finally, the total energies were recalculated for all internuclear distances around the equilibrium geometry with the optimized basis.

Table 6.1 shows the results of our calculations for the binding energies (D_e) of $TiCl_4$, $ZrCl_4$, $HfCl_4$ and $RfCl_4$. In the first column, the values of D_e obtained within

Table 6.3 Comparison of D_e and R_e for $RfCl_4$.

Method	R_e (a.u.)	D_e (eV)
RECP-KRHF	4.51	16.9
AREP-MP2	4.42	20.4
RECP-CCSD(T)	4.50	18.8
DFB	4.51	15.5
Our RLDA	4.46	21.4
Our RGGA	4.54	19.5

the RLDA are listed, while in the second column the values of D_e obtained by the RGGA are shown. All the values were calculated by subtracting the total energies of the atoms from the total energies of the corresponding molecules after the atomic DFT calculations in the same approximation were performed. The atomic values had then to be corrected by the multiplet splitting contributions listed in the third column of Table 6.1 (these atomic corrections are denoted as 'AC'). The latter were obtained by performing atomic multiconfiguration Dirac–Fock (MCDF) calculations. The final theoretical values of D_e are presented in the second to last column. They can be compared with the experimental dissociation energies (last column of Table 6.1), which were calculated via Born–Haber cycles for the formation enthalpies. The energetics of the compounds are shown to be improved within the RGGA, as compared with the RLDA.

Table 6.2 presents the values of bond distances R_e. A shift to larger values can be observed in going from the RLDA to RGGA, which is in agreement with results of calculations for diatomic molecules (Varga *et al.* 1999).

In Table 6.3, the values of D_e for $RfCl_4$ are compared with those obtained within various approximations using relativistic effective core potentials (RECP) Kramers-restricted Hartree–Fock (KRHF) (Han *et al.* 1999), averaged RECP including second-order Møller–Plesset perturbation theory (AREP-MP2) for the correlation part (Han *et al.* 1999), RECP coupled-cluster single double (triple) [CCSD(T)] excitations (Han *et al.* 1999), and a Dirac–Fock–Breit (DFB) method (Malli and Styszynski 1998). The AREP-MP2 calculation of D_e gives 20.4 eV, while the RECP-CCSD(T) method with correlation leads to 18.8 eV. Our value of D_e of 19.5 eV is just between these calculated values.

Thus, it was demonstrated that the present fully relativistic DFT can be applied successfully to calculations of binding energies of compounds containing superheavy elements. In the present publication, the use of the DFT method will, however, be restricted to calculations of electronic density distribution data of heavy-element complexes in terms of the Mulliken numbers (Mulliken 1955). Since we were dealing within the REHE project mostly with the solution chemistry of the very heavy elements, optimization of geometry of large, often negatively charged complexes did not seem to be feasible or economically efficient, so that geometries and bond lengths were chosen on the basis of a careful analysis of numerous experimental data for the lighter homologues of the transactinides. The bond lengths for transactinide complexes were then estimated taking into account bond lengths optimized within the DFT and other approximations for simpler gas-phase compounds. To predict stability of complexes and free-energy changes of complex formation reactions, some special models were used, as described in Section 6.4.1.

6.3 Experiment

The transactinide elements are being synthesized in nuclear fusion reactions with heavy-ion projectiles. Details of the production and decay of these elements can be found in Münzenberg and Hofmann (1999) and Kratz (1999). The production rates rapidly decrease from about 1 atom per minute for element 104 to 1 atom per several days for the heaviest artificial elements. Half-lives decreasing from about 1 min to the order of 1 ms for the longest-lived isotopes of these elements present an additional challenge. Low production rates and short half-lives lead to the situation in which, on the average, each synthesized atom has decayed before a new one is made. The consequences for chemical studies with one atom at a time have been discussed in Kratz (1999). Cold fusion reactions in which ^{50}Ti, ^{54}Cr, ^{58}Fe, $^{62,64}Ni$ and ^{70}Zn projectiles are fused with ^{208}Pb and ^{209}Bi targets tend to give the highest possible cross-sections; however, the neutron deficient product nuclei in these reactions have half-lives too short for chemical investigation. More neutron rich and, hence, longer-lived product nuclei are obtained in hot fusion reactions of ^{18}O, ^{22}Ne, ^{26}Mg and ^{34}S projectiles with actinide targets. Typical experimental conditions are heavy-ion beam currents of 3×10^{12} particles per second and a maximum useful target thickness of about 900 $\mu g\ cm^{-2}$. The studies referred to in this report used 34 s ^{262}Db produced in the $^{249}Bk(^{18}O, 5n)$ reaction to chemically characterize element 105 (dubnium) and 7.4 s ^{265}Sg produced in the $^{248}Cm(^{22}Ne,5n)$ reaction to characterize element 106 (seaborgium). These are produced with cross-sections of 6 nb and 240 pb, respectively.

6.3.1 Target and transport systems

A schematic of a target- and recoil-chamber arrangement is shown in Figure 6.1. Heavy-ion beams pass through a vacuum isolation window, a volume of nitrogen cooling gas, and a target backing before interacting with the target material. Reaction products recoiling out of the target are thermalized in a volume of He gas loaded with aerosol particles of 10–200 nm in size to which the reaction products attach. At a flow rate of about $2\ l\ min^{-1}$ the transport gas with the aerosols is transported through capillary tubes ($\sim$1.5 mm inner diameter) to the chemistry apparatus where it deposits the reaction products. He–aerosol jets allow for transportation over distances of several tens of metres with yields of about 50% (Schädel *et al.* 1988; Trautmann 1995). Transport times are of the order of 2–5 s. Aerosol materials are selected so as to minimize their influence on the chemical procedures. Separations in the aqueous phase often use KCl as an aerosol.

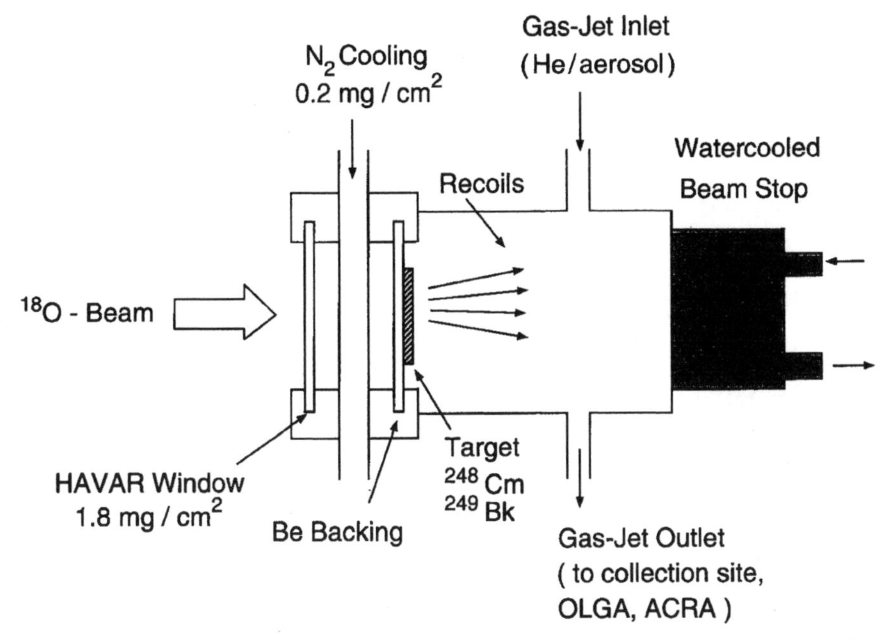

Figure 6.1 Schematic representation of a target-and-recoil chamber arrangement with He–aerosol jet. Reproduced with permission from Trautmann (1995). © 1995 R. Oldenbourg Verlag.

The experiments within the frame of the REHE project were performed in the aqueous phase in a discontinuous, batch-wise manner. It was necessary, in order to get a statistically significant result, to repeat the same experiment several hundred or even several thousand times with a cycle time of typically 45 s. These studies were performed with the Automatic Rapid Chemistry Apparatus (ARCA) II (Schädel *et al.* 1989), a computer-controlled apparatus for fast, repetitive high-performance liquid chromatography (HPLC) separations. A schematic of the ARCA II components is shown in Figure 6.2.

ARCA II consists of a central catcher-chemistry part incorporating the sliders SL1–SL3, and two movable magazines containing 20 of the chromatographic columns C1, C2 (1.6×8 mm^2) each, and peripheral components, i.e. three chemically inert HPLC pumps, P_1–P_3, and a number of pneumatically driven four-way slider valves, S_1–S_3. Each pump pumps one eluent—in the case of the separations of element 105 in HCl solutions (Paulus *et al.* 1999), one 10 M HCl, another 6 M HCl, and the third 6 M HNO_3/0.015 M HF—through Teflon tubing of 0.3 mm inner diameter to the central catcher-chemistry unit. The He(KCl) gas jet deposits the transported reaction products continuously onto one of two frits F. After 1 min collection, the frit is moved on top of one of the microcolumns C1, washed with 10 M HCl, whereby the reaction products are dissolved, complexed and extracted into the organic phase (the columns are filled with Teflon grains coated with the quaternary ammonium salt Aliquat 336(Cl^-),

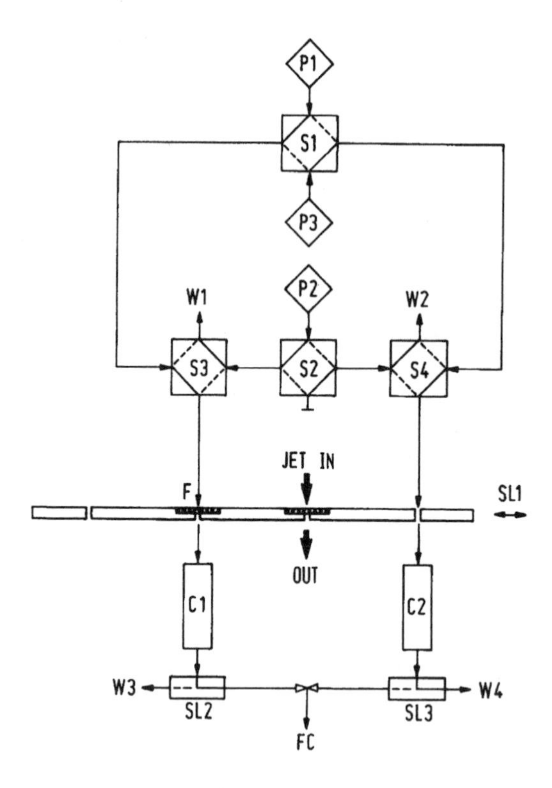

Figure 6.2 Schematic of the computer-controlled HPLC-system ARCA II; for details see text. Reproduced with permission from Schädel *et al.* (1989). © 1989 R. Oldenbourg Verlag.

while the nonextractable species (notably the actinides) run through into the waste, W3. The column is then washed with 6 M HCl, and the effluent (containing the lighter homologue Ta) is directed through SL2 to the fraction collector FC, where it is collected on a Ta disc and quickly evaporated to dryness by intense IR light and hot He gas. Next, the column is stripped with 6 M HNO_3/0.015 M HF (containing Nb, Pa and element 105), which is collected on a Ta disc and evaporated to dryness. The Ta discs are inserted into vacuum chambers where they are assayed for α activity by silicon detectors starting 60 s after the end of collection. Simultaneously, the next 1 min collection on the twin frit is complete. That frit is moved on top of the column C2 contained in the opposite magazine, and the next separation cycle is carried out. After each separation, the magazine is moved by one step, thus introducing a new column into the elution position. In this way, the time-consuming reconditioning of used columns and cross-contaminations from previous separations are avoided. After 40 min of continuous collection and separation cycles, the program is stopped, the used magazines are removed, two new magazines are introduced, and another 40 cycles are started. More than 7800 of such and similar ARCA II experiments have so far been conducted in the study of element 105.

6.4 Element 105

6.4.1 Theoretical predictions of complex formation of element 105 in aqueous acidic solutions

Experimentally, complex formation is studied by solvent extraction or anion exchange separations. By studying complex formation of the transactinides along with their lighter homologues, we obtain information about similarity or differences in complexing ability and ionic radii of the extracted metals. Results of previous experimental investigations of complex formation (Kratz *et al.* 1989) of element 105 in aqueous HCl/HF solutions have shown that complex formation and extraction do not simply continue trends within the chemical group found for the lighter homologues. Thus, element 105, Db, has revealed a nontantalum-like behaviour and similarity to Nb and Pa in its extraction from mixed HCl/HF solutions by amines (Kratz *et al.* 1989). In a similar way, in group 4, Rf was extracted between Zr and Hf from 8 M HCl by TBP (Günther *et al.* 1998). In order to interpret the results of these experiments and to predict those of new experiments on extraction chromatography separations of element 105 and its homologues from pure HF, HCl and HBr solutions, the following theoretical study has been undertaken within the REHE project (Pershina 1998a,b; Pershina and Bastug 1999).

Hydrolysis of the transactinides

Since transition elements are known to be subject to strong hydrolysis, the latter influences both the complex formation and extraction.

Hydrolysis of cations. In studying hydrolysis, we have to distinguish between hydrolysis of cations and hydrolysis of complexes. For hydrolysis of cations, each step in the formation of a mononuclear species can be described as a successive loss of a proton

$$M(H_2O)_n^{z+} \rightleftharpoons MOH(H_2O)_{n-1}^{(z-1)+} + H^+ \tag{6.5}$$

with the hydrolysis constant being

$$\log K = -\Delta G^r/2.3RT. \tag{6.6}$$

Experimental data (Baes and Mesmer 1976) show that, within the entire range of HCl concentrations, hydrolysis of group 5 elements and Pa has the trend,

$$\text{Nb} > \text{Ta} \gg \text{Pa}.$$

The simple electrostatic model of hydrolysis (Baes and Mesmer 1976), which predicts log K to change linearly with the ratio of the ionic charge to ionic radius, does not explain, for example, the difference between Nb and Ta having the same value of the formal charge and ionic radius, as it does not explain the reversed order in hydrolysis of Zr relative to that of Hf. It was thought to be even less applicable to the transactinides, where covalent interactions are the strongest among the known transition elements. The following proposed model is free of those drawbacks.

Since calculations of total energies of large (often negative) complexes do not seem feasible, or efficient, the free-energy change of Reaction (6.5) was suggested to be calculated via differences in both the ionic and covalent contributions to the binding energy of reaction components separately. For a reaction component, in a fashion analogous to that of Kassiakoff and Harker (1938),

$$-\frac{\Delta G^f(u, v, w)}{2.3RT} = \sum a_i + \sum a_{ij} + \log P - \log(u!v!w!2^w) + (2u + v + 1)\log 55.5, \qquad (6.7)$$

where the first term on the right-hand side, $\sum a_i$, is a sum of nonelectrostatic contributions from M, O, OH and H_2O, and the next term, $\sum a_{ij}$, is a sum of each pairwise electrostatic interaction. P is the partition function representing the contribution of structural isomers and the last two terms are statistical: one is a correction for the indistinguishable configurations of the species, and the other is a conversion to the molar scale of concentration.

The second term on the right-hand side of Equation (6.7) can be calculated as

$$2.3RTa_{ij} = E^{\mathrm{C}} = -BQ_iQ_j/d_{ij}\epsilon, \qquad (6.8)$$

where Q_i is the effective charge, d_{ij} is the metal–ligand distance, ϵ is the dielectric constant and B transforms the energy in eV. The free-energy change of a reaction is then

$$-\Delta G^r/2.3RT = \Delta E^{\mathrm{C}} + K\Delta OP + \Delta S, \qquad (6.9)$$

where OP is the difference in overlap populations for complexes on the right- and left-hand sides of Reaction (6.5). Both ΔE^{C} and ΔOP are then obtained as a result of the Mulliken electronic density distribution analysis implemented in the DFT method described in Section 6.2.

With this aim in mind, calculations of the electronic structure of the hydrated, $M(H_2O)_6^{5+}$, and hydrolysed, $M(OH)_6^-$, complexes of Nb, Ta, Db and Pa, have been performed using the DFT method (Pershina 1998a,b). The calculations have shown E^{C} to be the predominant type of the metal–ligand interaction, so that by calculating only ΔE^{C}, correct trends in the complex formation can be predicted for all the elements under discussion (see Table 6.4). This electrostatic interaction must, however, be defined on the basis of the real (relativistic) electronic density distribution in the considered systems.

Thus, the following trend in the hydrolysis of group 5 elements has been defined:

$$\mathrm{Nb} > \mathrm{Ta} > \mathrm{Db} \gg \mathrm{Pa}.$$

Table 6.4 E^C and ΔE^C (in eV) for reaction $M(H_2O)_6^{5+} \Leftrightarrow M(OH)_6^-$, where M = Nb, Ta, Db and Pa (Pershina 1998a).

E^C	Nb	Ta	Db	Pa
$M(OH)_6^-$	−21.74	−23.33	−21.48	−19.53
$M(H_2O)_6^{5+}$	−21.92	−25.38	−25.37	−29.71
ΔE^C	0.18	2.05	3.89	9.18

The way to define hydrolysis constants is shown in Pershina (1998b). The theoretical results for Nb, Ta and Pa obtained here are in agreement with experiments on hydrolysis of Nb, Ta and Pa (Baes and Mesmer 1976), with the difference between Nb and Ta being reproduced. Thus, a weaker hydrolysis of Db with respect to that of Nb and Ta has been predicted.

Complex formation and extraction of group 5 elements

Complex formation and hydrolysis of complexes. In acidic solutions, hydrolysis of group 5 elements is competing with complex formation. For group 5 complexes, it is described by the following equilibrium:

$$M(OH)_y^{(z-y)-} + aH^+ + aX^- \rightleftharpoons M(OH)_{y-a}X_a^{(z-y)-} + aH_2O. \quad (6.10)$$

As a result, in HCl solutions the following complexes of Nb, Ta, Pa and, assuming the same stoichiometry, of Db are formed: $M(OH)_2Cl_4^-$, $MOCl_4^-$, $MOCl_5^{2-}$ and MCl_6^-.

To predict the stability of products of Reaction (6.10), the same technique has been applied as in the case of hydrolysis of group 5 cations (Equations (6.7)–(6.9)). With this aim in mind, calculations of the electronic structure of various complexes of Nb, Ta, Pa and Db (indicated above) have been performed using the DFT method (Pershina 1998a,b; Pershina and Bastug 1999). ΔE^C and relative free-energy changes of Equilibria (6.10) determined on their basis indicate the following trend in the complex formation of group 5 elements

$$Pa \gg Nb > Db > Ta. \quad (6.11)$$

The Sequence (6.11) for Pa, Nb and Ta is in agreement with experimental data showing that among analogous complexes, those of Pa are formed in much more dilute HCl solutions, while much higher acid concentrations are needed to form complexes of Ta (Gmelin 1970; Scherff and Herrmann 1966). The calculations have also confirmed the experimentally found sequence in the formation of the following types of complexes of the same element as a function of HCl concentrations

$$M(OH)_2Cl_4^- > MOCl_4^- > MCl_6^-.$$

We have also studied formation of the MF_6^- and MBr_6^- complexes (M = Nb, Ta, Db and Pa) in HF and HBr solutions (Pershina and Bastug 1999). The same trend in formation

Table 6.5 ΔE^{C} (in eV) for reaction $M(OH)_6^- \Leftrightarrow MX_6^-$, where M = Nb, Ta, Db and Pa; X = F, Cl and Br (Pershina and Bastug 1999).

Complex	F	Cl	Br
NbX_6^-	12.20	19.57	21.40
TaX_6^-	12.69	20.78	22.63
DbX_6^-	12.38	20.46	22.11
PaX_6^-	12.19	17.67	19.91

of each types of complexes as that obtained for MCl_6^- has been found. Calculations have reproduced well the sequence in the formation of the MX_6^- complexes (X = F, Cl and Br) of the same metal as a function of the ligand (Table 6.5): $MF_6^- > MCl_6^- > MBr_6^-$. Thus, MF_6^- are the strongest complexes formed in much more dilute HF solutions, while to form MCl_6^- or MBr_6^-, more concentrated HCl and HBr solutions should be used.

The analysis of all the factors contributing to the extraction process has shown the latter to be governed by the complex formation and ion transfer, so that the distribution of the group 5 complexes between an organic and aqueous (above 4 M) HCl and HBr phases has the following trend (Pershina 1998b):

$$\text{Pa} \gg \text{Nb} \geqslant \text{Db} > \text{Ta}. \tag{6.12}$$

Thus, the trends in the complex formation and extraction known for group 5 elements Nb and Ta turned out to be reversed in going to Db. This could not be predicted by any straightforward extrapolation of these properties within the group, but came out as a result of considering the real chemical equilibria and calculating relativistically the electronic structure of their components.

6.4.2 Experimental results

The first studies of the aqueous phase chemistry of element 105 were conducted by Gregorich *et al.* (1988). Like Nb and Ta, Db was adsorbed on glass surfaces upon fuming with nitric acid. In 801 manually performed experiments, 24 α events due to the decay of 34 s ^{262}Db or its 3.9 s ^{258}Lr daughter including five $\alpha\alpha$ mother–daughter correlations were observed. In an attempt to study the extraction of the dubnium fluoride complex from 3.8 M HNO_3/1.1 M HF into methyl isobutyl ketone (MIBK), no decays attributable to ^{262}Db could be observed. Under these conditions, Ta extracts while Nb does not. From an extrapolation in group 5 it was expected that Db would behave more like Ta than Nb but, surprisingly, Db apparently did not extract. The nontantalum-like behaviour of Db might indicate that Db forms polynegative anions like DbF_7^{2-} under the chosen conditions. The higher charge would then prevent extraction even into solvents with a relatively high dielectric constant such as MIBK.

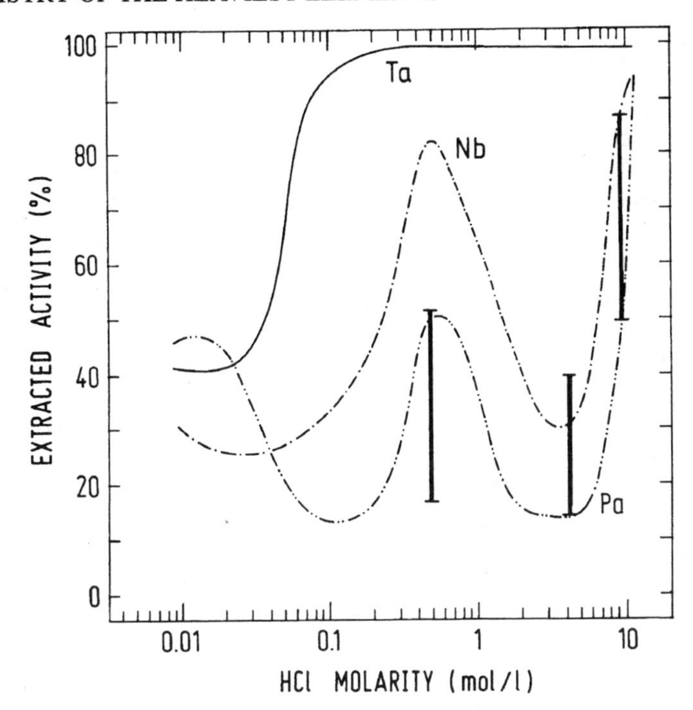

Figure 6.3 Percentage extracted activity of Nb, Ta and Pa tracers (curves) as a function of HCl concentration in the system TiOA–HCl/0.03 M HF. The bold bars encompass the upper and lower limits deduced for the Db extraction from the elution positions in the chromatography experiments (Kratz *et al.* 1989; Zimmermann *et al.* 1993). The complete extraction of Db into TiOA from 12 M HCl/0.02 M HF is not indicated for clarity. Reproduced with permission from Zimmermann *et al.* (1993). © 1993 R. Oldenbourg Verlag.

To investigate this unexpected finding and more facets of dubnium chemistry, a large number of automated separations were conducted with the ARCA II (Schädel *et al.* 1989). In the first experiments, extraction chromatography separations with the liquid anion exchanger triisooctylamine (TiOA) on an inert support (Kratz *et al.* 1989) were performed. TiOA extracts all group 5 elements including Pa, irrespective of the formation of mono- or polynegative anions, from HCl solutions above 10 M. At lower concentrations, selective back extractions allowed to distinguish between the chemical behaviour of Nb, Ta, Pa and Db. Small amounts of HF (typically 0.02 M) were added to the HCl solutions as this is recommended in the literature (Korkisch 1989) to prevent hydrolysis and 'to maintain reproducible solution chemistry' of the group 5 elements.

Element 105 was shown (Kratz *et al.* 1989) to extract into the TiOA on the chromatographic columns in ARCA II from 12 M HCl/0.02 M HF as do Nb, Ta and Pa, due to the formation of anionic halide complexes. In subsequent elutions, the elution positions of element 105 relative to those of Nb, Ta and Pa were determined (Kratz *et al.* 1989) in 10 M HCl/0.025 M HF, in 4 M HCl/0.02 M HF and in 0.5 M HCl/0.01 M

HF (Zimmermann *et al.* 1993). 2198 collection and separation cycles on a 1 min timescale (Kratz *et al.* 1989; Zimmermann *et al.* 1993) were necessary to obtain the results shown in Figure 6.3.

It is seen that element 105 shows a striking nontantalum-like behaviour and that it follows, at all HCl concentrations below 12 M, the behaviour of its lighter homologue Nb and that of its pseudohomologue Pa. The electronic structure of the group 5 anionic complexes MCl_6^-, $MOCl_4^-$, $M(OH)_2Cl_4^-$ and $MOCl_5^{2-}$ was calculated with the DS-DVM code (Pershina *et al.* 1994). By applying Born's theory of ion transfer between the aqueous phase and the organic phase (mixed HCl/HF solutions and TiOA (Kratz *et al.* 1989)), the extraction sequence Pa $\gg$ Db $\geqslant$ Nb was found theoretically (Pershina *et al.* 1994) that was the inverse sequence compared with that found experimentally (Kratz *et al.* 1989).

Due to the complicated situation in mixed HCl/HF solutions with the possibility of forming mixed chloride/fluoride or fluoride complexes, it was advisable to repeat the experiments in the pure HCl system (Pershina *et al.* 1994).

In the frame of the REHE project, the amine extractions of the group 5 elements were systematically revisited by Paulus *et al.* (1999). Pershina (1998b), by considering the competition between hydrolysis and halide complex formation, predicted the extraction sequence

$$\text{Pa} \gg \text{Nb} \geqslant \text{Db} > \text{Ta},$$

as described in Section 6.4.1. Distribution coefficients (K_d values) for Nb, Ta and Pa were measured in new batch extraction experiments with the quaternary ammonium salt Aliquat 336 and pure HF, HCl and HBr solutions (Paulus *et al.* 1999). Based on these results, new chromatographic column separations with ARCA II were elaborated to study separately the fluoride and chloride complexation of element 105. As an example, the separation of Eu-, Ta-, Nb- and Pa-tracers in the Aliquat 336 (Cl^-)/HCl system is shown in Figure 6.4. After feeding of the activities onto the column in the 10 M HCl (whereby the disturbing actinides modelled by trivalent Eu run through the column), Ta is eluted in 6 M HCl, Nb in 4 M HCl and Pa in 0.5 M HCl (Paulus *et al.* 1999).

1307 experiments were conducted with element 105 with a cycle time of 50 s. In the system Aliquat 336/HCl, after feeding of the activity onto the column in 10 M HCl, a Ta fraction was eluted in 6 M HCl, as shown in Figure 6.4. This was followed by stripping of a combined Nb, Pa fraction from the column in 6 M HNO_3/0.015 M HF. From the distribution of α-decays between the Ta fraction and the Nb, Pa fraction, a K_d value of 438^{+532}_{-166} for element 105 in 6 M HCl was deduced, which is close to that of Nb and differs from the values for Pa and Ta; see Figure 6.5. Thus, the extraction sequence Pa > Nb $\geqslant$ Db > Ta is established *exactly as theoretically predicted* (see Section 6.4.1).

In the system Aliquat 336/HF, the reaction products were loaded onto the column in 0.5 M HF. In elutions with 4 M HF (Pa fraction) and with 6 M HNO_3/0.015 M HF (combined Nb, Ta fraction) all α-decay events of element 105 were observed in the Nb, Ta fraction. This results in a lower limit for the K_d value of element 105 in 4 M

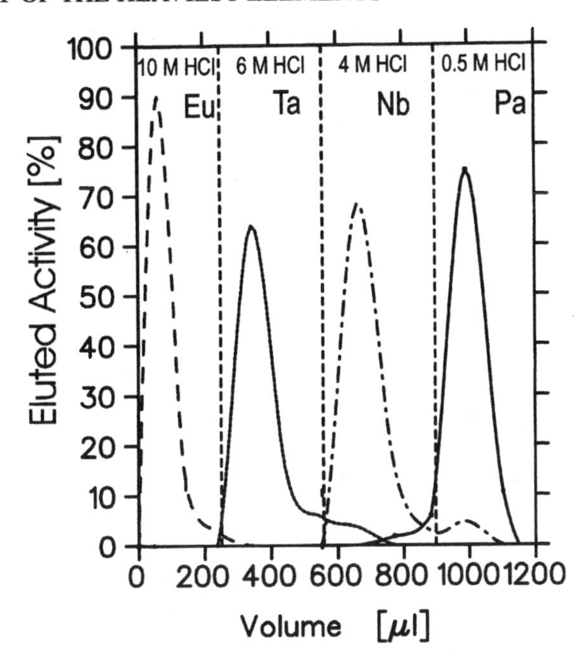

Figure 6.4 Elution curves for trivalent cations (Eu) and for Ta, Nb and Pa from Aliquat 336/Voltalef columns ($1.6 \times 8\ mm^2$) in ARCA II. The activities are fed onto the column in 10 M HCl. This is followed by separate elutions of a Ta fraction in 6 M HCl, of a Nb fraction in 4 M HCl and of a Pa fraction in 0.5 M HCl. Reproduced with permission from Paulus *et al.* (1999). © 1999 R. Oldenbourg Verlag.

HF of more than 570, which is close to that of Nb and Ta ($\geqslant 10^3$) and differs markedly from that of Pa ($\sim$10).

It is satisfying to see that not only is the extraction sequence in the system Aliquat 336/HCl correctly predicted by theory (Pershina 1998b), but the calculated free-energy changes of the reactions of complex formation are of the order of 12 eV for the fluorides, 20 eV for the chlorides and 22 eV for the bromides (see Table 6.5 and (Pershina and Bastug 1999)) (not taking into account the free enthalpy of formation of H_2O, which is 3 eV), which, again, is in agreement with the experimental findings. For fluorides, extractable complexes are formed even at low HF concentrations; for chlorides, it takes more than 3 M HCl to form extractable chloride complexes; and for bromides, the threshold is shifted to above 6 M HBr.

To conclude, the amine extraction behaviour of dubnium halide complexes is always close to that of its lighter homologue Nb, in agreement with the predicted *inversion of the trend of the properties when going from the* 5d *to the* 6d *elements* (Pershina 1998b; Pershina *et al.* 1994). In pure HF solutions, it differs mostly from the behaviour of Pa. In pure HCl solutions, it differs considerably from both Pa and Ta. In mixed HCl/HF solutions, it differs markedly from the behaviour of Ta. The studies of the halide complexing and amine extraction of the group 5 elements both theoretically (Pershina

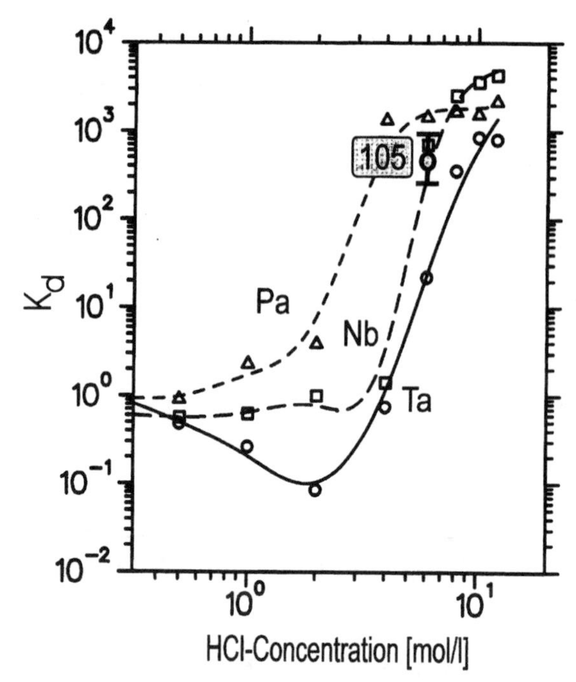

Figure 6.5 Distribution coefficients of Pa, Nb and Ta in the system Aliquat 336/HCl. The K_d for Db in 6 M HCl is also indicated. The system shows the inverse extraction sequence as compared with the TiOA/HCl/HF system as theoretically predicted. Reproduced with permission from Paulus *et al.* (1999). © 1999 R. Oldenbourg Verlag.

1998a; Pershina *et al.* 1994) and experimentally (Paulus *et al.* 1999) demonstrate that enormous progress has been made in understanding detailed facets of the chemistry of the transactinide elements.

6.5 Element 106

6.5.1 Theoretical predictions

Estimates of redox potentials of Sg

Knowledge of the stable oxidation states of an element is very important since many other properties depend on these states. It is also important to know about the relative stability of oxidation states, i.e. redox potentials, for a chemical application. Trends in their values can also provide information about similarities or differences between the transactinides and their lighter homologues. Thus, for example, the stability of the maximum oxidation state is known to increase within transition element groups. It is therefore of great interest to investigate whether transactinides fall within this trend: those at the beginning of the 6d row were expected to be stabilized in lower oxidation states (+3 for Db or +4 for Sg) due to the relativistically stabilized $7s^2$ electronic ground-state configuration. Thus, aiming at achieving a general knowledge of trends in the stability of oxidation states as well as predicting redox potentials for a future experiment on the reduction of Sg (Strub *et al.* 1999), the following theoretical study has been undertaken.

To characterize an oxidation state, the oxidation–reduction potential E^0 is of crucial importance. For the reaction

$$M^{z+n} + ne \longrightarrow M^{z+}, \tag{6.13}$$

E^0 is given by

$$E^0 = -\Delta G^\circ / nF, \tag{6.14}$$

where ΔG° is the standard Gibbs energy change for Reaction (6.13) and F is the Faraday constant. The change in the free energy of the redox Reaction (6.13) can be expressed as

$$\Delta G^\circ = -(\mathrm{IP} + \Delta G^\circ_{\mathrm{hydr}}), \tag{6.15}$$

where IP $= I \pm \Delta E$, I is the ionization energy $M^{z+} \longrightarrow M^{z+n}$ and ΔE is the energy needed to reconstruct the electronic configuration of the metal ion when going from the $z+$ to the $(z+n)$ ionized state. I and $\Delta G^\circ_{\mathrm{hydr}}$ are usually smooth functions of the atomic number. Thus, ΔE correlates linearly with redox potentials and defines all the changes in their values. If IP includes the changes in electronic configurations, $E^0(M^{z+n}/M^{z+})$ is directly proportional to IP.

In Ionova *et al.* (1992), redox potentials for Db in aqueous solutions have been estimated on the basis of the above-mentioned linear correlations between E^0 and multiple IPs calculated using the MCDF method (Johnson *et al.* 1990). In the present work, the IP for Sg and those experimentally unknown for Mo and W were determined using the same approach (Pershina *et al.* 1999). For that purpose, multiple IPs for element 106, Sg and its lighter homologues, Cr, Mo and W, were calculated using the MCDF method (Johnson *et al.* 1999). Using the calculated MCDF IP and, where available, experimental redox potentials, the unknown values of E^0 were defined for group 6 elements on the basis of the linear correlations (Equations (6.14) and (6.15)). As an example, one of those plots for $E^0(MO_2/M^{3+})$ is shown in Figure 6.6. From this figure, we can see that the 4+ oxidation state of Sg is less stable than the 4+ oxidation state of Mo and W. The total scheme of obtained redox potentials is shown in Figure 6.7.

The data of Figure 6.7 indicate that the stability of the maximum oxidation state, 6+, increases in the group, while that of the 5+, 4+ and 3+ states decreases. In contrast to expectations, the 4+ state of Sg will be less stable than that of Mo and W. This fact can be explained by the step-wise ionization scheme of Sg (see Figure 10 from Pershina *et al.* (1999)) showing that the 4+ state of Sg has the $6d^2$ electronic configuration, by analogy with Mo and W, and not the $7s^2$. Since the 6d orbitals of Sg

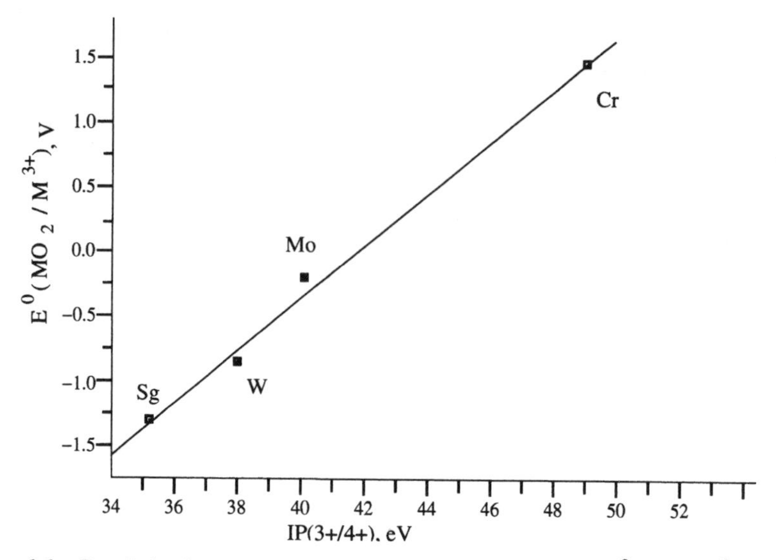

Figure 6.6 Correlation between IP(3+/4+) and standard potentials $E^0(MO_2/M^{3+})$, where M = Cr, Mo, W and Sg. Reproduced with permission from Pershina *et al.* (1999). © 1999 American Chemical Society.

are more destabilized than those of Mo and W (due to an indirect relativistic effect), the 4+ state of Sg is less stable than those of Mo and W.

An analysis of the influence of relativistic effects on redox potentials has shown the stability of the maximum oxidation state to increase with increasing Z in groups 4–6 as a result of relativity. Nonrelativistic IP corresponding to transitions from the neutral to the maximum oxidation state would give a less stable maximum oxidation state of the 6d elements than those of the 4d and 5d homologues. A correlation between relativistic and nonrelativistic energies of the charge-transfer transitions and E^0 for some compounds (see Figure 6 in Pershina and Fricke (1999)) clearly shows that this is a pure relativistic effect. The calculations have also shown the increase in stability of the maximum oxidation state in the groups to become less pronounced in going from group 4 to group 6: for W and Sg the stability of the 6+ state is nearly the same. Thus, along the transactinide series, the stability of the maximum oxidation state decreases: $Lr^{3+} > Rf^{4+} > Db^{5+} > Sg^{6+}$. The 7+ oxidation state of element 107 will probably not be attainable in solutions. Calculations of MCDF IP for elements 107 and 108 and their homologues allowing for predictions of redox potentials have just been finished (Jacob *et al.* 2000).

Hydrolysis and complex formation of Sg

Results of the first aqueous chemistry experiments on Sg showed that it formed neutral or anionic oxyfluorides in 0.1 M $HNO_3/5 \times 10^{-4}$ M HF (Schädel *et al.* 1997a). The next experiments on the ion exchange of Sg from pure 0.1 M HNO_3 solutions (Schädel *et al.* 1998) showed that Sg was not eluted from the cation exchange column in contrast

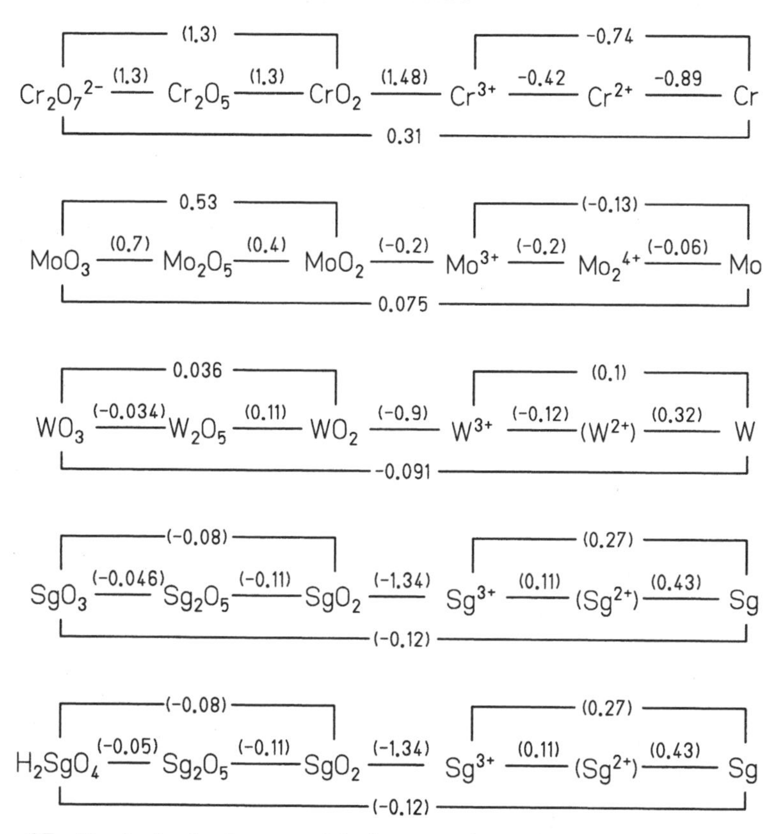

Figure 6.7 Standard reduction potentials for group 6 crystalline and aqueous phase compounds. Reproduced with permission from Pershina *et al.* (1999). © 1999 American Chemical Society.

to Mo and W. This nontungsten-like behaviour of Sg was tentatively attributed to its lower tendency to hydrolyse compared with that of W. To interpret the behaviour of Sg in these experiments and to predict its complex formation in HF/HCl solutions, the following theoretical study was undertaken.

For group 6 elements, Mo and W, the protonation scheme is known to be as that shown in Figure 6.8. Equilibrium constants (Baes and Mesmer 1976) indicate that hydrolysis of Mo is stronger than that of W. The question arises whether Sg will follow the trend found for the lighter homologues.

The model used for the present study was the same as that used for predictions of hydrolysis of group 5 elements. For that purpose, calculations of the electronic structure of all types of complexes for Mo, W and Sg indicated in Figure 6.8 were performed using the DFT method. As in the previous case, ΔE^C and ΔOP were defined for each reaction, with the results for ΔE^C shown in Table 6.6 and in Pershina and Kratz (2001). As in the case of group 5 complexes, changes in the Coulomb part of the metal–ligand interaction, ΔE^C, were shown to define ΔG° of the processes.

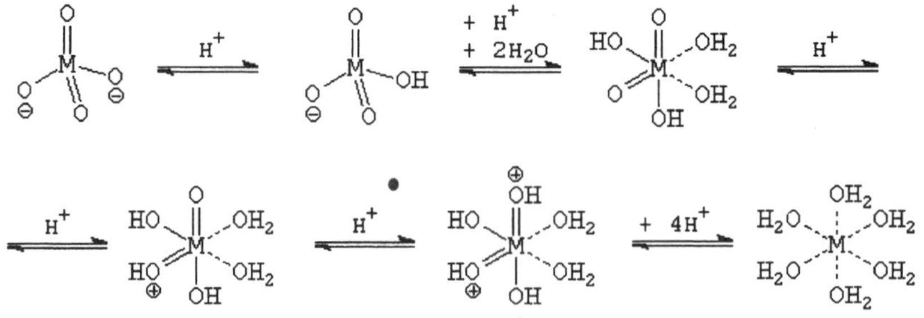

Figure 6.8 The protonation scheme for group 6 oxocomplexes, where M = Mo, W and Sg. Reproduced with permission from Pershina and Kratz (2001). © 2001 American Chemical Society.

Table 6.6 ΔE^{C} (in eV) for the stepwise protonation of MO_4^{2-} (M = Mo, W and Sg).

Reaction	ΔE^{C}		
	Mo	W	Sg
$MO_4^{2-} + H^+ \rightleftharpoons MO_3(OH)^-$	−12.98	−13.13	−12.96
$MO_3(OH)^- + H^+ \rightleftharpoons MO_2(OH)_2(H_2O)_2$	−21.43	−22.08	−21.61
$MO_2(OH)_2(H_2O)_2 + H^+ \rightleftharpoons MO(OH)_3(H_2O)_2^+$	−5.84	−6.35	−6.65
$MO(OH)_3(H_2O)_2^+ + H^+ \rightleftharpoons M(OH)_4(H_2O)_2^{2+}$	−0.43	−0.76	−1.23
$M(OH)_4(H_2O)_2^{2+} + H^+ \cdots \rightleftharpoons M(H_2O)_6^{6+}$	41.97	38.71	37.11

Thus, for the first two protonation steps, the trend is W > Sg > Mo. For the further processes, the trend is Sg > W > Mo. It was also possible to give absolute values of the protonation constants for Sg using the results of the present calculations and the experimental $\log K$ for Mo and W. Thus, with the use of Equations (6.7)–(6.9), the values of $\log K$ for the known protonation steps are

$$\log K_1 = k\Delta \mathrm{OP} + B\Delta E^{\mathrm{C}} - 1.142,$$
$$\log K_2 = k\Delta \mathrm{OP} + B\Delta E^{\mathrm{C}} + 2BD_{\mathrm{H_2O}} - 2.464,$$
$$\log K_3 = k\Delta \mathrm{OP} + B\Delta E^{\mathrm{C}} - 0.176,$$

where B and k correct not only for an unknown dielectric constant in solution, but also for the fact that the values of E^{C} are calculated for the species in vacuum. B and k, in turn, can be defined by solving a couple of equations for Mo and W, such as, for example,

$$\log K_1(\mathrm{Mo}) = 0.16k + 12.98B - 1.142 = 3.7,$$
$$\log K_1(\mathrm{W}) = 0.12k + 13.13B - 1.142 = 3.8.$$

Using, then, ΔE^{C} from Table 6.6 and $\Delta\mathrm{OP}$ from Pershina and Kratz (2001),

$$\log K_1(\mathrm{W}) = 0.11k + 12.95B - 1.142 = 3.74.$$

The values of log K obtained for Sg and the experimentally unknown log K for W are given in Table 1 of Pershina and Kratz (2001). They confirm the sequence in the hyrolysis/protonation obtained here just on the basis of the ΔE^{C} data. Thus, hydrolysis of the neutral species with formation of negative oxo-complexes has a reversed trend in the chemical group: Mo > Sg > W. An analogous trend was observed for complex formation of group 5 elements in HCl and HBr solutions: Nb > Db > Ta (Paulus *et al.* 1999). For positively charged complexes, the trend in hydrolysis from Mo to W continues further to Sg so that Mo > W > Sg. This theoretical result is in agreement with experimental data on hydrolysis/protonation of Mo and W (Tytko and Glemser 1976) and, recently, on Sg (Schädel *et al.* 1998). The present calculations have again clearly shown the decisive factor in the complex formation (hydrolysis/protonation) process to be predominant changes in the electrostatic metal–ligand interaction energy.

Complex formation of Sg in aqueous HF solutions

The first experiments on the chemical identification (the valence state and possible form of complexes) of Sg have been carried out by cation exchange separations (Schädel *et al.* 1997a). Sg was eluted from cation exchange columns with dilute nitric/hydrofluoric acid together with Mo and W, thus confirming that it formed probably $SgO_2F_3^-$, SgO_3F^- or SgO_2F_2 by analogy with Mo and W. The values of the distribution coefficient, K_d were, however, not defined in that experiment. In future experiments, K_d values are planned to be determined on an anion exchanger in HF/0.1 M HNO_3 solutions using on-line chromatography with the multicolumn technique (Pfrepper *et al.* 1997). To predict an outcome of the chemical experiments, the following theoretical study has been started, with some preliminary results shown here.

The group 6 elements Mo and W are known to form the $MO_2F_3(H_2O)^-$ complexes at lower HF concentration and MOF_5^- at higher HF concentration (Caletka and Krivan 1990). In solutions at pH = 1 (0.1 M HNO_3), neutral complexes are in equilibrium with monocharged ones, so that the reactions with formation of each type of the fluoro-complexes could be of the following types, starting from the monocharged and neutral complexes, respectively,

$$MO(OH)_3(H_2O)_2^+ + HF \rightleftharpoons MOF_5^- \text{ or } MO_2F_3(H_2O)^-, \qquad (6.16)$$
$$MO_2(OH)_2(H_2O)_2 + HF \rightleftharpoons MOF_5^- \text{ or } MO_2F_3(H_2O)^-. \qquad (6.17)$$

Table 6.7 E^C and ΔE^C(in eV) for reactions: (1) $MO_2(OH)_2(H_2O)_2 \Leftrightarrow MO_2F_3(H_2O)^-$ and (2) $MO_2(OH)_2(H_2O)_2 \Leftrightarrow MOF_5^-$, where M = Mo, W and Sg.

E^C	Mo	W	Sg
$MO_2(OH)_2(H_2O)_2$	−29.32	−30.68	−28.97
$MO_2F_3(H_2O)^-$	−14.80	−16.77	−14.86
MOF_5^-	−13.11	−15.33	−14.46
ΔE^C (1)	14.52	13.90	14.11
ΔE^C (2)	16.21	15.35	14.51

To calculate the free energies of those equilibria, we have calculated the electronic structure of $MO_2F_3(H_2O)^-$ and MOF_5^- for M = Mo, W and Sg using the DFT method. Using the E^C for the neutral and monocharged hydrolysed complexes of Mo, W and Sg from Pershina and Kratz (2001), we have calculated ΔE^C for Reactions (6.16) and (6.17) according to the model described by Equations (6.7)–(6.9). The work is still in progress and some preliminary results are shown in Table 6.7.

The data of Table 6.7 show that at pH = 1, in the area of low HF concentrations, the complex formation in group 6 has a reversed trend, W > Sg > Mo, while in the area of high HF concentrations, Sg follows the trend found for Mo and W, so that Sg > W > Mo. The obtained sequences for Mo and W agree with the experimental sequence in the values of K_d for the sorption of Mo and W from HF solutions by an anion exchange resin (Caletka and Krivan 1990). Thus, Sg will be extracted below or above W depending on the HF concentrations. Considerations of further reactions at much lower HF concentrations with formation of the MO_3F^- and $MO_2F_2(H_2O)_2$ complexes of Mo, W and Sg are in progress.

6.5.2 Experimental results

The ARCA II that has been successful in studying chemical properties of element 105 (Kratz 1999; Kratz *et al.* 1989; Paulus *et al.* 1999; Zimmermann *et al.* 1993), and recently also of element 104 (Günther *et al.* 1998; Strub *et al.* 2000) in aqueous solutions, was foreseen to also perform the first aqueous chemistry with element 106, seaborgium.

Several chemical systems were tested with the fission products ^{93}Y, ^{97}Zr, ^{99}Mo and W isotopes produced in the $^{20}Ne + ^{152}Gd$ reaction (Brüchle *et al.* 1992) at the PSI Philips cyclotron. α-hydroxyisobutyric acid solutions of 5×10^{-2} M, pH = 2.65 or pH = 5 used to elute W in a rapid, one-stage separation from cation exchange columns provided a good separation from Hf and Lu (Brüchle *et al.* 1992). Likewise (Brüchle *et al.* 1992), solutions with 0.1 M HCl and various HF concentrations between 10^{-4} M and 10^{-2} M were eluting W rapidly while Hf was safely retained on the column below

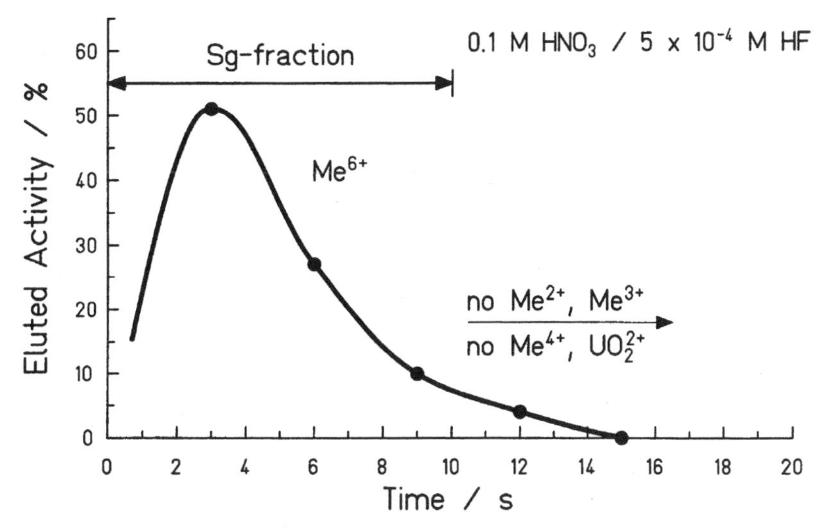

Figure 6.9 Elution curve for W-tracer modelling the seaborgium separation in ARCA II using a solution of 0.1 M $HNO_3/5 \times 10^{-4}$ M HF with a flow rate of 1 ml min^{-1}. The $1.6 \times 8\ mm^2$ columns are filled with the cation exchange resin Aminex A6. Reproduced with permission from Schädel *et al.* (1997b). © 1997 R. Oldenbourg Verlag.

10^{-3} M HF. Hf was observed to be partially eluted for not less than 2.8×10^{-3} M HF in 0.1 M HCl. Finally, the decision was made to use 0.1 M $HNO_3/5 \times 10^{-4}$ M HF to elute a seaborgium fraction from cation exchange columns (Günther *et al.* 1995) in order to avoid the formation of mixed chloride–fluoride complexes, which are difficult to model. $MO_2F_3(H_2O)^-$ is a likely form of the complexes that are eluted, but neutral species such as MO_2F_2 cannot be excluded. Some problems were encountered with adsorption of the activities on the slider in ARCA II. Among the various materials tested, titanium showed the lowest losses of W and Hf due to adsorption. Figure 6.9 shows the elution curve for short-lived W isotopes from the reaction of ^{20}Ne with enriched ^{152}Gd.

The activity was transported to ARCA II with a He(KCl)-jet within about 3 s and deposited on a titanium slider, dissolved and washed through the $1.6 \times 8\ mm^2$ chromatographic column (filled with the cation exchange resin Aminex A6, 17.5 ± 2 μm) at a flow rate of 1 ml min^{-1} with 0.1 M $HNO_3/5 \times 10^{-4}$ M HF. 85% of the W are eluted within 10 s. No di- or trivalent metal ions and no group 4 ions are eluted within the first 15 s. Also, uranium, in the form of UO_2^{2+}, is completely retained on the column.

In the seaborgium experiments (Schädel *et al.* 1997b), a 950 μg cm^{-2} ^{248}Cm target was bombarded with 3×10^{12} ^{22}Ne ions per second at 121 MeV. 3900 identical separations were conducted with a collection and cycle time of 45 s and a total dose of 5.48×10^{17} ^{22}Ne ions. The transport efficiency of the He(KCl) jet was 45%. On average, counting of the samples started 38 s after the end of collection. The overall chemical yield was 80%. Three correlated $\alpha\alpha$ mother–daughter decays were

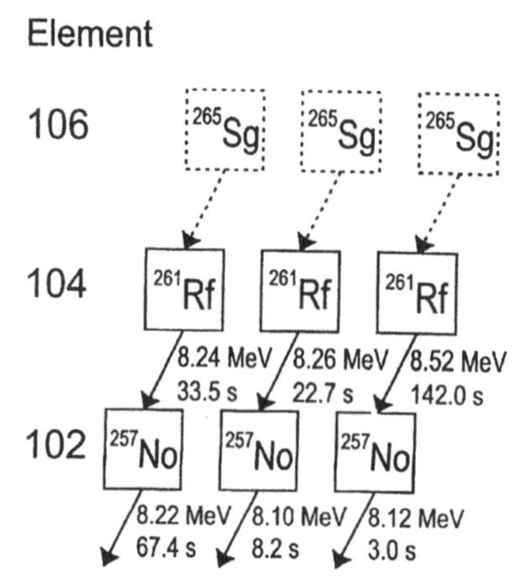

Figure 6.10 Nuclear decay chains originating with ^{265}Sg after chemical separation with ARCA II. The α-decay energies are given in MeV and the observed lifetimes in seconds. Reproduced with permission from Schädel *et al.* (1997b). © 1997 *Nature* & Macmillan Publishers Ltd.

observed that are assigned to the decay of ^{261}Rf and ^{257}No as the decay products of ^{265}Sg (see Figure 6.10). The three correlated events have to be compared with an expectation value of 0.27 for random correlations. This gives a probability of 0.24% that the three events are random correlations. As the mother decays were not observed, Figure 6.10, it is important to note that ^{261}Rf and ^{257}No can only be observed if ^{265}Sg passed through the column because group 4 elements and No are strongly retained on the cation exchange columns in ARCA II. Most likely, the decay of 7 s ^{265}Sg was not observed because it decayed in the time interval between the end-of-separation and the start-of-measurement, which was equivalent to four half-lives. That the columns really retained ^{261}Rf was demonstrated recently in an experiment where ^{261}Rf was produced directly in the ^{248}Cm(^{18}O,5n) reaction at the PSI Philips cyclotron (Strub *et al.* 2000), and processed in the seaborgium chemistry in 0.1 M $HNO_3/5 \times 10^{-4}$ M HF. ^{261}Rf did not elute from the column and was subsequently stripped from the column with 0.1 M $HNO_3/10^{-1}$ M HF.

From the observation of the three correlated α-decay chains of ^{265}Sg daughters, it was concluded that, for the first time, a chemical separation of seaborgium had been performed in aqueous solution. Seaborgium shows a behaviour typical for a hexavalent element located in group 6 of the Periodic Table and different from that of the pseudogroup-6 element uranium, which is fixed as UO_2^{2+} on the cation exchange column. Presumably, Sg forms $SgO_2F_3(H_2O)^-$ or the neutral species SgO_2F_2, but due to the low fluoride concentration used, the anionic SgO_4^{2-} ('seaborgate' in analogy to molybdate, MoO_4^{2-}, or tungstate, WO_4^{2-}) cannot be excluded.

In order to get experimental information on this latter question, a new series of seaborgium experiments with ARCA II was performed in which 0.1 M HNO_3 without HF was used as the mobile aqueous phase and Aminex A6 as stationary phase. If the 'seaborgate' ion was what was isolated in Schädel *et al.* (1997b), it was supposed to show up here again. A 690 μg cm^{-2} ^{248}Cm target containing 22 μg cm^{-2} enriched ^{152}Gd was bombarded with 123 MeV ^{22}Ne ions. The simultaneously produced ^{169}W served as a yield monitor. 45 s cycles were run in which the effluent was evaporated on thin (~500 μg cm^{-2}) Ti foils mounted on Al frames. These were thin enough to be counted in close geometry by pairs of PIPS detectors, thus increasing the efficiency for $\alpha\alpha$-correlations by a factor of four as compared with the 1995 experiment. A beam dose of 4.32×10^{17} beam particles was collected in 4575 separations. Only one $\alpha\alpha$-correlation attributable to the ^{261}Rf–^{257}No pair was observed. With an expected number of random correlations of 0.5 this is likely (the probability is 30%) to be a random correlation. From the beam integral and the overall yield as measured simultaneously for ^{169}W (27% on the average), a total of five correlated events was to be expected. This tends to indicate that, in the absence of fluoride ion, there is sorption of seaborgium on the cation exchange resin (Schädel *et al.* 1998).

This nontungsten-like behaviour of seaborgium under the given conditions may be attributed to its weaker tendency to hydrolyse (see the equations in Table 6.6). For Mo and W, the sequence of subsequent hydrolysis reactions in diluted HNO_3 reaches the neutral species $MO_2(OH)_2$. A weaker tendency to hydrolyse for seaborgium would stop this sequence earlier, e.g. with $MO(OH)_3(H_2O)_2^+$, which sorbs on a cation exchange resin. Calculations of the electronic structure of hydrolysed species of Mo, W and Sg and application of the theoretical model for hydrolysis lead indeed to the prediction of the sequence of hydrolysis (Section 6.5.1)

$$\mathrm{Mo} > \mathrm{W} > \mathrm{Sg}$$

in acidic solutions, as indicated by the experiment (Schädel *et al.* 1998).

In the presence of fluoride ions having a strong tendency to replace OH–ligands, the formation of neutral or anionic fluoride species is favoured:

$$\left.\begin{aligned} MO_2(OH)_2(H_2O)_2 + 2HF &\rightleftharpoons MO_2F_2 + 4H_2O, \\ MO_2F_2 + F^- &\rightleftharpoons MO_2F_3^-. \end{aligned}\right\} \qquad (6.18)$$

Thus, in the early experiments with seaborgium in the presence of fluoride ions, neutral or anionic fluoride complexes, e.g. MO_2F_2 or $MO_2F_3^-$, were likely to be formed and were eluted from the cation exchange columns. Experiments using the on-line three-column technique (Pfrepper *et al.* 2000) are in preparation to determine the K_d value of seaborgium oxofluorides on an anion exchange resin relative to the respective values for Mo and W. The theoretical predictions (Section 6.5.1) provide a guideline for these experiments.

6.6 Summary

We have shown here how the chemical properties of the very heavy elements can be studied both experimentally and theoretically, each complementing the other and leading to an understanding of the chemistry of the very exotic short-lived elements. The study has also shown that theoretical models are adequate to describe real processes and are able to predict the results of chemical experiments.

The investigations have established trends, as well as revealed new unexpected features in the properties of the heavy elements in relation to the well-studied lighter homologues. Theoretical predictions of redox potentials in solutions and complex formation using the DFT method seem to be especially promising. The present investigations have also demonstrated that linear extrapolations of properties are no more reliable in the area of the very heavy elements, where relativistic effects are very strong, and that fully relativistic calculations are indispensable for defining trends in the correct way.

The combination of fully relativistic calculations with some additional physicochemical schemes is still presently the best way of reliably predicting the outcome of sophisticated experiments.

Acknowledgments

We thank S. Varga from the Institut für Kernchemie, Universität Mainz, for help in preparing the manuscript. V.P. thanks Professor B. Fricke in whose group the latest version of the DFT program has been developed. Financial support from the Deutsche Forschungsgemeinschaft under contracts Kr 1458/5-1, Kr 1458/4-2 and Kr 1458/2-2 is gratefully acknowledged. We also appreciate support from the BMBF under contract 06MZ864 and from Gesellschaft für Schwerionenforschung under contract MZKRAK.

7 Experimental Probes for Relativistic Effects in the Chemistry of Heavy d and f Elements

Detlef Schröder, Martin Diefenbach, Helmut Schwarz
Institut für Chemie,
Technische Universität Berlin

Annette Schier, Hubert Schmidbaur
Anorganisch-chemisches Institut,
Technische Universität München

7.1 Introduction

In the design of experimental probes for relativistic effects in the chemistry of heavy d- and f-metals, gold is a key element because relativistic effects reach a local maximum (Bartlett 1998; Kaltsoyannis 1997). The two research groups involved in this collaborative study follow different approaches. The Berlin group aims to examine the gas-phase ion chemistry of heavy elements with respect to the role of relativistic effects on metal–ligand binding in general and chemical reactivity in particular. Based on the expertise in the preparation of 5d metal compounds, the Munich group is predominantly oriented towards the investigation of the influence of relativity on molecular and supramolecular structures. Common to both projects is the attempt to experimentally probe predictions made by theoreticians as well as to guide computational studies towards problems of chemical relevance. These efforts are summarized in this chapter, which focuses on some more recent, representative studies rather than covering all our research projects carried out in this context.

7.2 Gas-Phase Ion Chemistry of Heavy Elements

The starting point for the involvement of the Berlin group in the topic of relativistic effects was the experimental detection of the elusive gold(I) fluoride, whose existence had been conjectured by theory, but had not so far been seen experimentally. In a collaborative effort with Klapötke and co-workers we were able to unambiguously establish the existence of AuF as a long-lived molecule in the gas phase by mass spectrometric means (Schröder *et al.* 1994a,b). Almost simultaneously, Schwerdtfeger

Relativistic Effects in Heavy-Element Chemistry and Physics. Edited by B. A. Hess

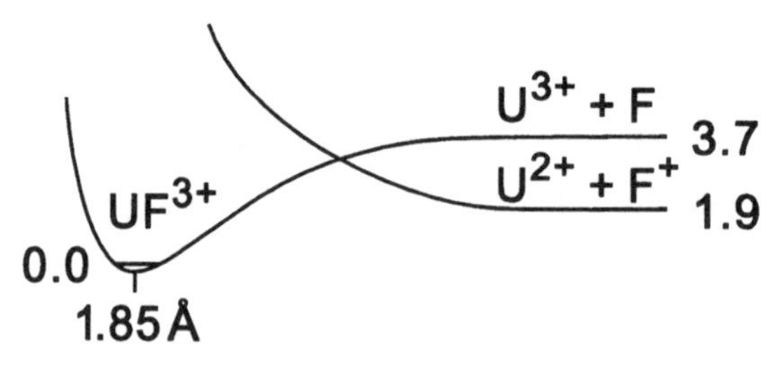

Figure 7.1 Stability diagram (in eV) of the diatomic UF^{3+} trication.

et al. (1994) also provided evidence for the formation of gaseous AuF; this molecule continues to receive considerable attention (Andreev and DelBruno 2000; Evans and Gerry 2000; Schwerdtfeger *et al.* 1995b).

7.2.1 Thermochemistry

There exist numerous reports on the influence of relativistic effects on calculated bond strengths and redox properties of heavy elements and their compounds (Hess 1997). While quantitative predictions require explicit consideration, some general patterns and trends evolve clearly. As far as transition metals are concerned, the relativistic contraction of s orbitals is most significant because of their large contribution to chemical bonding. For many metal atoms, the low-lying s orbitals result in both enlarged electron affinities (EAs) and ionization energies (IEs). For example, EA(Au) = 2.31 eV and IE(Au) = 9.23 eV of the neutral gold atom having an s^1d^{10} configuration are significantly larger than those of the lighter 3d and 4d congeners, i.e. copper and silver with EA(Cu) = 1.23 eV, EA(Ag) = 1.30 eV, IE(Cu) = 7.73 eV and IE(Ag) = 7.58 eV. These trends translate to the compounds of 5d-block elements. Of course, the effect vanishes when s occupation is not involved in ionization and even reverses for f-block elements because the f orbitals are destabilized by relativistic effects. Indeed, the IEs of lanthanide and actinide atoms are rather low, and some of them are close to those of alkali metals, for example, IE(Pr) = 5.47 eV and IE(Ac) = 5.17 eV compared with IE(Na) = 5.12 eV. Among the numerous examples of relativistic effects on bond and redox energies, let us consider three specific cases.

The relativistically lowered ionization energies of lanthanides and actinides permit the generation of several multiply charged cations in oxidation and charge states which are difficult to reach with other elements. For example, the diatomic cations UF^{n+} with n up to 3 can be generated by electron ionization of uranium hexafluoride. Moreover, not only the mono- and dications, but also the diatomic trication UF^{3+}, is thermochemically stable in that the Coulomb explosion is endothermic (Figure 7.1) (Schröder *et al.* 1999a,b). By comparison, while the diatomic trication TiF^{3+} exists in a local minimum, it is unstable with respect to the charge separation of Ti^{2+} + F^+ by more than 5 eV (Schröder *et al.* 1998a). Choosing heavier group 4 elements does not lead to thermochemical stability, because the metals' IEs do not decrease monotonically. Thus, while the ionization energies of titanium, IE(Ti) = 6.82 eV, IE(Ti^+) = 13.57 eV and IE(Ti^{2+}) = 27.47 eV, exceed those of zirconium, IE(Zr) = 6.64 eV, IE(Zr^+) = 13.13 eV, IE(Zr^{2+}) = 22.98 eV, those of hafnium slightly increase again due to relativistic contraction of the 6s shell: IE(Hf) = 6.83 eV, IE(Hf^+) = 14.9 eV and IE(Hf^{2+}) = 23.2 eV.

While none of the bare M^+ cations of the 3d and 4d series is capable of activating methane at ambient temperatures, pioneering work of Irikura and Beauchamp (1991a,b) demonstrated that several 5d metals can give rise to efficient dehydrogenation of methane according to Reaction (7.1) leading to the exothermic formation of cationic metal–carbene complexes (Schwarz and Schröder 2001):

$$M^+ + CH_4 \longrightarrow MCH_2^+ + H_2. \tag{7.1}$$

This dramatic difference in reactivity can be attributed to the significantly stronger M^+–CH_2 bonds of the 5d metals (Heinemann *et al.* 1995d). For example, the dissociation energies $D(Ni^+–CH_2)$ = 3.17 eV and $D(Pd^+–CH_2)$ = 2.95 eV are much lower than for the 5d congener, $D(Pt^+–CH_2)$ = 4.80 eV (see Zhang *et al.* 2001, and references cited therein); thermal occurrence of Reaction (7.1) requires $D(M^+–CH_2) >$ 4.7 eV. High-level *ab initio* calculations with explicit consideration of relativistic effects have demonstrated that the large bond strength of $PtCH_2^+$ is to a notable extent caused by relativity (Heinemann *et al.* 1995e, 1996b; Rakowitz *et al.* 2000). This effect is complemented by a significantly better overlap between the π-orbitals of the carbene fragment with the $5d_\pi$ orbitals of Pt compared with the $3d_\pi$ and $4d_\pi$ counterparts of Ni and Pd. A related consequence is that high oxidation states are much more favourable for 5d compared with 3d and 4d elements.

A particularly intriguing case evolves from the consideration of the Mo–O bond strengths in MoO^+ and MoO_2^+. Thus, $D(Mo^+–O)$ = 3.78 eV computed nonrelativistically increases to 4.60 eV when relativity is included, whereas the nonrelativistic and relativistic values for the corresponding dioxide are almost identical, i.e. $D(OMo^+–O)$ = 5.21 versus 5.24 eV (Kretzschmar *et al.* 1997). Hence, relativity is of prime importance for MoO^+, but negligible for MoO_2^+. The origin of this unusually large difference is associated with the significant 5s occupation in MoO^+, while this is small in Mo^+ (^6S) and MoO_2^+ ($^2A_1$). The lowering of the 5s orbital due to relativity therefore results in a differential stabilization which is most pronounced for MoO^+. Notable is the significant effect of relativity on the thermochemistry, although molybdenum is a 4d element for which relativistic effects are often assumed to be negligible.

A final system mentioned here demonstrates the strength of contemporary *ab initio* methods. During a study of metal-mediated C–C coupling processes, it was discovered that thermalized Pd^+ reacts with methyl iodide to yield a $PdCH_2I^+$ cation (Schwarz *et al.* 1996a). As no obvious intermediates were observed, occurrence of Reaction (7.2) was assumed:

$$Pd^+ + CH_3I \longrightarrow PdCH_2I^+ + H. \tag{7.2}$$

Reaction (7.2) is quite surprising because the strong C–H bond is selectively activated in the presence of the considerably weaker C–I bond. Hence, the $PdCH_2I^+$ product seems to experience some particular stabilization. In fact, *ab initio* studies reveal a notable interaction between palladium and iodine, giving rise to a Pd–C–I angle of only 78°. Nevertheless, none of the rather demanding theoretical methods applied led to $D(Pd^+–CH_2I) > 4.5$ eV, which is required to render Reaction (7.2) exothermic. While we may speculate about an additional stabilization of $PdCH_2I^+$ due to synergistic interaction of two heavy elements being involved in the bonding, advanced theoretical investigations enforced a reconsideration of the experiments. Upon more detailed analysis, the true route for the formation of $PdCH_2I^+$ was found to involve two elementary steps, Reactions (7.3) and (7.4):

$$Pd^+ + CH_3I \longrightarrow PdCH_3^+ + I, \tag{7.3}$$

$$PdCH_3^+ + CH_3I \longrightarrow PdCH_2I^+ + CH_4. \tag{7.4}$$

Failure to observe the $PdCH_3^+$ intermediate in the first study is due to a delicate balance of the associated rate constants leading to an extremely low steady-state concentration of the $PdCH_3^+$ intermediate. The sequence via Reactions (7.3) and (7.4) begins with C–I bond activation and is consistent with all experimental and computational findings (Schwarz *et al.* 1996b). The lesson from this study is that only the consequent improvement of the theoretical methods leads to a correct interpretation of the experiments.

7.2.2 Coordination chemistry

Gold(I) is particularly suited to probing the effects of relativity on the coordination chemistry of closed-shell ligands L for several reasons. At first, relativistic effects are particularly pronounced for gold; this has also been referred to as the 'gold maximum' (Pyykkö 1988). Next, the $6s^05d^{10}$ configuration of the valence space facilitates the treatment of gold(I) compounds at appropriate levels of theory. Further, many ligated gold(I) compounds are easily accessible in the gas phase (see below), thereby allowing a reasonable flexibility in the choice of the binding partners. Last but not least, in the gas phase Au^+ shows little chemical reactivity—except electron transfer (ET) and reactions with highly reactive substrates (Wesendrup *et al.* 1995)—such that skeletal rearrangements of the ligands, competing bond cleavages, etc., do not severely disturb the examination of the coordination properties.

The first challenge concerned the exploration of a route which allows for a flexible synthesis of various gold(I) compounds in the gas phase. The equivalent of gold(I) in these studies is the bare metal cation Au^+. Due to the high ionization energy of atomic gold (IE = 9.23 eV), direct association with organic ligands is inappropriate in many cases because ET as well as anion abstractions compete (Chowdhury and Wilkins 1987; Ho and Dunbar 1999; Weil and Wilkins 1985). Another difficulty in preparing $Au(L)^+$ complexes in the diluted gas phase is that, despite the attractive potentials between Au^+ and L, direct associations in two-body collisions are unlikely to occur for ligands of small to moderate sizes. This is because the energy gained upon complexation is, of course, also sufficient for redissociation; only species with sufficient lifetimes can experience radiative and/or collisional stabilization. For example, although the dissociation energy $D(Au^+–C_2H_4)$ amounts to about 3 eV, bimolecular association of Au^+ and ethene is almost too slow to be measured in the low-pressure regime of the experiments (typically less than 10^{-6} mbar).

A versatile route for the gas-phase synthesis of various gold(I) complexes is provided by the reaction of Au^+ with hexafluorobenzene. While $IE(C_6F_6) = 9.91$ eV is large enough to prevent ET, C_6F_6 has a sufficient number of rovibronic states to allow for efficient formation of the $Au(C_6F_6)^+$ complex via radiative stabilization in the low-pressure regime according to Reaction (7.5) (Schröder *et al.* 1995):

$$Au^+ + C_6F_6 \longrightarrow Au(C_6F_6)^+, \qquad (7.5)$$

$$Au(C_6F_6)^+ + C_6F_6 \longrightarrow Au(C_6F_6)_2^+. \qquad (7.6)$$

In an excess of hexafluorobenzene, the adduct continues to react yielding the bisligated ion in Reaction (7.6); complexes of gold(I) with more than two ligands were not observed in these and similar experiments conducted in the low-pressure regime (Taylor *et al.* 1997). By pulsing the C_6F_6 reactant into the mass spectrometer and appropriately adjusting pulse lengths and heights, yields of $Au(C_6F_6)^+$ up to 70% (based on Au^+) can be achieved in a rapid and well-reproducible manner. The key aspect is that hexafluorobenzene adds efficiently to Au^+ according to Reaction (7.5), while at the same time it is a poor ligand due to the electron deficient π-system. In fact, theoretical studies indicate that coordination of the gold cation to one of the fluorine atoms is energetically almost equivalent to binding at the π-system (Schröder *et al.* 1998c). Consequently, C_6F_6 can be exchanged by other equally well or more strongly bound ligands L according to Reaction (7.7):

$$Au(C_6F_6)^+ + L \longrightarrow Au(L)^+ + C_6F_6, \qquad (7.7)$$

$$Au(L)^+ + L' \longrightarrow Au(L')^+ + L. \qquad (7.8)$$

One of these ligands is water, hardly an avoidable impurity in high-vacuum devices, and formation of $Au(H_2O)^+$ was indeed observed in all experiments conducted with $Au(C_6F_6)^+$. The same approach can then be applied using other ligands L′ in subsequent reactions such as (7.8). Under thermal gas-phase conditions, Reaction (7.8) can only occur when the reaction enthalpy $\Delta_r G_{298}(7.8)$ is close to or lower than zero (Bouchoux *et al.* 1996). Thus, ligand exchange is only observed if $D(Au^+–L')$ approaches or exceeds $D(Au^+–L)$; in some cases, notable entropic contributions to $\Delta_r G_{298}$ need to be considered as well (Schröder *et al.* 1998c). In addition to these qualitative aspects, the experiments permit quantitative analysis of the reaction kinetics if the internal energy deposited in radiative association is adequately acknowledged (Schröder *et al.* 2000a). Fortunately in some cases, equilibrium constants K_{eq} can be determined that provide accurate thermochemical information using the Gibbs–Helmholtz equation $\Delta_r G = -RT \ln K_{eq}$. In the present case, a series of systematic studies led to an order of relative Au^+ affinities of several prototype ligands (Schröder

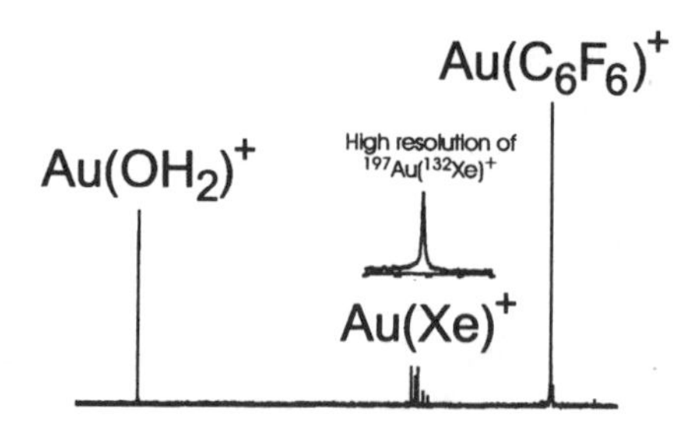

Figure 7.2 Generation of $Au(Xe)^+$ upon reacting $Au(C_6F_6)^+$ with xenon. Note the competitive formation of $Au(H_2O)^+$ due to reactions with background water.

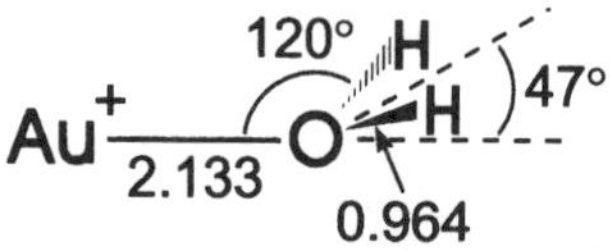

Figure 7.3 Computed structure of the $Au(H_2O)^+$ cation (bond lengths in angstroms and angles in degrees).

et al. 1995, 1998c), i.e. $Xe < C_6F_6 < H_2O < CO < H_2S < CH_3CN \approx C_2H_4 \approx C_6H_6 \approx NH_3 \approx CH_3NC < CH_3SCH_3 < PH_3$. Various experimental and theoretical studies, including a detailed examination of equilibrium isotope effects for $Au(C_2X_4)^+$ (X = H, D) (Schröder *et al.* 2000c), were used to determine relevant properties of gaseous gold(I) complexes of which we would like to discuss three systems in more detail here.

First, the experimental studies conducted by the Berlin group did confirm Pyykkö's prediction that xenon is a suitable ligand for gold(I) (Pyykkö 1995). Thus, atomic Xe is able to replace the arene ligand in $Au(C_6F_6)^+$ according to Reaction (7.7) to afford $Au(Xe)^+$ as revealed by the characteristic isotope pattern as well as high-resolution mass analysis (Figure 7.2) (Schröder *et al.* 1998c). At longer reaction times, the bisligated cation $Au(Xe)_2^+$ is also formed, again confirming the theoretical prediction (Pyykkö 1995). These results suggest that liquid xenon may be used as a suitable solvent in the synthesis of gold compounds in the condensed phase (see also Seidel and Seppelt 2000).

Next, the coordination geometry of the $Au(H_2O)^+$ complex is noteworthy. Unlike all other monocationic complexes $M(H_2O)^+$ of main-group or transition metals M with water, $Au(H_2O)^+$ is a nonplanar molecule (Figure 7.3, Hrušák *et al.* 1994). This deviation from planarity is basically caused by relativistic effects which energetically lower the 6s orbital, such that Au^+ in $Au(H_2O)^+$ behaves like a proton in H_3O^+: it prefers covalent to predominantly electrostatic bonding (Hrušák *et al.* 1995). Theoretical studies of the corresponding benzene complex $Au(C_6H_6)^+$ predict a similar deviation from usual coordination geometries: unlike most other monocationic

Table 7.1 Bond dissociation energies $D(M^+–C_2H_4)$ (in eV) of $M(C_2H_4)^+$ complexes for transition metals M (taken from Schröder *et al.* 1998c).

	D		D		D
Sc	1.39	Y	1.43	La	1.99
Ti	1.34	Zr	1.52	Hf	1.56
V	1.21	Nb	1.60	Ta	1.91
Cr	1.30	Mo	1.08	W	2.17
Mn	0.87	Tc	1.13	Re	1.30
Fe	1.50	Ru	1.30	Os	1.95
Co	1.86	Rh	1.34	Ir	2.43
Ni	1.82	Pd	1.21	Pt	2.39
Cu	1.95	Ag	1.47	Au	2.99

metal–benzene complexes, that of gold trades off C_{6v} geometry in that η^1- and η^2-type coordinations are strongly preferred (Dargel *et al.* 1999; Hertwig *et al.* 1995; Schröder *et al.* 2000a). Again, this behaviour of Au^+ resembles that of a proton interacting with benzene (Bouchoux *et al.* 1999; Glukhovtsev *et al.* 1995; Mason *et al.* 1995), and the facile interconversion of the η^1- and η^2-structures of $Au(C_6H_6)^+$ finds its equivalent in the hydrogen ring-walk of protonated arenes (Kuck 1990).

The third aspect concerns the strengths of Au^+–L interactions in general. A suitable comparison can be made with L = ethene for which sufficient data are available (Table 7.1). Among all $M(C_2H_4)^+$ complexes studied so far, $D(Au^+–C_2H_4)$ is outstandingly large, being in fact almost twice that of the lower congeners copper and silver (Hertwig *et al.* 1996; Hrušák *et al.* 1995). The bond strength is even sufficient for ethene replacing the formally covalent iodine ligand in the AuI^+ cation (Reaction (7.9), Schröder *et al.* 1995).

$$AuI^+ + C_2H_4 \longrightarrow Au(C_2H_4)^+ + I. \tag{7.9}$$

This large bond strength can be attributed to the bonding scheme of $Au(C_2H_4)^+$, which is best described as that of a metallacyclopropane, rather than a simple metal–olefin complex. Thus, an atoms-in-molecules analysis (Bader 1994) reveals primarily T-shaped types of electrostatic bonding for $Cu(C_2H_4)^+$ and $Ag(C_2H_4)^+$, whereas $Au(C_2H_4)^+$ clearly shows bond critical points between gold and carbon (Hertwig *et al.* 1996).

7.2.3 Reactivity

Gas-phase reactivity studies form the key issue in the research of the Berlin group. Note that relativity is not explicitly addressed here because reactivity is determined by thermochemical and kinetic aspects, and nonrelativistic considerations of reaction kinetics are almost meaningless (see also Section 7.4). Whereas bare Au^+ shows

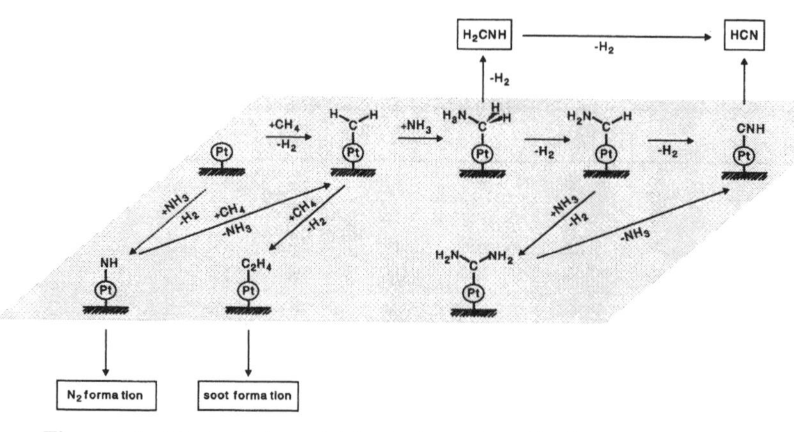

Figure 7.4 Mechanistic model for the Pt-catalysed HCN synthesis from CH_4 and NH_3 in the Degussa process.

little chemical reactivity, a rather diverse and fascinating chemistry is exhibited by gaseous platinum ions. For example, Pt^+ is not only capable of activating methane according to Reaction (7.1) (Heinemann *et al.* 1995f; Irikura and Beauchamp 1991a,b; Schwarz and Schröder 2001), the resulting $PtCH_2^+$ cation shows interesting reactivities towards methane (Irikura and Beauchamp 1991b; Wesendrup *et al.* 1994a,b), oxygen (Pavlov *et al.* 1997; Wesendrup *et al.* 1994a,b), as well as various nucleophiles, such as water, ammonia, hydrogen sulphide, several alcohols and amines (Brönstrup *et al.* 1999). The most instructive system in this context is $Pt^+/CH_4/NH_3$ (Aschi *et al.* 1998a,b; Diefenbach *et al.* 1999), which provides a gas-phase model for the synthesis of hydrogen cyanide in the Degussa process according to Reaction (7.10):

$$CH_4 + NH_3 \longrightarrow HCN + 3H_2. \quad (7.10)$$

In the gas phase, Pt^+ gives rise to a sequence of reactions. The first step is the activation of methane according to Reaction (7.1) to yield the platinum carbene cation $PtCH_2^+$; in contrast, bare Pt^+ does not activate ammonia. Subsequently, $PtCH_2^+$ undergoes efficient C–N coupling with ammonia whereas successive reactions with methane are of minor importance. Interestingly, two kinds of coupling products are formed: the aminocarbene complex $PtCH(NH_2)^+$ and the aminomethyl cation $CH_2NH_2^+$, namely protonated formimine. Examination of the consecutive reactions, collisional activation studies, labelling experiments, and *ab initio* calculations allowed us to suggest a mechanistic scenario of the Degussa process (Figure 7.4) (Diefenbach *et al.* 1999).

Thus, HCN is proposed to be formed in two competing routes, a surface-bound pathway via platinum aminocarbenes and a gas-phase channel involving successive dehydrogenation of formimine. Common to both routes is the Pt-mediated activation of methane as the first step. A crucial aspect is the remarkable selectivity observed, i.e. bare Pt^+ only reacts with methane while the resulting $PtCH_2^+$ preferentially reacts with ammonia. Exploratory studies of other transition metals indicate that platinum is most efficient in fulfilling both tasks simultaneously, the activation of methane and the subsequent C–N coupling with ammonia (Diefenbach *et al.* 1999). Mimicking the related Andrussow process, in which oxygen is added to the CH_4/NH_3 mixture, reveals that the Pt/O species formed only affect oxidation of methane (Brönstrup *et al.* 2001; Pavlov *et al.* 1997; Wesendrup *et al.* 1994a,b), while not intervening in the C–N coupling that is crucial for HCN formation.

Another reaction to be mentioned briefly is the Ta^+-mediated coupling of methane and carbon dioxide to afford ketene; the latter can serve as a precursor for acetic acid (Wesendrup and Schwarz 1995a,b). The rather attractive, yet hypothetical, coupling Reaction (7.11) is endothermic, however. In the mass spectrometric model study, the thermochemical driving force is provided by the oxidation of tantalum yielding the TaO_2^+ cation as the final product (Sändig and Koch 1998). Therefore, the coupled activation of methane and carbon dioxide so far remains stoichiometric:

$$CH_4 + CO_2 \longrightarrow CH_3COOH. \tag{7.11}$$

In addition to the chemistry of the d-block elements, several chemical transformations promoted by lanthanides (Cornehl *et al.* 1995), actinides (Heinemann *et al.* 1995a) and their oxides (Cornehl *et al.* 1996a,b, 1997b; Heinemann and Schwarz 1995) have been examined, such as C–C bond couplings (Heinemann *et al.* 1994), C–F bond activations (Cornehl *et al.* 1996c; Heinemann *et al.* 1995b,c), olefin oxidation (Cornehl *et al.* 1997a; Heinemann *et al.* 1996a) and alkadiene oligomerization (Cornehl *et al.* 1997b). In this context, an elegant technology developed by Gibson is noteworthy as it permits us to investigate the gas-phase ion chemistry of 'hot' elements such as plutonium (see Gibson 2001, and references therein).

In the course of the research project, it turned out that a major aspect of relativity for chemical processes is associated with reaction dynamics, rather than mere energetics. Moreover, this effect seems to be more relevant for lighter than for heavier elements. Numerous reactions of organometallic fragments in the gas phase—among them important processes such as hydrocarbon oxidation—involve formally spin-forbidden steps. In the reactions of heavy elements, for example, in the rich chemistry of gaseous platinum, we may assume that spin is not a meaningful quantum number. Instead, spin–orbit coupling predominates and spin rules do not play a crucial role any more. For lighter elements, however, spin constraints can affect chemical reactivity and may even become a decisive factor. For example, several hydroxylations of hydrocarbons by transition-metal oxides involve formally spin-forbidden steps (Fiedler *et al.* 1994; Shaik *et al.* 1998). Yet, the reactions occur at ambient temperature via surface crossings mediated by spin–orbit coupling (Yarkony 2001). This behaviour has also been termed two-state reactivity (TSR) (Shaik *et al.* 1995). In brief, the key feature of TSR is a change of spin multiplicity along the reaction coordinate from the reactants to the rate-determining transition structure (Schröder *et al.* 2000b). In addition to oxidation reactions (see Shiota and Yoshizawa 2000, and references therein) evidence for the mechanistically decisive role of TSR in main-group (Aschi *et al.* 1998c; Harvey *et al.* 1998; Schröder *et al.* 1998b) and transition-metal chemistry (Kretzschmar *et al.*

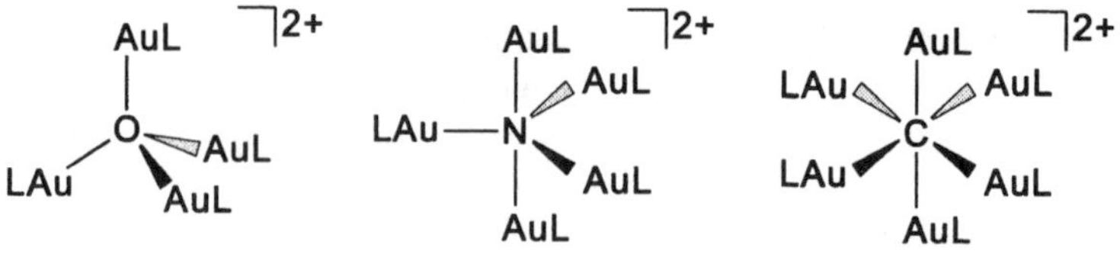

Figure 7.5 Multiply aurated oxygen, nitrogen and carbon.

1998; Rue *et al.* 1999), as well as in surface reactions (Triguero *et al.* 1998), has been provided.

7.3 Structural Chemistry of Gold Compounds in the Condensed Phase

In recent years, extensive preparative, structural, spectroscopic and theoretical studies, which have been oriented towards a better understanding of relativistic phenomena in the chemistry of gold, have been carried out by the Munich group (Schmidbaur 2000). Not only atomic and ionic radii are influenced by relativity, but also the stability of the oxidation states and coordination numbers are heavily affected, with consequences ranging from relatively small to supramolecular ensembles.

7.3.1 AuL^+: a big proton?

At first sight, the structures of gold(I) compounds appear to be quite simple. Thus, the s^0d^{10} configuration of Au^+ leads to bisligated complexes with perfectly linear arrangement of the ligands irrespective of the net charge of the compounds, for example, (OC)AuCl, $Au(NH_3)_2^+$ and $Au(CN)_2^-$ show L–Au–L angles of 180°. However, gold(I) compounds exhibit a fascinating structural chemistry when multiply ligated to elements with more than one valency. Most notable is the formation of several hypercoordinated, multiply charged complexes which give insight into some quite unexpected molecular architectures (Schmidbaur 1995). For example, when counterions and stoichiometries are chosen appropriately, compounds with tetracoordinated oxygen (Schmidbaur *et al.* 1995), pentacoordinated nitrogen (see Schier *et al.* 2000, and references therein) and even hexacoordinated carbon (Scherbaum *et al.* 1988a,b) can be synthesized as bulk substances and were fully characterized by X-ray diffraction (Figure 7.5).

The formation of hypercoordinated compounds is not restricted to carbon, nitrogen and oxygen, but also occurs with their heavier congeners (Preisenberger *et al.* 1999a), and the phosphine ligand can be replaced by others, for example, arsines (Tripathi *et al.* 1998a). These multiply charged cations can be regarded as analogues of the elusive dications H_4O^{2+}, NH_5^{2+} and CH_6^{2+}, which it has been suggested exist in the gas phase (Lammertsma *et al.* 1989a,b) as well as condensed media (Olah 1993a,b,

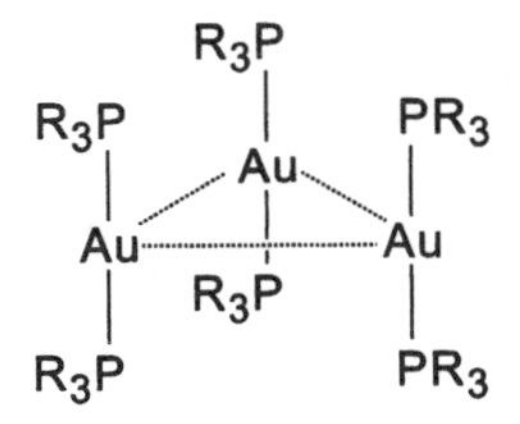

Figure 7.6 Sketch of a gold triangle resulting from aurophilic interaction.

1995a,b). Hence, while the coordination chemistry of gold(I) does indeed resemble that of a proton (see also Section 7.2.2), the evidence for the existence of the hydrogen compounds is merely indirect, whereas the corresponding gold(I) complexes can be isolated in crystalline form and in appreciable yields, for example, a 53% yield (Preisenberger *et al.* 1999a) in reaction

$$[(Ph_3P)Au]_4N^+ \, BF_4^- + (Ph_3P)Au^+ \, BF_4^- \longrightarrow [(Ph_3P)Au]_5N^{2+} \cdot 2\,BF_4^- .$$

An important reason for the stability of these hypercoordinated complexes is the aurophilic interaction between the gold atoms linked to the core element, as discussed in the next section.

7.3.2 Aurophilicity

The research conducted by the Munich group *inter alia* contributed to the uncovering of a new type of bonding: the attractive interaction between gold(I) centres for which the term 'aurophilicity' has been coined and widely accepted (Schmidbaur 2000). From a classical point of view, this interaction is quite unexpected because the perfect-pairing s^0d^{10} configuration of Au^+ does not imply an obvious bonding scheme; in fact, we might anticipate repulsion of the positively charged gold(I) centres. The interaction between the closed-shell d^{10} configuration is based largely on electron correlation, somewhat similar to van der Waals interactions. Though the closed-shell interaction is not inherently relativistic, inclusion of relativity increases the attractive interaction by an appreciable amount (Pyykkö 1997). In contrast to the weak van der Waals type bonding, the strengths of aurophilic interactions are comparable with hydrogen bonding. Thus, the $Au \cdots Au$ contacts do not impose major barriers for molecular motion at ambient temperatures, but can crucially determine the structural chemistry of gold compounds. For example, a fluoroboryl-bridged organophosphinite ligand leads to the formation of a triangular arrangement of three gold atoms joined by aurophilic bonding (Figure 7.6) (Hollatz *et al.* 1998, 1999b).

An ongoing topic in the Munich group is the search for a corresponding argentophilic interaction. While $Ag \cdots Ag$ bonding is predicted to be much weaker than for gold, some results indicate the occurrence of a similar phenomenon in silver compounds. The shortest Ag–Ag distance realized so far occurs in a binuclear silver(I) complex with two tridentate phosphine ligands (Zank *et al.* 1999) and amounts to

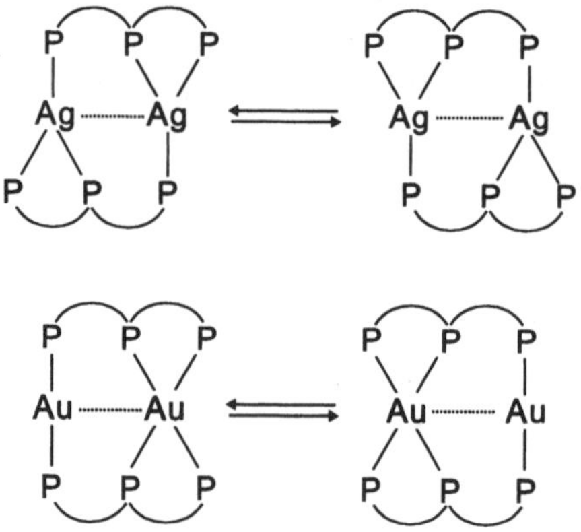

Figure 7.7 Dynamics of the dinuclear $[M_2(TP)_2][BF_4]_2$ complexes of silver and gold. TP = bis[2-(diphenylphosphino)phenyl]phenylphosphine, M = Ag, Au.

$r_{AgAg} = 2.857$ Å compared with $r_{AgAg} = 2.889$ Å in bulk silver. Likewise, attention is paid to other metallophilic interactions including those between different metal atoms, for example, Ag $\cdots$ Au.

7.3.3 Ligand design

Considering that the strength of aurophilic interaction is comparable with hydrogen bonding, its occurrence crucially depends on the steric requirements of the ligands (Hollatz *et al.* 1999a; Monge Oroz *et al.* 1999; Tripathi *et al.* 1998b). An interesting case is the comparison of the dinuclear complexes of a tailored tridentate phosphine with silver(I) and gold(I), respectively (Figure 7.7) (Zank *et al.* 1998, 1999). In solution, dynamic NMR experiments reveal fluxional behaviour for both complexes, yet the patterns differ greatly. In the silver complex, both Ag atoms undergo degenerate displacements of the ligands to afford two identical, tricoordinated metal centres. While degeneracy also occurs for the gold compound, the metal atoms exhibit different coordination numbers in the equilibrium geometry, i.e. one gold atom is coordinated to two and the other to four phosphorus atoms. This notable difference is ascribed to the strength of the metallophilic interaction of gold compared with that of silver. Nevertheless, in the silver complex the Ag–Ag distance is also remarkably short (see above). Complexes with chiral ligands have also been examined and provided deeper insight into the rather complicated organization of chiral units at three- and four-coordinate gold(I) centres (Bayler *et al.* 1998). An interesting competition comes into play with several sulphur-containing ligands. Because coordination of gold to sulphur is particularly favourable, it can in fact override the aurophilicity, resulting in a switch from Au $\cdots$ Au to Au $\cdots$ S contacts (Preisenberger *et al.* 1999b; Tzeng *et al.* 1999b).

Like hydrogen bonding, the aurophilic interaction can give rise to di- and tridimensional structures in which the gross architecture is determined by weak but significant

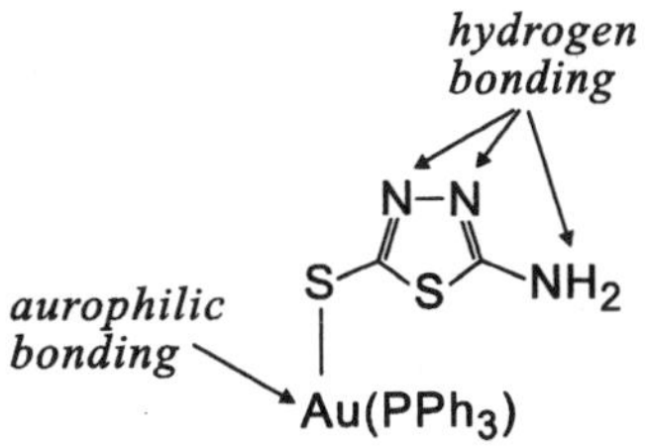

Figure 7.8 Supramolecular binding sites of gold(I) thiazolate.

forces (Schmidbaur 2000). For example, dithiophosphate complexes of gold show a chain-like arrangement in the solid which is determined by Au $\cdots$ Au contacts (Preisenberger *et al.* 1998). Particularly fascinating is the combination of aurophilic interaction with hydrogen bonding found for thiazolate complexes of gold(I) (Tzeng *et al.* 1999a). Here, the supramolecular architecture of the solid is determined by Au $\cdots$ Au interactions as well as N $\cdots$ H bridges of the thiazole ligands with themselves as well as interstitial solvents such as methanol. These coordination modes are sketched in Figure 7.8 (see Tzeng *et al.* (1999a) for the complete structures). Other options arise from Au $\cdots$ S coordination (Tzeng *et al.* 1999b), and a combination of these effects gives rise to several interesting supramolecular structures (Schmidbaur 2000).

7.4 Conclusions

The chemistry of transition metals, lanthanides and actinides is significantly influenced by relativistic effects. Qualitatively, these effects become apparent in the comparison of certain structural properties or reactivity patterns for a group of metals, for example, trends in the chemistry of copper, silver and gold. Quantification of relativistic effects can, however, only be achieved by relating the experimental findings to the results of adequate *ab initio* studies. Reference to theory is required because nonrelativistic properties cannot be probed directly. Thus, elements behave relativistically in any kind of experiment, whether one deals with the spectrum of H_2^+ or the properties of transuranium compounds.

As far as thermochemistry, ligand bonding and molecular structures in the gaseous as well as condensed phase are concerned, the case studies described above demonstrate the fruitful interplay of experiment and theory. For example, an understanding of the unusual stability of the gaseous $PtCH_2^+$ cation can only be achieved upon inclusion of relativistic effects. Likewise, the bond strengths of $Au(L)^+$ complexes can only be rationalized if relativity is considered explicitly, and the same applies to the structural chemistry of gold(I) compounds in the condensed phase, for example, aurophilic interaction. Somewhat different is the situation as far as reactivity is concerned, because nonrelativistic descriptions of chemical reactivity are somewhat artificial and cannot be probed at all. If, for example, the thermochemistry of gold(I) compounds is already poorly described by nonrelativistic approaches, it is simply inadequate to use these results for any conclusion with respect to reaction kinetics. Likewise, the search for extreme relativistic effects should be taken with care. Depending on the choice of the reference systems, we may in fact construct cases in which relativistic effects are either extremely large or small, which are nothing other than accumulations or cancellations of errors, however.

From an applied point of view, there is another, rather important aspect to consider. Nowadays, there exist quite reliable theoretical methods to describe heavy elements and their compounds within chemical accuracy as far as minima are concerned. The appropriate inclusion of spin–orbit coupling in the quantum chemical description of complete potential-energy surfaces is, however, usually limited to systems comprising only a few atoms (Rakowitz *et al.* 2000, e.g. $PtCH_2^+$). Unfortunately, this is particularly true for the description of transition structures as well as the crossing points between surfaces of different spin multiplicities. The ability to handle reaction barriers and crossing points in polyatomic systems is, however, of prime importance for the understanding of the chemical reactivity of heavy transition metals—the thermochemical properties of the minima are at best a prerequisite. Improvement of the theoretical tools would *inter alia* require the development of gradients for geometry optimizations while appropriately acknowledging spin–orbit coupling. In this respect the performance of contemporary *ab initio* methods is still too limited to really *predict* the chemical behaviour of heavy elements, which is good news for experimentalists and constitutes an ongoing challenge to theory.

Acknowledgments

Our studies are based on the contributions of all co-workers and collaborating partners mentioned in the references, and we very much acknowledge their input. Financial support was provided by the Deutsche Forschungsgemeinschaft, the Volkswagen-Stiftung, the European Commission (RTN1-1999-254), the Fonds der Chemischen Industrie, and the Gesellschaft von Freunden der Technischen Universität Berlin. Furthermore, the Konrad–Zuse Zentrum, Berlin, is appreciated for generous allocation of computer time. Particularly valuable were the insightful discussions with leading R&D scientists from the Bayer AG and Degussa–Hüls AG. Finally, Professor P. B. Armentrout is thanked for providing a preprint (Zhang *et al.* 2001) before its publication.

Appendix A

This volume is based on the work and the reports of the following members of the 'Schwerpunkt' of the German Science Foundation (Deutsche Forschungsgemeinschaft) on 'Relativistic Effects in Heavy-Element Chemistry and Physics'.

Jürgen Braun	Theoretische Physik, Universität Osnabrück
Helmut Bross	Theoretische Physik, Universität München
Robert J. Buenker	Theoretische Chemie, Universität Wuppertal
Peter H. Dederichs	Forschungszentrum Jülich
Michael Dolg	Theoretische Chemie, Universität Bonn
Reiner Dreizler	Theoretische Physik, Universität Frankfurt/Main
Eberhard Engel	Theoretische Physik, Universität Frankfurt/Main
Hubert Ebert	Physikalische Chemie, Universität München
Farhad H. M. Faisal	Theoretische Physik, Universität Bielefeld
Gregor-Martin Fehrenbach	Theoretische Physik, Universität München
Burkhard Fricke	Theoretische Physik, Universität Kassel
Lothar Fritsche	Theoretische Physik, Universität Clausthal-Zellerfeld
Stephan Fritzsche	Theoretische Physik, Universität Kassel
Walter Greiner	Theoretische Physik, Universität Frankfurt/Main
Stefan Grimme	Organische Chemie, Universität Münster
Eberhard K. U. Gross	Theoretische Physik, Universität Würzburg
Bernd Hess	Theoretische Chemie, Universität Erlangen
Jürgen Hinze	Theoretische Chemie, Universität Bielefeld
Martin Kaupp	Organische Chemie, Universität Würzburg
Dietmar Kolb	Theoretische Physik, Universität Kassel
Jens Volker Kratz	Kernchemie, Universität Mainz
Jürgen Kübler	Festkörperphysik, Technische Hochschule Darmstadt
Janos Ladik	Theoretische Chemie, Universität Erlangen
Christel Marian	Theoretische Chemie, Universität Düsseldorf
Valeria Pershina	GSI Darmstadt
Günter Plunien	Theoretische Physik, Universität Dresden
Notker Rösch	Theoretische Chemie, Technische Universität München
Leonid Sandratskii	Festkörperphysik, Technische Hochschule Darmstadt

Werner Scheid	Theoretische Phsyik, Universität Gießen
Hubert Schmidbaur	Anorganische Chemie, Technische Universität München
Detlef Schröder	Organische Chemie, Technische Universität Berlin
Gisela Schütz	Experimentalphysik, Universität Würzburg
Helmut Schwarz	Organische Chemie, Technische Universität Berlin
Eugen Schwarz	Theoretische Chemie, Universität Siegen
Heinz Siedentop	Mathematik, LMU München
Herrmann Stoll	Theoretische Chemie, Universität Stuttgart
Gerhard Soff	Theoretische Physik, Universität Dresden
Günther Wunner	Theoretische Physik, Universität Stuttgart

References

Abragam, A. and Pryce, M. H. L. (1951) *Proc. R. Soc. Lond.* A **205**, 135.

Ackermann, B. (1985) Relativistische Theorie der Photoemission und Streuung langsamer Elektronen von ferromagnetischen Oberflachen (Relativistic theory of photoemission and low energy electron scattering from ferromagnetic surfaces). PhD thesis, University of Duisburg.

Ackermann, B., Feder, R. and Tamura, E. (1984) *J. Phys.* F **14**, L173.

Aerts, P. J. C. (1986) Towards relativistic quantum chemistry—on the *ab initio* calculation of relativistic electron wave functions for molecules in the Hartree–Fock–Dirac approximation. PhD thesis, Rijksuniversiteit te Groningen, Netherlands.

Aerts, P. J. C. and Nieuwpoort, W. C. (1986) *Int. J. Quant. Chem. Quant. Chem. Symp.* **19**, 267–277.

Ahlers, D. (1998) Magnetic EXAFS—an experimental and theoretical investigation. PhD thesis, University of Würzburg.

Ahlers, D. and Schütz, G. (1998) *Phys. Rev.* B **57**, 3466.

Ahlers, D., Attenkofer, K. and Schütz, G. (1998) *J. Appl. Phys.* **83**, 7085.

Akai, H. and Kotani, T. (1999) *Hyperfine Interactions* **120–121**, 3.

Akhiezer, I. A. and Peletminskii, S. V. (1960) *Zh. Eksp. Teor. Fiz.* **38**, 1829. (*Sov. Phys. JETP* **11**, 1316 (1960).)

Albert, K., Neyman, K. M., Nasluzov, V. A., Ruzankin, S. P., Yeretzian, C. and Rösch, N. (1995) *Chem. Phys. Lett.* **245**, 671–678.

Albert, K., Neyman, K. M., Paccioni, G. and Rösch, N. (1996) *Inorg. Chem.* **35**, 7370–7376.

Alekseyev, A., Liebermann, H.-P., Boustani, I., Hirsch, G. and Buenker, R. J. (1993) *Chem. Phys. Lett.* **173**, 333.

Alekseyev, A. B., Liebermann, H.-P., Buenker, R. J., Hirsch, G. and Li, Y. (1994a) *J. Chem. Phys.* **100**, 8956–8968.

Alekseyev, A., Buenker, R. J., Liebermann, H.-P. and Hirsch, G. (1994b) *J. Chem. Phys.* **100**, 2989.

Almbladh, C.-O. and von Barth, U. (1985) *Phys. Rev.* B **31**, 3231.

Andersen, O. K. (1975) *Phys. Rev.* B **12**, 3060–3083.

Andrae, D. (1997) *J. Phys.* B **30**, 4435–4451.

Andrae, D. (2000) *Phys. Rep.* **336**, 413–525.

Andrae, D. (2001) *Mol. Phys.* **99**, 327–334.

Andrae, D., Häußermann, U., Dolg, M., Stoll, H. and Preuß, H. (1990) *Theor. Chim. Acta* **77**, 123.

Andrae, D., Reiher, M. and Hinze, J. (2000a) *Int. J. Quant. Chem.* **76**, 473–499.

Andrae, D., Reiher, M. and Hinze, J. (2000b) *Chem. Phys. Lett.* **320**, 457–468.

Andreev, O. Y., Labzowsky, L. N., Plunien, G. and Soff, G. (2001) *Phys. Rev.* A **64**, 042513.

Andreev, S. and DelBruno, J. J. (2000) *Chem. Phys. Lett.* **329**, 490.

Antoniewicz, P. R. and Kleinman, L. (1985) *Phys. Rev.* B **31**, 6779.

Antropov, V. P., Liechtenstein, A. I. and Harmon, B. N. (1995) *J. Magn. Magn. Mater.* **140–144**, 1161.

Arima, A. and Horie, H. (1955) *Prog. Theor. Phys.* **11**, 509.

Artemyev, A. N., Shabaev, V. M. and Yerokhin, V. A. (1995) *Phys. Rev.* A **52**, 1884.

Artemyev, A. N., Beier, T., Plunien, G., Shabaev, V. M., Soff, G. and Yerokhin, V. A. (1999) *Phys. Rev.* A **60**, 45.

Artemyev, A. N., Beier, T., Plunien, G., Shabaev, V. M., Soff, G. and Yerokhin, V. A. (2000) *Phys. Rev.* A **62**, 022116.

Aschi, M., Brönstrup, M., Diefenbach, M., Harvey, J. N., Schröder, D. and Schwarz, H. (1998a) *Angew. Chem.* **110**, 858.

Aschi, M., Brönstrup, M., Diefenbach, M., Harvey, J. N., Schröder, D. and Schwarz, H. (1998b) *Angew. Chem. Int. Ed. Engl.* **37**, 829.

Aschi, M., Harvey, J. N., Schalley, C. A., Schröder, D. and Schwarz, H. (1998c) *J. Chem. Soc. Chem. Commun.*, p. 531.

Aucar, G. A., Saue, T., Visscher, L. and Jensen, H. J. A. (1999) *J. Chem. Phys.* **110**, 6208–6218.

Auth, T. (1999) Master's thesis, University of Frankfurt.

Autschbach, J. and Schwarz, W. H. E. (2000) *Theor. Chem. Acc.* **104**, 82–88.

Avgoustoglou, E., Johnson, W. R., Liu, Z. W. and Sapirstein, J. (1995) *Phys. Rev.* A **51**, 1196–1208.

Aziz, R. A. and Slaman, M. J. (1991) *J. Chem. Phys.* **94**, 8047.

Bach, V., Barbaroux, J. M., Helffer, B. and Siedentop, H. (1998) *Doc. Math.* (electronic) **3**, 353.

Bach, V., Barbaroux, J. M., Helffer, B. and Siedentop, H. (1999) *Commun. Math. Phys.* **201**, 445.

Bachelet, G. B., Hamann, D. R. and Schlüter, M. (1982) *Phys. Rev.* B **26**, 4199.

Bader, R. F. (1994) *Atoms in Molecules*. Clarendon, Oxford.

Baes, C. F. J. and Mesmer, R. E. (1976) *The Hydrolysis of Cations*. Wiley.

Bagno, P., Jepsen, O. and Gunnarsson, O. (1989) *Phys. Rev.* B **40**, 1997.

Bakasov, A. and Quack, M. (1999) *Chem. Phys. Lett.* **203**, 547–557.

Bakasov, A., Ha, T.-K. and Quack, M. (1998) *J. Chem. Phys.* **109**, 7263–7285.

Balasubramanian, B. and Pitzer, K. S. (1983) *J. Chem. Phys.* **78**, 2928.

Balasubramanian, K. (1988) *J. Chem. Phys.* **89**, 5731.

Balasubramanian, K. (1998) in Schleyer (1998), p. 2471.

Balian, R. and Werthamer, N. R. (1963) *Phys. Rev.* **131**, 1553.

Ballard, C. C., Hada, M., Kaneko, H. and Nakatsuji, H. (1996) *Chem. Phys. Lett.* **254**, 170–178.

Baltz, A. J. (1997) *Phys. Rev. Lett.* **78**, 1231.

Baltz, A. J. and McLerran, L. (1998) *Phys. Rev.* C **58**, 1679.

Baltz, A. J., Gelis, F., McLerran, L. and Peshier, A. (2001) Coulomb corrections to e^+e^--production in ultrarelativistic nuclear collisions (nucl-th/0101024).

Band, I. M. and Trzhaskovskaya, M. B. (1997) *J. Phys.* B **30**, 5185–5195.

Banhart, J. and Ebert, H. (1995) *Europhys. Lett.* **32**, 517.

Banhart, J., Ebert, H. and Vernes, A. (1997) *Phys. Rev.* B **56**, 10165.

Banhart, J., Ebert, H., Voitländer, J. and Winter, H. (1986) *J. Magn. Magn. Mater.* **61**, 221.

Banhart, J., Ebert, H., Weinberger, P. and Voitländer, J. (1994) *Phys. Rev.* B **50**, 2104.

Banhart, J., Vernes, A. and Ebert, H. (1996) *Solid State Commun.* **98**, 129.

Bansmann, J., Lu, L., Getzlaff, M., Fluchtmann, M. and Braun, J. (2000) *Surf. Sci.* **454–456**, 686–691.

Barandiarán, Z. and Seijo, L. (1992) *Can. J. Chem.* **70**, 409.

Barandiarán, Z. and Seijo, L. (1994) *J. Chem. Phys.* **101**, 4049.

Bardsley, J. (1974) *Case Stud. Atom. Phys.* **4**, 299.

Barnett, R. N. and Landman, U. (1993) *Phys. Rev.* B **48**, 2081.

Bartlett, N. (1998) *Gold Bull.* **31**, 22.

Barysz, M., Sadlej, A. J. and Snijders, J. G. (1997) *Int. J. Quant. Chem.* **65**, 225–239.

Bastug, T., Sepp, W.-D., Kolb, D., Fricke, B., Baerends, E. J. and Velde, G. T. (1995) *J. Phys.* B **28**, 2325–2331.

Bastug, T., Rashid, K., Sepp, W.-D., Kolb, D. and Fricke, B. (1997a) *Phys. Rev.* A **55**, 1750.

Bastug, T., Kürpick, P., Meyer, J., Sepp, W.-D., Fricke, B. and Rosén, A. (1997b) *Phys. Rev.* B **55**, 5015.

Battocletti, M. (1997) Relativistische Berechnung der Hyperfeinfelder in Übergangsmetallsystemen und Berücksichtigung von Orbitalpolarisationseffekten. PhD thesis, University of München.

Battocletti, M. and Ebert, H. (1996) Relativistic influences on the hyperfine fields of magnetic metals. In *Proc. 10th Int. Conf. on Hyperfine Interactions, Part II, Balzer, Basel* (ed. M. Rots, A. Vantomme, J. Dekoster, R. Coussement and G. Langouche), p. 5.

Baudelet, F., Odin, S., Giorgetti, C., Dartyge, E., Itie, J. P., Polian, A., Pizzini, S., Fontaine, A. and Kappler, J. P. (1997) *J. Phys. (Paris)* **7**, C2 441.

Bayler, A., Schier, A. and Schmidbaur, H. (1998) *Inorg. Chem.* **37**, 4353.

Bayram, S. B., Havey, M., Rosu, M., Sieradzan, A., Derevianko, A. and Johnson, W. R. (2000) *Phys. Rev.* A **61**, 050502–1–4.

Beck, D. R. and Datta, D. (1995) *Phys. Rev.* A **52**, 2436–2438.

Beck, D. R. and Norquist, P. L. (2000) *Phys. Rev.* A **61**, 044504–1–3.

Becke, A. (1992) *J. Chem. Phys.* **96**, 2155.

Becke, A. D. (1988a) *Phys. Rev.* A **38**, 3098.

Becke, A. D. (1988b) *J. Chem. Phys.* **88**, 2547.

Becke, A. D. and Dickinson, R. M. (1988) *J. Chem. Phys.* **89**, 2993.

Becker, U., Grün, N. and Scheid, W. (1986) *J. Phys.* B **19**, 1347.

Becker, U., Grün, N., Momberger, K. and Scheid, W. (1987) *Physics of Strong Fields.* NATO ASI Series B: Physics vol. 153, p. 609. Plenum, New York.

bei der Kellen, S. and Freeman, A. J. (1996) *Phys. Rev.* B **54**, 11 187–11 198.

Beier, T. (2000) *Phys. Rep.* **339**, 79.

Beier, T. and Soff, G. (1988) *Z. Phys.* D **8**, 129.

Beier, T., Mohr, P. J., Persson, H., Plunien, G., Greiner, M. and Soff, G. (1997a) *Phys. Lett.* A **236**, 329.

Beier, T., Plunien, G., Greiner, M. and Soff, G. (1997b) *J. Phys.* B **30**, 2761.

Beier, T., Mohr, P. J., Persson, H. and Soff, G. (1998) *Phys. Rev.* A **58**, 954.

Beiersdorfer, P., Osterheld, A. L., Scofield, J. H., López-Urrutia, J. R. C. and Widmann, K. (1998) *Phys. Rev. Lett.* **80**, 3022.

Beitel, F. P. and Pugh, E. M. (1958) *Phys. Rev.* **112**, 1516.

Belkacem, A., Gould, H., Feinberg, B., Bossingham, R. and Meyerhof, W. E. (1993) *Phys. Rev. Lett.* **71**, 1514.

Belkacem, A., Gould, H., Feinberg, B., Bossingham, R. and Meyerhof, W. E. (1997) *Phys. Rev.* A **56**, 2806.

Belling, T., Grauschopf, T., Krüger, S., Mayer, M., Nörtemann, F., Staufer, M., Zenger, C. and Rösch, N. (1999a) Quantum chemistry on parallel computers: Concepts and results of a density functional method. In *High Performance Scientific and Engineering Computing, Proc. 1st Int. FORTWIHR Conf., Munich, 1998* (ed. H.-J. Bungartz, F. Durst and C. Zenger), pp. 439–453. Springer.

Belling, T., Grauschopf, T., Krüger, S., Nörtemann, F., Staufer, M., Mayer, M., Nasluzov, V. A., Birkenheuer, U., Hu, A., Mateev, A. V. and Rösch, N. (1999b) ParaGauss V 2.1.

Benkowitsch, J. and Winter (1983) *J. Phys.* F **13**, 991.

Bennett, S. C. and Wieman, C. E. (1999) *Phys. Rev. Lett.* **82**, 2484–2487.

Berger, R. and Quack, M. (2000a) *J. Chem. Phys.* **112**, 57–60.

Berger, R. and Quack, M. (2000b) *Chem. Phys. Chem.* **1**, 3148–3158.

Berger, R., Quack, M. and Tschumper, G. S. (2000) *Helv. Chim. Acta* **83**, 1919–1950.

Bergner, A., Dolg, M., Küchle, W., Stoll, H. and Preuss, H. (1993) *Mol. Phys.* **80**, 1431.

Berning, A., Schweizer, M., Werner, H.-J., Knowles, P. J. and Palmieri, P. (2000) *Mol. Phys.* **98**, 1823–1833.

Bertulani, C. A. and Baur, G. (1988) *Phys. Rep.* **163**, 299.

Best, C., Greiner, W. and Soff, G. (1992) *Phys. Rev.* A **46**, 261.

Bethe, H. and Salpeter, E. (1957) *Quantum Mechanics of One- and Two-Electron Atoms.* Springer.

Beyer, H. F. (1995) *IEEE Trans. Instrum. Meas.* **IM44**, 510.

Beyer, H. F., Liesen, D., Bosch, F., Finlayson, K. D., Jung, M., Klepper, O., Moshammer, R., Beckert, K., Eickhoff, H., Franzke, B., Nolden, F., Spädtke, P., Steck, M., Menzel, G. and Deslattes, R. D. (1994) *Phys. Lett.* A **184**, 435.

Biémont, E., Froese Fischer, C., Godefroid, M. R., Palmeri, P. and Quinet, P. (2000) *Phys. Rev.* A **62**, 032512.

Bieroń, J. (1999) *Phys. Rev.* A **59**, 4295–4299.

Bieroń, J. and Grant, I. P. (1998) *Phys. Rev.* A **58**, 4401–4405.

Bieroń, J., Froese Fischer, C. and Ynnerman, A. (1994) *J. Phys.* B **27**, 4829–4834.

Bieroń, J., Jönsson, P. and Froese Fischer, C. (1996) *Phys. Rev.* A **53**, 2181–2188.

Bieroń, J., Grant, I. P. and Froese Fischer, C. (1997) *Phys. Rev.* A **56**, 316–321.

Bieroń, J., Jönsson, P. and Froese Fischer, C. (1999) *Phys. Rev.* A **60**, 3547–3557.

Bieroń, J., Parpia, F. A., Froese Fischer, C. and Jönsson, P. (1995) *Phys. Rev.* A **51**, 4603–4610.

Blaha, P., Schwarz, K., Dufek, P. and Augustyn, R. (1995) WIEN95, Technical University of Vienna. (Improved and updated Unix version of the original copyrighted WIEN-code, by P. Blaha, K. Schwarz, P. Sorantin and S. B. Trickey. *Comput. Phys. Commun.* **59** (1990), 399.)

Blasco, A. J., Plaja, L., Roso, L. and Faisal, F. H. M. (2001) *Phys. Rev.* E **64**, 026505.

Blase, A. (2001) PhD thesis, Universität Bielefeld.

Blaudeau, J.-P. and Curtiss, L. A. (1997) *Int. J. Quant. Chem.* **61**, 943.

Blügel, S. and Dederichs, P. (1989) *Europhys. Lett.* **9**, 597.

Blügel, S., Drittler, B., Zeller, R. and Dederichs, P. (1989) *Appl. Phys.* A **49**, 547.

Blundell, S. (1993) *Phys. Rev.* A **47**, 1790.

Blundell, S. A. and Snyderman, N. J. (1991) *Phys. Rev.* A **44**, 1427.

Blundell, S., Mohr, P., Johnson, W. R. and Sapirstein, J. (1993) *Phys. Rev.* A **48**, 2615.

Blundell, S. A., Cheng, K. T. and Sapirstein, J. (1997) *Phys. Rev.* A **55**, 1857.

Bodwin, G. T. and Yennie, D. R. (1988) *Phys. Rev.* D **37**, 498.

Boerrigter, P. M., Velde, G. T. and Baerends, E. J. (1988) *Int. J. Quant. Chem.* **33**, 87.

Boettger, J. C. (1998a) *Phys. Rev.* B **57**, 8743–8746.

Boettger, J. C. (1998b) *Phys. Rev.* B **57**, 8743–8746.

Bohr, A. (1951) *Phys. Rev.* **81**, 331.

Bohr, A. and Weisskopf, V. F. (1950) *Phys. Rev.* **77**, 94.

Bonner, W. A. (2000) *Chirality* **12**, 114–126.

Borneis, S., Dax, A., Engel, T., Holbrow, C., Huber, G., Kühl, T., Marx, D., Merz, P., Quint, W., Schmitt, F., Seelig, P., Tomaselli, M., Beckert, H. W. K., Franzke, B., Nolden, F., Reich, H. and Steck, M. (2000) *Hyperfine Interactions* **127**, 305.

Bosselmann, P., Staude, U., Horn, D., Schartner, K.-H., F. Folkmann, F., Livingston, A. E. and Mokler, P. H. (1999) *Phys. Rev.* A **59**, 1874.

Boucard, S. and Indelicato, P. (2000) *Eur. Phys. J.* D **8**, 59–73.

Bouchoux, G., Salpin, J. Y. and Leblanc, D. (1996) *Int. J. Mass Spectrom. Ion Processes* **153**, 37.

Bouchoux, G., Yáñez, M. and Mó, O. (1999) *Int. J. Mass Spectrom.* **185–187**, 241.

Brako, R. and Şokçević, D. (1998) *Surf. Sci.* **401**, L388.

Braun, J. (1996) *Rep. Prog. Phys.* **59**, 1267.

Breit, G. (1928) *Nature* **122**, 649.

Brewer, L. (1977) Technical Report LBL-3720 Rev., Lawrence Berkeley Laboratory, University of California, Berkeley.

Brönstrup, M., Schröder, D. and Schwarz, H. (1999) *Organometallics* **18**, 1939.

Brönstrup, M., Schröder, D., Kretzschmar, I., Schwarz, H. and Harvey, J. N. (2001) *J. Am. Chem. Soc.* **123**, 142.

Brooks, M. S. S. (1985) *Physica* B **130**, 6.

Brooks, M. S. S. and Kelly, P. J. (1983) *Phys. Rev. Lett.* **51**, 1708.

Brown, G. E., Langer, J. S. and Schaefer, G. W. (1959) *Proc. R. Soc. Lond.* A **251**, 92.

Brüchle, W., Schausten, B., Jäger, E., Schimpf, E., Schädel, M., Kratz, J. V., Trautmann, N., Zimmermann, H. P., Bruchertseifer, H. and Heller, N. (1992) GSI scientific report 1991, p. 315.

Brummelhuis, R., Röhrl, N. and Siedentop, H. (2001) *Doc. Math.* **6**, 1.

Brummelhuis, R., Siedentop, H. and Stockmeyer, E. (2002) *Doc. Math.* **7**, 167–182.

Bruna, P. J., Lushington, G. H. and Grein, F. (1997) *Chem. Phys.* **225**, 1.

Bruno, P. (1989) *Phys. Rev.* B **39**, 865.

Bruno, P. (1993) Magnetismus von Festkörpern und Grenzflächen. In *IFF-Ferienkurs* (ed. F. Jülich), p. 24.1.

Bucksbaum, P. H., Bashkansky, M. and McIlrath, T. J. (1987) *Phys. Rev. Lett.* **58**, 349.

Buenker, R. J., Alekseyev, A. B., Liebermann, H.-P., Lingott, R. M. and Hirsch, G. (1998) *J. Chem. Phys.* **108**, 3400–3408.

Buenker, R. J. and Peyerimhoff, S. D. (1974) *Theor. Chim. Acta* **35**, 33.

Bühl, M., Kaupp, M., Malkina, O. L. and Malkin, V. G. (1999) *J. Chem. Phys.* **110**, 3897–3902.

Bula, C., McDonald, K. T., Prebys, E. J., Bamber, C., Boege, S., Kotseroglou, T., Melissinos, A. C., Meyerhofer, D. D., Ragg, W., L.Burke, D., Field, R. C., Horton-Smith, G., Odian, A. C., Spencer, J. E., Walz, D., Berridge, S. C., Bugg, W. M., Shmakov, K. and Weidemann, A. W. (1996) *Phys. Rev. Lett.* **76**, 3116.

Bulliard, C., Allan, M., Smith, J. M., Hrovat, D. A., Borden, W. T. and Grimme, S. (1998) *Chem. Phys.* **225**, 153.

Busic, O., Grün, N. and Scheid, W. (1999a) *Physica Scr.* T **80**, 432.

Busic, O., Grün, N. and Scheid, W. (1999b) *Phys. Lett.* A **254**, 337.

Butler, W. H. (1985) *Phys. Rev.* B **31**, 3260.

Butler, W. H., Zhang, X. G., Nicholson, D. M. C. and MacLaren, J. M. (1995) *J. Magn. Magn. Mater.* **151**, 354.

Bylander, D. M. and Kleinman, L. (1995a) *Phys. Rev. Lett.* **74**, 3660.

Bylander, D. M. and Kleinman, L. (1995b) *Phys. Rev.* B **52**, 14 566.

Bylander, D. M. and Kleinman, L. (1996) *Phys. Rev.* B **54**, 7891.

Bylander, D. M. and Kleinman, L. (1997) *Phys. Rev.* B **55**, 9432.

Cabrera, B. and Peskin, M. E. (1989) *Phys. Rev.* B **39**, 6425.

Cabria, I., Deng, M. and Ebert, H. (2000) *Phys. Rev.* B **62**, 14 287.

Cabria, I., Perlov, A. and Ebert, H. (2001) *Phys. Rev.* B **63**, 104424.

Cabria, I., Nonas, B., Zeller, R. and Dederichs, P. (2002) *Phys. Rev.* B **65**, 054414.

Caletka, R. and Krivan, V. (1990) *J. Radioanal. Nucl. Chem.* **142**, 239–371.

Callaway, J. and Wang, C. S. (1973) *Phys. Rev.* B **7**, 1096.

Canal Neto, A., Librelon, P. R. and Jorge, F. E. (2000) *Chem. Phys. Lett.* **326**, 501–508.

Cao, X., Liao, M., Chen, X. and Li, B. (1998) *J. Comp. Chem.* **17**, 851–863.

Capelle, K. (2001) *Phys. Rev.* B **63**, 52 503.

Capelle, K. and Gross, E. K. U. (1995) *Phys. Lett.* A **198**, 261.

Capelle, K. and Gross, E. K. U. (1997a) *Phys. Rev. Lett.* **78**, 1872.

Capelle, K. and Gross, E. K. U. (1997b) *Int. J. Quant. Chem.* **61**, 325.

Capelle, K. and Gross, E. K. U. (1999a) *Phys. Rev.* B **59**, 7140.

Capelle, K. and Gross, E. K. U. (1999b) *Phys. Rev.* B **59**, 7155.

Capelle, K., Gross, E. K. U. and Gyorffy, B. L. (1997) *Phys. Rev. Lett.* **78**, 1872.

Capelle, K., Gross, E. K. U. and Gyorffy, B. L. (1998) *Phys. Rev.* B **58**, 473.

Capelle, K., Marques, M. and Gross, E. K. U. (2001) In *Condensed Matter Theories* (ed. S. Hernandez), vol. 16, Nova.

Carra, P., Thole, B. T., Altarelli, M. and Wang, X. (1993) *Phys. Rev. Lett.* **70**, 694.

Casarubios, M. and Seijo, L. (1998) *J. Mol. Struct.* **426**, 59.

Casarubios, M. and Seijo, L. (1999) *J. Chem. Phys.* **110**, 784.

Castro, M. and Salahub, D. R. (1994) *Phys. Rev.* B **49**, 11 842.

Cauchois, Y. and Manescu, I. (1940) *C. R. Acad. Sci.* **210**, 172.

Chaix, P. and Iracane, D. (1989) *J. Phys.* B **22**, 3791.

Chaix, P., Iracane, D. and Lions, P. L. (1989) *J. Phys.* B **22**, 3815.

Chang, A. and Pitzer, R. M. (1989) *J. Am. Chem. Soc.* **111**, 2500.

Chang, C., Pélissier, M. and Durand, P. (1986) *Physica Scr.* **34**, 394–404.

Chelikowsky, J. and Cohen, M. (1992) In *Handbook on Semiconductors* (ed. P. Landsberg), vol. 1, p. 59. Elsevier, Amsterdam.

Chen, J., Krieger, J. B., Li, Y. and Iafrate, G. J. (1996) *Phys. Rev.* A **54**, 3939.

Chen, M. H., Cheng, K. T. and Johnson, W. R. (1993) *Phys. Rev.* A **47**, 3692–3703.

Chen, M. H., Cheng, K. T., Johnson, W. R. and Sapirstein, J. (1995) *Phys. Rev.* A **52**, 266.

Chen, Z. and Goldman, S. P. (1993) *Phys. Rev.* A **48**, 1107–1113.

Chen, Z., Fonte, G. and Goldman, S. P. (1994) *Phys. Rev.* A **50**, 3838–3844.

Cheng, K. T., Johnson, W. R. and Sapirstein, J. (1991) *Phys. Rev. Lett.* **66**, 2960.

Cheng, K. T., Chen, M. H. and Sapirstein, J. (2000) *Phys. Rev.* A **62**, 054501.

Chernysheva, L. V. and Yakhontov, V. L. (1999) *Comp. Phys. Commun.* **119**, 232–255.

Chetty, N., Weinert, M., Rahman, T. S. and Davenport, J. W. (1995) *Phys. Rev.* B **52**, 6313.

Cheung, A. S. C., Gordon, R. M. and Merer, A. J. (1981) *J. Mol. Spectrosc.* **87**, 289.

Chevary, J. A. and Vosko, S. H. (1990) *Phys. Rev.* B **42**, 5320.

Chou, H.-S. (2000) *Phys. Rev.* A **62**, 042507.

Chowdhury, A. K. and Wilkins, C. L. (1987) *J. Am. Chem. Soc.* **109**, 5336.

Chriplovic, I. B. (1991) *Parity Nonconservation in Atomic Phenomena*. Gordon and Breach.

Christensen, N. E. and Seraphin, B. O. (1971) *Phys. Rev.* B **4**, 3321.

Christiansen, O., Gauss, J. and Schimmelpfennig, B. (2000) *Phys. Chem. Chem. Phys.* **2**, 965–971.

Christiansen, P. A. (1983) *J. Chem. Phys.* **79**, 2928.

Christiansen, P. A. (1984) *Chem. Phys. Lett.* **109**, 145.

Christiansen, P. A., Lee, Y. S. and Pitzer, K. S. (1979) *J. Chem. Phys.* **71**, 4445.

Christiansen, P., Ermler, W. and Pitzer, K. (1988) *Adv. Quant. Chem.* **19**, 139.

Chung, S., Krüger, S., Paccioni, G. and Rösch, N. (1995) *J. Chem. Phys.* **102**, 3695–3702.

Chung, S., Krüger, S., Ruzankin, S. P., Paccioni, G. and Rösch, N. (1996a) *Chem. Phys. Lett.* **248**, 109–115.

Chung, S., Krüger, S., Schmidbaur, H. and Rösch, N. (1996b) *Inorg. Chem.* **35**, 5387–5392.

Clementi, E. and Corongiu, G. (eds) (1995) *Methods and Techniques in Computational Chemistry METECC-95*. STEF, Cagliari.

Clementi, E. (ed.) (1993) *Methods and Techniques in Computational Chemistry: METECC-94*, vol. A. STEF, Cagliari.

Colle, R. and Salvetti, O. (1975) *Theoret. Chim. Acta (Berlin)* **37**, 329.

Collins, C. L., Dyall, K. G. and Schaefer III, H. F. (1995) *J. Chem. Phys.* **102**, 2024–2031.

Connerade, J.-P. and Keitel, C. H. (1996) *Phys. Rev.* A **53**, 2748.

Corkum, P. B., Burnett, N. H. and Brunel, F. (1992) In *Atoms in Intense Laser Fields* (ed. M. Gavirila). Academic.

Cornehl, H. H., Heinemann, C., Schröder, D. and Schwarz, H. (1995) *Organometallics* **14**, 992.

Cornehl, H. H., Heinemann, C., Marçalo, J., de Matos, A. P. and Schwarz, H. (1996a) *Angew. Chem.* **108**, 950.

Cornehl, H. H., Heinemann, C., Marçalo, J., de Matos, A. P. and Schwarz, H. (1996b) *Angew. Chem. Int. Ed. Engl.* **35**, 891.

Cornehl, H. H., Hornung, G. and Schwarz, H. (1996c) *J. Am. Chem. Soc.* **118**, 9960.

Cornehl, H. H., Wesendrup, R., Diefenbach, M. and Schwarz, H. (1997a) *Chem. Eur. J.* **3**, 1083.

Cornehl, H. H., Wesendrup, R., Harvey, J. N. and Schwarz, H. (1997b) *J. Chem. Soc., Perkin Trans. 2*, p. 2283.

Cortona, P. (1989) *Phys. Rev.* A **40**, 12 105.

Cortona, P., Doniach, S. and Sommers, C. (1985) *Phys. Rev.* A **31**, 2842.

Cowan, R. D. and Griffin, D. C. (1976) *J. Opt. Soc. Am.* **66**, 1010.

Cracknell, A. P. (1970) *Phys. Rev.* B **1**, 1261.

Cundari, T., Benson, M., Lutz, M. and Sommerer, S. (1996) *Rev. Comp. Chem.* **8**, 145.

Cundari, T. R. and Stevens, W. J. (1993) *J. Chem. Phys.* **98**, 5555.

Dahl, J. P. (1977) *Mat. Fys. Medd. Dan. Vid. Selsk.* **30**, 12.

Danovich, D., Marian, C. M., Neuheuser, T., Peyerimhoff, S. D. and Shaik, S. (1998) *J. Phys. Chem.* A **102**, 5923–5936.

Dargel, T., Hertwig, R. and Koch, W. (1999) *Mol. Phys.* **96**, 583.

Darwin, C. G. (1928) *Proc. R. Soc. Lond.* A **118**, 654–680.

Das, M. P., Ramana, M. V. and Rajagopal, A. K. (1980) *Phys. Rev.* A **22**, 9.

Da Silva, A. B. F., Malli, G. L. and Ishikawa, Y. (1993) *Chem. Phys. Lett.* **203**, 201–203.

Datta, D. and Beck, D. R. (1995) *Phys. Rev.* A **52**, 3622–3627.

Davidson, E. R., Hagstrom, S. A., Chakravorty, S. J., Umar, V. M. and Fischer, C. F. (1991) *Phys. Rev.* A **44**, 7071.

Davis, H., Bethe, H. A. and Maximon, L. C. (1954) *Phys. Rev.* **93**, 788.

de Gennes, P. G. (1966) *Superconductivity of Metals and Alloys*. Addison-Wesley, Reading.

de Jong, W. A. and Nieuwpoort, W. C. (1998) *Int. J. Quant. Chem.* **58**, 203–216.

de Jong, W. A., Styszynski, J., Visscher, L. and Nieuwpoort, W. C. (1998) *J. Chem. Soc.* **108**, 5177–5184.

de Jong, W. A., Visscher, L. and Nieuwpoort, W. C. (1997) *J. Chem. Phys.* **107**, 9046–9058.

de Jong, W. A., Visscher, L. and Nieuwpoort, W. C. (1999) *J. Mol. Struct. (THEOCHEM)* **458**, 41–52.

Deng, M. (2001) PhD thesis, University of Munich.

Deng, M., Freyer, H. and Ebert, H. (2000) *Solid State Commun.* **114**, 365.

Desclaux, J. P. (1973) *At. Data Nucl. Data Tables* **12**, 311–406.

Desclaux, J. P. (1975) *Comp. Phys. Commun.* **9**, 31–45.

Desclaux, J. P. (1993) In Clementi (1993), pp. 253–274.

Desclaux, J. P. and Pyykkö, P. (1974) *Chem. Phys. Lett.* **29**, 534–539.

Desiderio, A. M. and Johnson, W. R. (1971) *Phys. Rev.* A **3**, 1267.

Díaz-Megías, S. and Seijo, L. (1999) *Chem. Phys. Lett.* **299**, 613.

Diefenbach, M., Brönstrup, M., Aschi, M., Schröder, D. and Schwarz, H. (1999) *J. Am. Chem. Soc.* **121**, 10 614.

DiLabio, G. A. and Christiansen, P. A. (1998) *J. Chem. Phys.* **108**, 7527–7533.

Dirac, P. A. M. (1928) *Proc. R. Soc. Lond.* A **117**, 610–624.

Dirac, P. A. M. (1929) *Proc. R. Soc. Lond.* A **123**, 714–733.

Dixon, R. N. and Robertson, I. L. (1978) *Spec. Period. Rep., Theor. Chem., The Chemical Society, London* **3**, 100.

Dolbeault, J., Esteban, M. J. and Séré, E. (2000a) *J. Funct. Analysis* **174**, 208–226.

Dolbeault, J., Esteban, M. J. and Séré, E. (2000b) Variational methods in relativistic quantum mechanics: new approach to the computation of Dirac eigenvalues. In *Mathematical Models and Methods for Ab Initio Quantum Chemistry* (ed. M. Defranceschi and C. Le Bris), pp. 211–226. Springer.

Dolbeault, J., Esteban, M. J., Séré, E. and Vanbreugel, M. (2000c) *Phys. Rev. Lett.* **85**, 4020–4023.

Dolg, M. (1996a) *Theor. Chim. Acta* **93**, 141–156.

Dolg, M. (1996b) *Mol. Phys.* **88**, 1645–1655.

Dolg, M. (1996c) *J. Phys. Chem.* **104**, 4061–4067.

Dolg, M. (1996d) *Chem. Phys. Lett.* **250**, 75–79.

Dolg, M. (2000) In Grotendorst (2000), pp. 507 – 540.

Dolg, M. and Flad, H.-J. (1996a) *J. Phys. Chem.* **100**, 6147–6151.

Dolg, M. and Flad, H.-J. (1996b) *J. Phys. Chem.* **100**, 6147–6151.

Dolg, M. and Stoll, H. (1995) In *Handbook on the Physics and Chemistry of Rare Earths* (ed. K. A. Gschneidner Jr and L. Eyring), vol. 22. Elsevier.

Dolg, M., Wedig, U., Stoll, H. and Preuss, H. (1987) *J. Chem. Phys.* **86**, 866.

Dolg, M., Stoll, H. and Preuss, H. (1989a) *J. Chem. Phys.* **90**, 1730.

Dolg, M., Stoll, H., Savin, A. and Preuss, H. (1989b) *Theor. Chim. Acta* **75**, 173.

Dolg, M., Stoll, H. and Preuss, H. (1993a) *Theor. Chim. Acta* **85**, 441.

Dolg, M., Stoll, H., Preuss, H. and Pitzer, R. M. (1993b) *J. Phys. Chem.* **97**, 5852.

Dolg, M., Stoll, H., Seth, M. and Schwerdtfeger, P. (2001) *Chem. Phys. Lett.* **345**, 490.

Doniach, S. and Sommers, C. (1981) In *Valence Fluctuations in Solids* (ed. L. M. Falicov, W. Hanke and M. B. Maple), p. 349. North-Holland, Amsterdam.

Douglas, M. and Kroll, N. M. (1974) *Ann. Phys.* **82**, 89.

Dreizler, R. M. and Gross, E. K. U. (1990) *Density Functional Theory*. Springer.

Dufek, P., Blaha, P. and Schwarz, K. (1994) *Phys. Rev.* B **50**, 7279.

Durand, P. and Barthelat, J. C. (1975) *Theor. Chim. Acta* **38**, 283.

Düsterhöft, C., Yang, L., Heinemann, D. and Kolb, D. (1994) *Chem. Phys. Lett.* **229**, 667–670.

Düsterhöft, C., Heinemann, D. and Kolb, D. (1998) *Chem. Phys. Lett.* **296**, 77–83.

Dyall, K. G. (1992) *J. Chem. Phys.* **96**, 1210–1217.

Dyall, K. G. (1993a) *J. Chem. Phys.* **98**, 9678–9686.

Dyall, K. G. (1993b) *J. Chem. Phys.* **98**, 2191–2197.

Dyall, K. G. (1994a) *Chem. Phys. Lett.* **224**, 186–194.

Dyall, K. G. (1994b) *J. Chem. Phys.* **100**, 2118–2127.

Dyall, K. G. (1994c) In Malli (1994), pp. 17–58.

Dyall, K. G. (1999) *Mol. Phys.* **96**, 511–518.

Dyall, K. G. and Enevoldsen, T. (1999) *J. Chem. Phys.* **111**, 10 000–10 007.

Dyall, K. G. and Fægri Jr, K. (1996) *Theor. Chim. Acta* **94**, 39–51.

Dyall, K. G., Grant, I. P. and Wilson, S. (1984) *J. Phys.* B **17**, 493–503.

Dyall, K. G., Grant, I. P., Johnson, C. T., Parpia, F. A. and Plummer, E. P. (1989) *Comp. Phys. Commun.* **55**, 425–456.

Dyall, K. G., Fægri, K. and Taylor, P. R. (1991a) Polyatomic Molecular Dirac–Hartree–Fock Calculations with Gaussian Basis Sets. In Wilson *et al.* (1991), pp. 167–184.

Dyall, K. G., Taylor, P. R., Fægri Jr, K. and Partridge, H. (1991b) *J. Chem. Phys.* **95**, 2583–2594.

Dyck, R. S. V., Schwinberg, P. B. and Dehmelt, H. G. (1987) *Phys. Rev. Lett.* **59**, 26.

Eberly, J. H. (1969) *Progress in Optics* (ed. E. Wolf). North-Holland, Amsterdam.

Ebert, H. (1988) *Phys. Rev.* B **38**, 9390.

Ebert, H. (1989) *J. Phys.: Condens. Matter* **1**, 9111.

Ebert, H. (1995) (Unpublished.)

Ebert, H. (1996) *Rep. Prog. Phys.* **59**, 1665.

Ebert, H. (2000) Fully relativistic band structure calculations for magnetic solids—formalism and application. In *Electronic Structure and Physical Properties of Solids* (ed. H. Dreyssé), p. 191. Lecture Notes in Physics, vol. 535. Springer.

Ebert, H. and Akai, H. (1993) *Int. J. Mod. Phys.* B **7**, 922.

Ebert, H. and Battocletti, M. (1996) *Solid State Commun.* **98**, 785.

Ebert, H. and Gyorffy, B. L. (1988) *J. Phys.* F **18**, 451.

Ebert, H. and Schütz, G. (eds) (1996) *Spin–Orbit Influenced Spectroscopies of Magnetic Solids.* Lecture Notes in Physics, vol. 466. Springer.

Ebert, H., Abart, J. and Voitländer, J. (1984) *J. Phys.* F **14**, 749.

Ebert, H., Strange, P. and Gyorffy, B. L. (1988a) *J. Phys.* F **18**, L135.

Ebert, H., Strange, P. and Gyorffy, B. L. (1988b) *J. Appl. Phys.* **63**, 3055.

Ebert, H., Strange, P. and Gyorffy, B. L. (1988c) *Z. Phys.* B **73**, 77.

Ebert, H., Zeller, R., Drittler, B. and Dederichs, P. H. (1990) *J. Appl. Phys.* **67**, 4576.

Ebert, H., Akai, H., Maruyama, H., Koizumi, A., Yamazaki, H. and Schütz, G. (1993) *Int. J. Mod. Phys.* B **7**, 750.

Ebert, H., Freyer, H., Vernes, A. and Guo, G.-Y. (1996a) *Phys. Rev.* B **53**, 7721.

Ebert, H., Vernes, A. and Banhart, J. (1996b) *Phys. Rev.* B **54**, 8479.

Ebert, H., Battocletti, M. and Gross, E. K. U. (1997a) *Europhys. Lett.* **40**, 545.

Ebert, H., Freyer, H. and Deng, M. (1997b) *Phys. Rev.* B **56**, 9454.

Ebert, H., Vernes, A. and Banhart, J. (1997c) *Solid State Commun.* **104**, 243.

Ebert, H., Popescu, V. and Ahlers, D. (1999) *Phys. Rev.* B **60**, 7156.

Ebert, H., Minár, J., Popescu, V., Sandratskii, L. M. and Mavromaras, A. (2000) *AIP Conf. Proc.* **514**, 110.

Eichinger, M., Tavan, P., Hutter, J. and Parinello, M. (1999) *J. Chem. Phys.* **110**, 10 452.

Eichler, J. and Meyerhof, W. E. (1995) *Relativistic Atomic Collisions*, Academic Press, San Diego.

Eichler, R., Brüchle, W., Dressler, R., Düllmann, C. E., Eichler, B., Gäggeler, H. W., Gregorich, K. E., Hoffman, D. C., Hübener, S., Jost, D. T., Kirbach, U., Laue, C. A., Lavanchy, V. M., Nitsche, H., Patin, J. B., Piguet, D., Schädel, M., Shaughnessy, A., Strellis, D. A., Taut, S., Tobler, L., Tsyganov, Y., Türler, A., Vahle, A., Wilk, P. A. and Yakushev, A. B. (2000) *Nature* **407**, 63.

Eichmann, U. (2000) PhD Thesis, JW Goethe-Universität Frankfurt.

Eichmann, U., Reinhardt, J., Schramm, S. and Greiner, W. (1999) *Phys. Rev.* A **59**, 1223.

Eichmann, U., Reinhardt, J. and Greiner, W. (2000a) *Phys. Rev.* A **61**, 62 710.

Eichmann, U., Reinhardt, J. and Greiner, W. (2000b) *Phys. Rev.* C **61**, 64 901.

Eliav, E. and Kaldor, U. (1996) *Chem. Phys. Lett.* **248**, 405–408.

Eliav, E., Kaldor, U. and Ishikawa, Y. (1994a) *Int. J. Quant. Chem. Quant. Chem. Symp.* **28**, 205–214.

Eliav, E., Kaldor, U. and Ishikawa, Y. (1994b) *Phys. Rev.* A **49**, 1724–1729.

Eliav, E., Kaldor, U., Schwerdtfeger, P., Hess, B. A. and Ishikawa, Y. (1994c) *Phys. Rev. Lett.* **73**, 3203–3206.

Eliav, E., Kaldor, U. and Ishikawa, Y. (1995) *Phys. Rev. Lett.* **74**, 1079–1082.

Eliav, E., Kaldor, U. and Ishikawa, Y. (1996) *Phys. Rev.* A **53**, 3050–3056.

Eliav, E., Ishikawa, Y., Pyykkö, P. and Kaldor, U. (1997) *Phys. Rev.* A **56**, 4532–4536.

Eliav, E., Kaldor, U. and Hess, B. A. (1998a) *J. Chem. Phys.* **108**, 3409–3415.

Eliav, E., Kaldor, U. and Hess, B. A. (1998b) *J. Chem. Phys.* **108**, 3409–3415.

Eliav, E., Shmulyian, S., Kaldor, U. and Ishikawa, Y. (1998c) *J. Chem. Phys.* **109**, 3954–3958.

Eliav (Ilyabaev), E., Kaldor, U. and Ishikawa, Y. (1994) *Chem. Phys. Lett.* **222**, 82–87.

Ellingsen, K., Matila, T., Saue, T., Aksela, H. and Gropen, O. (2000) *Phys. Rev.* A **62**, 032502–1–6.

Enevoldsen, T., Visscher, L., Saue, T., Jensen, H. J. A. and Oddershede, J. (2000) *J. Chem. Phys.* **112**, 3493–3498.

Engel, E. and Dreizler, R. M. (1996) In *Density Functional Theory II* (ed. R. F. Nalewajski), p. 1. Topics in Current Chemistry, vol. 181. Springer.

Engel, E. and Dreizler, R. M. (1999) *J. Comput. Chem.* **20**, 31.

Engel, E. and Facco Bonetti, A. (2000) In *Quantum Systems in Theoretical Chemistry and Physics* (ed. A. Hernández-Laguna, J. Maruani, R. McWeeny and S. Wilson), p. 227, vol. 1 of *Basic Problems and Model Systems*. Kluwer, Dordrecht.

Engel, E. and Vosko, S. H. (1990) *Phys. Rev.* B **42**, 4940.

Engel, E. and Vosko, S. H. (1993) *Phys. Rev.* B **47**, 13 164.

Engel, E., Keller, S., Facco Bonetti, A., Müller, H. and Dreizler, R. M. (1995a) *Phys. Rev.* A **52**, 2750.

Engel, E., Müller, H., Speicher, C. and Dreizler, R. M. (1995b) In *Density Functional Theory* (ed. E. K. U. Gross and R. M. Dreizler), p. 65. NATO ASI Series B, vol. 337. Plenum, New York.

Engel, E., Keller, S. and Dreizler, R. M. (1996) *Phys. Rev.* A **53**, 1367.

Engel, E., Facco Bonetti, A., Keller, S., Andrejkovics, I. and Dreizler, R. M. (1998a) *Phys. Rev.* A **58**, 964.

Engel, E., Keller, S. and Dreizler, R. M. (1998b) In *Electronic Density Functional Theory: Recent Progress and New Directions* (ed. J. F. Dobson, G. Vignale and M. P. Das), p. 149. Plenum, New York.

Engel, E., Höck, A. and Dreizler, R. M. (2000a) *Phys. Rev.* A **61**, 032502.

Engel, E., Höck, A. and Dreizler, R. M. (2000b) *Phys. Rev.* A **62**, 042502.

Engel, E., Auth, T. and Dreizler, R. M. (2001a) *Phys. Rev.* B **64**, 235126.

Engel, E., Höck, A. and Varga, S. (2001b) *Phys. Rev.* B **63**, 125121.

Engel, E., Höck, A., Schmid, R. N., Dreizler, R. M. and Chetty, N. (2001c) *Phys. Rev.* B **64**, 125111.

Engström, M., Minaev, B., Vahtras, O. and Ågren, H. (1998) *Chem. Phys.* **237**, 149.

Eriksson, O., Johansson, B. and Brooks, M. S. S. (1989) *J. Phys. Cond. Matt.* **1**, 4005.

Eriksson, O., Brooks, M. S. S. and Johansson, B. (1990a) *Phys. Rev.* B **41**, 7311.

Eriksson, O., Johansson, B., Albers, R. C., Boring, A. M. and Brooks, M. S. S. (1990b) *Phys. Rev.* B **42**, 2707.

Eriksson, O., Nordström, L., Pohl, A., Severin, L., Boring, A. M. and Johansson, B. (1990c) *Phys. Rev.* B **41**, 11 807.

Eriksson, O., Albers, R. C. and Boring, A. M. (1991) *Phys. Rev. Lett.* **66**, 1350.

Ermler, W., Ross, R. and Christiansen, P. (1985) *Ann. Rev. Phys. Chem.* **36**, 407.

Ermler, W. C., Ross, R. B. and Christiansen, P. A. (1991) *Int. J. Quant. Chem.* **40**, 829.

Erskine, J. L. and Stern, E. A. (1975) *Phys. Rev.* B **12**, 5016.

Ertl, G., Neumann, M. and Streit, K. M. (1977) *Surf. Sci.* **64**, 393.

Eschrig, H. (1996) *The Fundamentals of Density Functional Theory*. B. G. Teubner Verlagsgesellschaft, Stuttgart, Leipzig.

Eschrig, H. and Servedio, V. D. P. (1999) *J. Comp. Chem.* **20**, 23.

Eschrig, H., Seifert, G. and Ziesche, P. (1985) *Solid State Commun.* **56**, 777.

Essen, L., Donaldson, R. W., Bangham, M. J. and Hope, E. G. (1971) *Nature* **229**, 110.

Esser, M. (1984a) *Chem. Phys. Lett.* **111**(1,2), 58–63.

Esser, M. (1984b) *Int. J. Quant. Chem.* **26**, 313–338.

Esser, M., Butscher, W. and Schwarz, W. H. E. (1981) *Chem. Phys. Lett.* **77**, 359.

Esteban, M. J. and Sere, E. (1999) *Commun. Math. Phys.* **203**, 499–530.

Evans, C. J. and Gerry, M. C. L. (2000) *J. Am. Chem. Soc.* **122**, 1560.

Evans, W. D., Perry, P. and Siedentop, H. (1996) *Commun. Math. Phys.* **178**, 733.

Facco Bonetti, A., Engel, E., Dreizler, R. M., Andrejkovics, I. and Müller, H. (1998) *Phys. Rev.* A **58**, 993.

Facco Bonetti, A., Engel, E., Schmid, R. N. and Dreizler, R. M. (2001) *Phys. Rev. Lett.* **86**, 2241.

Fægri Jr, K. and Saue, T. (2001) *J. Chem. Phys.* **115**, 2456–2464.

Fægri Jr, K. and Visscher, L. (2001) *Theor. Chem. Acc.* **105**, 265–267.

Faisal, F. H. M. and Radozycki, T. (1993) *Phys. Rev.* A **48**, 554.

Fano, U. (1969a) *Phys. Rev.* **178**, 131.

Fano, U. (1969b) *Phys. Rev.* **184**, 250.

Faulkner, J. S. (1977) *J. Phys.* C **10**, 4661.

Faulkner, J. S. (1982) *Prog. Mater. Sci.* **27**, 3.

Faulkner, J. S. and Stocks, G. M. (1980) *Phys. Rev.* B **21**, 3222.

Feder, R. (ed.) (1985) *Polarized Electrons in Surface Physics*. World Scientific, Singapore, p. 125.

Feder, R. and Henk, J. (1996) In Ebert and Schütz (1996), p. 85.

Feder, R., Rosicky, F. and Ackermann, B. (1983) *Z. Phys.* B **52**, 31.

Fehrenbach, G. M. and Schmidt, G. (1997) *Phys. Rev.* B **55**, 6666–6669.

Feili, D., Bosselmann, P., Schartner, K.-H., Folkmann, F., Livingston, A. E., Träbert, E., Ma, X. and Mokler, P. H. (2000) *Phys. Rev.* A **62**, 022501.

Ferray, M., L'Huillier, A., Li, X. F., Lompre, L. A., Mainfray, G. and Manus, C. (1988) *J. Phys. B* **21**, L31.

Fiedler, A., Schröder, D., Shaik, S. and Schwarz, H. (1994) *J. Am. Chem. Soc.* **116**, 10 734.

Finkbeiner, M., Fricke, B. and Kühl, T. (1993) *Phys. Lett.* A **176**, 113.

Fischer, P., Eimüller, T., Schütz, G., Schmahl, G., Guttmann, P. and Bayreuther, G. (1999) *J. Magn. Magn. Mater.* **198**, 624.

Flad, H.-J. and Dolg, M. (1996a) *J. Phys. Chem.* **100**, 6152–6155.

Flad, H.-J. and Dolg, M. (1996b) *J. Phys. Chem.* **100**, 6152–6155.

Flad, H.-J., Dolg, M. and Shukla, A. (1997) *Phys. Rev.* A **55**, 4183–4195.

Fleig, T., Olsen, J. and Marian, C. M. (2001) *J. Chem. Phys.* **114**, 4775–4790.

Fluchtmann, M., Bei der Kellen, S., Braun, J. and Borstel, G. (1998) *Surf. Sci.* **402–404**, 663–668.

Fluchtmann, M., Braun, J. and Borstel, G. (1995) *Phys. Rev.* B **52**, 9564.

Fluchtmann, M., der Kellen, S. B., Braun, J. and Borstel, G. (1999) *Surf. Sci.* **432**, 291–296.

Foldy, L. L. and Wouthuysen, S. A. (1950) *Phys. Rev.* **78**, 29–36.

Forstreuter, J., Steinbeck, L., Richter, M. and Eschrig, H. (1997) *Phys. Rev.* B **55**, 9415.

Foucrault, M., Millie, P. and Daudey, J. P. (1992) *J. Chem. Phys.* **96**, 1297.

Fournier, R., Andzelm, J. and Salahub, D. R. (1989) *J. Chem. Phys.* **90**, 6371.

Freitas, P. P. and Berger, L. (1988) *Phys. Rev.* B **37**, 6079.

Frenking, G., Antes, I., Böhme, M., Dapprich, S., Ehlers, A., Jonas, V., Neuhaus, A., Otto, M., Stegmann, R., Veldkamp, A. and Vyboishchikov, S. (1996) *Rev. Comp. Chem.* **8**, 63.

Freyer, H., Deng, M. and Ebert, H. (1999) *Physica Status Solidi* (b) **215**, 833.

Fricke, B., Sepp, W.-D., Bastug, T., Varga, S., Schulze, K., Anton, J. and Pershina, V. (1997) *Adv. Quant. Chem.* **29**, 109.

Fritzsche, S. (1997) *Comp. Phys. Commun.* **103**, 51–73.

Fritzsche, S. (2000) *J. Electr. Spectra Rel. Phen.* **114–116**, 1155–1164.

Fritzsche, S. and Anton, J. (2000) *Comp. Phys. Commun.* **124**, 353–355.

Fritzsche, S., Froese Fischer, C. and Fricke, B. (1998a) *At. Data Nucl Data Tables* **68**, 149–179.

Fritzsche, S., Varga, S., Geschke, D. and Fricke, B. (1998b) *Comp. Phys. Commun.* **111**, 167–184.

Fritzsche, S., Fricke, B., Geschke, D., Heitmann, A. and Sienkiewicz, J. E. (1999) *Astrophys. J.* **518**, 994–1001.

Fritzsche, S., Dong, C. Z. and Gaigalas, G. (2000a) *At. Data Nucl. Data Tables* **76**, 155–175.

Fritzsche, S., Froese Fischer, C. and Dong, C. Z. (2000b) *Comp. Phys. Commun.* **124**, 340–352.

Froben, F. W., Schulze, W. and Kloss, U. (1983) *Chem. Phys. Lett.* **99**, 500.

Froese Fischer, C., He, X. and Jönsson, P. (1998) *Eur. Phys. J.* D **4**, 285–289.

Froitzheim, H., Hopster, H., Ibach, H. and Lehwald, S. (1977) *Appl. Phys.* **63**, 147.

Fuchs, M., Bockstedte, M., Pehlke, E. and Scheffler, M. (1998) *Phys. Rev.* B **57**, 2134.

Fuentealba, P., Preuss, H., Stoll, H. and Szentpály, L. (1982) *Chem. Phys. Lett.* **89**, 418.

Fujikawa, T., Yanagisawa, R., Yiwata, N. and Ohtani, K. (1997) *J. Phys. Soc. Jpn* **66**, 257.

Fujimura, N. and Matsuoka, O. (1992) *Int. J. Quant. Chem.* **42**, 751–759.

Furry, W. H. (1951) *Phys. Rev.* **81**, 115.

Gabbaï, P., Chung, S., Schier, A., Krüger, S., Rösch, N. and Schmidbaur, H. (1997) *Inorg. Chem.* **36**, 5699–5705.

Gäggeler, H. W. (1994) *J. Radioanal. Nucl. Chem.* **183**, 261.

Gagliardi, L., Willetts, A., Skylaris, C., Handy, N. C., Spencer, S., Ioannou, A. G. and Simper, A. M. (1998) *J. Am. Chem. Soc.* **120**, 11 727.

Gaigalas, G. and Fritzsche, S. (2001) *Comp. Phys. Commun.* **134**, 86–96.

Gavrila, M. (1992) *Adv. At. Mol. Phys. Suppl.* **1**, 453.

Geipel, N. J. M. and Hess, B. A. (1997) *Chem. Phys. Lett.* **273**, 62–70.

Gell-Mann, M. and Low, F. (1951) *Phys. Rev.* **54**, 350.

Gerlach, A., Matzdorf, R. and Goldmann, A. (1998) *Phys. Rev.* B **58**, 10 969–10 974.

Geschke, D., Fritzsche, S., Sepp, W.-D., Fricke, B., Varga, S. and Anton, J. (2000) *Phys. Rev.* B **62**, 15 439.

Gibson, J. K. (2001) *Int. J. Mass Spectrom.* **36**, 284.

Girichev, G. V., Petrov, G. M., Giricheva, N. I., Utkin, A. N. and Petrova, V. N. (1981) *Zh. Struk. Khim.*, p. 65.

Głowacki, L., Stanek, M. and Migdalek, J. (2000) *Phys. Rev.* A **61**, 064501–1–4.

Glukhovtsev, M., Pross, A., Nicolaides, A. and Radom, L. (1995) *J. Chem. Soc. Chem. Commun.* p. 2347.

Gmelin (1970) *Handbuch der Anorganischen Chemie*. Verlag Chemie.

Godby, R. W., Schlüter, M. and Sham, L. J. (1988) *Phys. Rev.* B **37**, 10 159.

Goldstone, J. (1957) *Proc. R. Soc. Lond.* A **239**, 267.

Gonis, A. (1992) *Green Functions for Ordered and Disordered Systems*. North-Holland, Amsterdam.

Görling, A. and Levy, M. (1994) *Phys. Rev.* A **50**, 196.

Grabo, T. and Gross, E. K. U. (1995) *Chem. Phys. Lett.* **240**, 141.

Gräf, P., Ebert, H., Akai, H. and Voitländer, J. (1993) *Hyperfine Interactions* **78**, 1011.

Grant, I. P. (1994) *Adv. At. Mol. Opt. Phys.* **32**, 169–186.

Grant, I. P. and Quiney, H. M. (1988) *Adv. At. Mol. Phys.* **23**, 37–86.

Grant, I. P. and Quiney, H. M. (2000a) *Phys. Rev.* A **62**, 022508–1–14.

Grant, I. P. and Quiney, H. M. (2000b) *Int. J. Quant. Chem.* **80**, 283–297.

Greenwood, D. A. (1958) *Proc. Phys. Soc.* **71**, 585.

Gregorich, K. E., Henderson, R. A., Lee, D. M., Nurmia, M., Chasteler, R. M., Hall, H. L., Bennett, D. A., Gannett, C. M., Chadwick, R. B., Leyba, J. D., Hoffman, D. C. and Herrmann, G. (1988) *Radiochim. Acta* **43**, 223.

Greiner, W. and Gupta, R. K. (eds) (1999) *Heavy Elements and Related New Phenomena*. World Scientific.

Griesemer, M. and Siedentop, H. (1999) *J. Lond. Math. Soc.* **60**, 490–500.

Griesemer, M., Lewis, R. T. and Siedentop, H. (1999) *Doc. Math.* **4**, 275–283.

Grimme, S. (1996) *Chem. Phys. Lett.* **259**, 128.

Grimme, S. and Waletzke, M. (1999) *J. Chem. Phys.* **111**, 5645–5655.

Grimme, S., Pischel, I., Laufenberg, S. and Vögtle, F. (1998) *Chirality* **10**, 147.

Grochmalicki, J., Lewenstein, M., Wilkens, M. and Rzazewski, K. (1990) *J. Opt. Soc. Am.* **7**, 607.

Gropen, O. (1988) In *Methods in Computational Chemistry* (ed. S. Wilson), vol. 2, p. 109. Plenum, New York.

Grotendorst, J. (ed.) (2000) *NIC Series*, vol. 3. John von Neumann Institute for Computing (NIC), Jülich, Germany.

Gukasov, A., Wiśniewski, P. and Henkie, Z. (1996) *J. Phys. Cond. Matt.* **8**, 10 589.

Gunnarsson, O. (1976) *J. Phys.* F **6**, 587.

Günther, R., Paulus, W., Posledni, A., Kratz, J. V., Schädel, M., Brüchle, W., Jäger, E., Schimpf, E., Schausten, B., Schumann, D. and Binder, R. (1995) *Jahresbericht 1994, IKMz 95-1*. Institut für Kernchemie, Universität Mainz.

Günther, R., Paulus, W., Kratz, J. V., Seibert, A., Thörle, P., Zauner, S., Brüchle, W., Jäger, E., Pershina, V., Schädel, M., Schausten, B., Schumann, D., Eichler, B., Gäggeler, H. W., Jost, D. T. and Türler, A. (1998) *Radiochim. Acta* **80**, 121.

Gyorffy, B. L. and Stott, M. J. (1973) In *Band Structure Spectroscopy of Metals and Alloys* (ed. D. J. Fabian and L. M. Watson), p. 385. Academic.

Häberlen, O. D. and Rösch, N. (1992) *Chem. Phys. Lett.* **199**, 491–496.

Häberlen, O. D., Chung, S., Stener, M. and Rösch, N. (1997) *J. Chem. Phys.* **106**, 5189–5201.

Häffner, H., Beier, T., Hermanspahn, N., Kluge, H.-J., Quint, W., Stahl, S., Verdu, J. and Werth, G. (2000) *Phys. Rev. Lett.* **85**, 5308.

Hafner, P. and Schwarz, W. H. E. (1979) *Chem. Phys. Lett.* **65**, 537.

Hamacher, P. and Hinze, J. (1991) *Phys. Rev.* A **44**, 1705–1711.

Hamann, D. R., Schlüter, M. and Chiang, C. (1979) *Phys. Rev. Lett.* **43**, 1494.

Hammer, B., Morikawa, Y. and Nørskov, J. K. (1996) *Phys. Rev. Lett* **76**, 2141.

Hamzić, A., Senoussi, S., Campbell, I. A. and Fert, A. (1978) *J. Phys.* F **8**, 1947.

Han, Y.-K. and Hirao, K. (2000) *Chem. Phys. Lett.* **328**, 453.

Han, Y.-K., Son, S.-K., Choi, Y. J. and Lee, Y. S. (1999) *J. Phys. Chem.* **103**, 9109.

Hartemann, F. V. and Kerman, A. K. (1996) *Phys. Rev. Lett.* **76**, 624.

Hartemann, F. V. and Luhmann, N. C. (1995) *Phys. Rev. Lett.* **74**, 1107.

Hartemann, F. V., Fuchs, S. N., Sage, G. P. L., Luhmann, N. C., Woodworth, J. G., Perry, M. D., Chen, Y. J. and Kerman, A. K. (1995) *Phys. Rev.* E **51**, 4833.

Hartley, A. C. and Sandars, P. G. H. (1991) Relativistic calculations of parity non-conserving effects in atoms. In Wilson *et al.* (1991), pp. 67–81.

Harvey, J. N., Aschi, M., Schwarz, H. and Koch, W. (1998) *Theor. Chem. Acc.* **99**, 95.

Häussermann, U., Dolg, M., Stoll, H., Preuss, H. and Schwerdtfeger, P. (1993) *Mol. Phys.* **80**, 1211.

Hay, P. J. (1983) *J. Chem. Phys.* **79**, 5469.

Hay, P. J. and Martin, R. L. (1998) *J. Chem. Phys.* **109**, 3875.

Hay, P. J. and Wadt, W. R. (1985a) *J. Chem. Phys.* **82**, 270.

Hay, P. J. and Wadt, W. R. (1985b) *J. Chem. Phys.* **82**, 299.

Hedin, L. and Lundqvist, S. (1969) *Solid State Phys.* **23**, 1.

Heinemann, C. and Schwarz, H. (1995) *Chem. Eur. J.* **1**, 7.

Heinemann, C., Schröder, D. and Schwarz, H. (1994) *Chem. Ber.* **127**, 1807.

Heinemann, C., Cornehl, H. H. and Schwarz, H. (1995a) *J. Organomet. Chem.* **501**, 201.

Heinemann, C., Goldberg, N., Tornieporth-Oetting, I. C., Klapötke, T. M. and Schwarz, H. (1995b) *Angew. Chem.* **107**, 225.

Heinemann, C., Goldberg, N., Tornieporth-Oetting, I. C., Klapötke, T. M. and Schwarz, H. (1995c) *Angew. Chem. Int. Ed. Engl.* **34**, 213.

Heinemann, C., Hertwig, R. H., Wesendrup, R., Koch, W. and Schwarz, H. (1995d) *J. Am. Chem. Soc.* **117**, 495.

Heinemann, C., Koch, W. and Schwarz, H. (1995e) *Chem. Phys. Lett.* **245**, 509.

Heinemann, C., Cornehl, H. H., Schröder, D., Dolg, M. and Schwarz, H. (1996a) *Inorg. Chem.* **35**, 2463.

Heinemann, C., Schwarz, H., Koch, W. and Dyall, K. G. (1996b) *J. Chem. Phys.* **104**, 4642.

Heinemann, C., Wesendrup, R. and Schwarz, H. (1995f) *Chem. Phys. Lett.* **239**, 75.

Heinzmann, U., Jost, K., Kessler, J. and Ohnemus, B. (1972) *Z. Phys.* **251**, 354.

Heitmann, A., Varga, S., Sepp, W.-D., Bastug, T. and Fricke, B. (2001). (Unpublished.)

Heiz, U., Vayloyan, A., Schumacher, E., Yeretzian, C., Stener, M., Gisdakis, P. and Rösch, N. (1995) *J. Chem. Phys.* **105**, 5774–5585.

Helgaker, T., Jørgensen, P. and Olsen, J. (2000) *Molecular Electronic-Structure Theory*. Wiley.

Hellwig, H., Vessot, R. F. C., Levine, M. W., Zitzewitz, P. W., Allan, D. W. and Glaze, D. J. (1970) *IEEE Trans. Instrum. Meas.* **IM19**, 200.

Hencken, K., Trautmann, D. and Baur, G. (1995) *Phys. Rev.* A **51**, 998.

Henk, J., Scheunemann, T. and Feder, S. V. H. R. (1996) *J. Phys. Cond. Matt.* **8**, 47.

Hermanspahn, N., Häffner, H., Kluge, H.-J., Quint, W., Stahl, S., Verdú, J. and Werth, G. (2000) *Phys. Rev. Lett.* **84**, 427.

Hertwig, R. H., Hrušák, J., Schröder, D. Koch, W. and Schwarz, H. (1995) *Chem. Phys. Lett.* **236**, 194.

Hertwig, R. H., Koch, W., Schröder, D., Schwarz, H., Hrušák, J. and Schwerdtfeger, P. (1996) *J. Phys. Chem.* **100**, 12 253.

Hess, B. A. (1986) *Phys. Rev.* A **33**, 3742–3748.

Hess, B. A. (1997) *Ber. BunsenGes. Phys. Chem.* **101**, 1.

Hess, B. A. and Kaldor, U. (2000) *J. Chem. Phys.* **112**, 1809–1813.

Hess, B. A. and Marian, C. M. (2000) Relativistic effects in the calculation of electronic energies. In *Computational Molecular Spectroscopy* (ed. P. Jensen and P. R. Bunker), pp. 169–219. Wiley.

Hess, B. A., Marian, C. M. and Peyerimhoff, S. D. (1995) *Ab initio* calculation of spin–orbit effects in molecules. In *Modern Electronic Structure Theory* (ed. D. R. Yarkony), pp. 152–278. World Scientific.

Hess, B. A., Marian, C. M., Wahlgren, U. and Gropen, O. (1996) *Chem. Phys. Lett.* **251**, 365–371.

Hess, B. A., Marian, C. M., Chandra, P., Hutter, S., Rakowitz, F. and Samzow, R. (2000) BNSOC is a spin–orbit coupling program package developed at the University of Bonn, Germany.

Heully, J. L., Lindgren, I., Lindroth, E., Lundquist, S. and Mårtensson-Pendrill, A. M. (1986) *J. Phys.* B **19**, 2799–2815.

Hibbert, A. (1982) *Adv. Atom. Mol. Phys.* **18**, 309.

Hjortstam, O., Trygg, J., Wills, J. M., Johansson, B. and Eriksson, O. (1996) *Phys. Rev.* B **53**, 9204.

Ho, Y.-P. and Dunbar, R. (1999) *Int. J. Mass Spectrom.* **182/183**, 175.

Hoare, F. E., Matthews, J. C. and Walling, J. C. (1953) *Proc. R. Soc. Lond.* A **216**, 502.

Höck, A. and Engel, E. (1998) *Phys. Rev.* A **58**, 3578.

Hoever, G. and Siedentop, H. (1999) *Math. Phys. Electron. J.* **5** (Paper 6), 11.

Hofmann, S. (1996) *Z. Phys.* A **354**, 229.

Hohenberg, P. and Kohn, W. (1964) *Phys. Rev.* B **136**, 864.

Hollatz, C., Schier, A. and Schmidbaur, H. (1998) *Inorg. Chem. Commun.* **1**, 115.

Hollatz, C., Schier, A. and Schmidbaur, H. (1999a) *Z. Naturf.* B **54**, 30.

Hollatz, C., Schier, A., Riede, J. and Schmidbaur, H. (1999b) *J. Chem. Soc., Dalton Trans.*, p. 111.

Hörmandinger, G. and Weinberger, P. (1992) *J. Phys. Cond. Matt.* **4**, 2185.

Horodecki, P., Kwela, J. and Sienkiewicz, J. E. (1999) *Eur. Phys. J.* D **6**, 435–440.

Hrušák, J., Schröder, D. and Schwarz, H. (1994) *Chem. Phys. Lett.* **225**, 416.

Hrušák, J., Schröder, D., Hertwig, R. H., Koch, W., Schwerdtfeger, P. and Schwarz, H. (1995) *Organometallics* **14**, 1284.

Hu, A., Otto, P. and Ladik, J. (1998) *Chem. Phys. Lett.* **293**, 277–283.

Hu, A., Otto, P. and Ladik, J. (1999) *J. Comp. Chem.* **20**, 655–664.

Hu, A., Otto, P. and Ladik, J. (2000) *Chem. Phys. Lett.* **320**, 6–7.

Huber, K. P. and Herzberg, G. L. (1950) *Molecular Spectra and Molecular Structure. I.* Van Nostrand Reinhold, New York.

Hughes, V. W. and Kinoshita, T. (1999) *Rev. Mod. Phys.* **71**, 133.

Huhne, T. (2000) PhD thesis, University of Munich.

Huhne, T. and Ebert, H. (1999) *Phys. Rev.* B **60**, 12 982.

Huhne, T. and Ebert, H. (2002) *Phys. Rev.* B **65**, 205125
Huhne, T., Zecha, C., Ebert, H., Dederichs, P. H. and Zeller, R. (1998) *Phys. Rev.* B **58**, 10 236.
Hulme, H. R. (1932) *Proc. R. Soc. Lond.* A **135**, 237.
Hundertmark, D., Röhrl, N. and Siedentop, H. (2000) *Commun. Math. Phys.* **211**, 629.
Hurley, M. M., Pacios, L. F., Christiansen, P. A., Ross, R. B. and Ermler, W. C. (1986) *J. Chem. Phys.* **84**, 6840.
Hussein, M. S., Pato, M. P. and Kerman, A. K. (1992) *Phys. Rev.* A **46**, 3562.
Hutter, S. J. (1994) Methodologische Aspekte der Behandlung der Spin-Bahn-Kopplung zweiatomiger Moleküle. PhD thesis, University of Bonn.
Huzinaga, S. (1991) *J. Mol. Struct.* **234**, 51.
Huzinaga, S. (1995) *Can. J. Chem.* **73**, 619.
Ilyabaev, E. and Kaldor, U. (1992a) *J. Chem. Phys.* **97**, 8455–8458.
Ilyabaev, E. and Kaldor, U. (1992b) *Chem. Phys. Lett.* **914**, 95–98.
Ilyabaev, E. and Kaldor, U. (1993) *Phys. Rev.* A **47**, 137–142.
Indelicato, P. (1991) *Theor. Chim. Acta* **80**, 207.
Indelicato, P. (1995) *Phys. Rev.* A **51**, 1132–1145.
Indelicato, P. (1996) *Phys. Rev. Lett.* **77**, 3323–3326.
Indelicato, P. and Desclaux, J. P. (1990) *Phys. Rev.* A **42**, 5139.
Indelicato, P. and Desclaux, J. P. (1993) *Physica Scr.* T **46**, 110–114.
Indelicato, P. and Lindroth, E. (1992) *Phys. Rev.* A **46**, 2426–2436.
Indelicato, P. and Mohr, P. J. (1992) *Phys. Rev.* A **46**, 172.
Indelicato, P. and Mohr, P. J. (1998) *Phys. Rev.* A **58**, 165.
Ionescu, D. C. and Eichler, J. (1993) *Phys. Rev.* A **48**, 1176.
Ionova, G. V., Pershina, V., Johnson, E., Fricke, B. and Schädel, M. (1992) *J. Phys. Chem.* **96**, 11 096.
Irikura, K. K. and Beauchamp, J. L. (1991a) *J. Am. Chem. Soc.* **113**, 2769.
Irikura, K. K. and Beauchamp, J. L. (1991b) *J. Phys. Chem.* **95**, 8344.
Ishikawa, H., Yamamoto, K., Fujima, K. and Iwasawa, M. (1999) *Int. J. Quant. Chem.* **72**, 509.
Ishikawa, Y. (1990a) *Int. J. Quant. Chem., Quant. Chem. Symp.* **24**, 383–391.
Ishikawa, Y. (1990b) *Chem. Phys. Lett.* **166**, 321–325.
Ishikawa, Y. and Koc, K. (1994) *Phys. Rev.* A **50**, 4733.
Ishikawa, Y. and Koc, K. (1997) *Int. J. Quant. Chem.* **65**, 545–554.
Ishikawa, Y. and Quiney, H. M. (1993) *Phys. Rev.* A **47**, 1732–1739.
Ishikawa, Y., Koc, K. and Schwarz, W. H. E. (1997) *Chem. Phys.* **225**, 239–246.
Ishikawa, Y., Malli, G. L. and Stacey, A. J. (1992) *Chem. Phys. Lett.* **188**, 145–148.
Ishikawa, Y., Quiney, H. M. and Malli, G. L. (1991) *Phys. Rev.* A **43**, 3270–3278.
Ishikawa, Y., Vilkas, M. J. and Koc, K. (2000) *Int. J. Quant. Chem.* **77**, 433–445.
Ismail, N., Heully, J.-L., Saue, T., Daudey, J.-P. and Marsden, C. J. (1999) *Chem. Phys. Lett.* **300**, 296–302.
Ivanov, D. Y., Schiller, A. and Serbo, V. G. (1999) *Phys. Lett.* B **454**, 155.
Jacob, T., Inghoff, T., Fricke, B., Fritzsche, S., Johnson, E. and Pershina, V. (2000) GSI annual report.
Jansen, G. and Hess, B. A. (1989) *Phys. Rev.* A **39**, 6016–6017.
Jansen, H. J. F. (1988) *J. Appl. Phys.* **64**, 5604.
Jaoul, O., Campbell, I. A. and Fert, A. (1977) *J. Magn. Magn. Mater.* **5**, 23.

Jáuregui, R., Bunge, C. F. and Ley-Koo, E. (1997) *Phys. Rev.* A **55**, 1781–1784.

Jeffrey, A.-M., Elmquist, R. E., Lee, L. H. and Dziuba, R. F. (1997) *IEEE Trans. Instrum. Meas.* **IM46**, 264.

Jen, S. U. (1992) *Phys. Rev.* B **45**, 9819.

Jen, S. U., Chen, T. P. and Chao, B. L. (1993) *Phys. Rev.* B **48**, 12 789.

Jenkins, A. C. and Strange, P. (1994) *J. Phys. Cond. Matt.* **6**, 3499.

Jensen, H. J. A., Dyall, K. G., Saue, T. and Fægri Jr, K. (1996) *J. Chem. Phys.* **104**, 4083–4097.

Jin, J., Khemliche, H. and Prior, M. H. (1996) *Phys. Rev.* A **53**, 615.

Johnson, B. G., Gill, P. M. and Pople, J. A. (1993a) *J. Chem. Phys.* **98**, 5612.

Johnson, E., Fricke, B., Keller Jr, O. L., Nestor Jr, C. W. and Tucker, T. C. (1990) *J. Chem. Phys.* **93**, 8041.

Johnson, E., Krause, M. O. and Fricke, B. (1996) *Phys. Rev.* A **54**, 4783–4788.

Johnson, E., Pershina, V. and Fricke, B. (1999) *J. Phys. Chem.* A **103**, 8458–8462.

Johnson, W. R. and Soff, G. (1985) *At. Data Nucl. Data Tables* **33**, 405.

Johnson, W. R., Blundell, S. A. and Sapirstein, J. (1988) *Phys. Rev.* A **37**, 307–315.

Johnson, W. R., Plante, D. R. and Sapirstein, J. (1995) *Adv. At. Mol. Phys.* **35**, 255–329.

Johnson, W. R., Sapirstein, J. and Blundell, S. A. (1993b) *Physica Scr.* T **46**, 184–192.

Jong, W. A. D. and Nieuwpoort, W. C. (1996) *Int. J. Quant. Chem.* **58**, 203.

Jönsson, P. and Froese Fischer, C. (1997) *J. Phys.* B **30**, 5861–5875.

Jönsson, P. and Froese Fischer, C. (1998) *Phys. Rev.* A **57**, 4967–4970.

Jönsson, P., Froese Fischer, C. and Träbert, E. (1998) *J. Phys.* B **31**, 3497–3511.

Jorge, F. E. and da Silva, A. B. F. (1996a) *J. Chem. Phys.* **105**, 5503–5509.

Jorge, F. E. and da Silva, A. B. F. (1996b) *J. Chem. Phys.* **104**, 6278–6285.

Jorge, F. E. and da Silva, A. B. F. (1997) *Z. Phys.* D **41**, 235–238.

Jorge, F. E. and da Silva, A. B. F. (1998) *Chem. Phys. Lett.* **289**, 469–472.

Kahn, L. (1984) *Int. J. Quant. Chem.* **25**, 149.

Kaldor, U. and Hess, B. A. (1994) *Chem. Phys. Lett.* **230**, 1–7.

Kaldor, U. (ed.) (1989) *Lecture Notes in Chemistry*, vol. 52. Springer.

Källén, A. O. G. (1958) *Handbuch der Physik*, vol. V, part 1. Springer.

Kaltsoyannis, N. (1997) *J. Chem. Soc., Dalton Trans.*, p. 1.

Karshenboim, S. G. (1999) *Can. J. Phys.* **77**, 241.

Karwowski, J. (2001) *Comp. Phys. Commun.* **138**, 10–17.

Kassiakoff, A. and Harker, D. (1938) *J. Am. Chem. Soc.* **60**, 2047.

Kaupp, M., Malkina, O. L. and Malkin, V. G. (1998a) In Schleyer (1998), p. 1857.

Kaupp, M., Malkina, O. L., Malkin, V. G. and Pyykkö, P. (1998b) *Chem. Eur. J.* **4**, 118–126.

Kaupp, M., Aubauer, C., Engelhardt, G., Klapötke, T. and Malkina, O. L. (1999) *J. Comput. Chem.* **20**, 91–105.

Keitel, C. H. and Knight, P. L. (1995) *Phys. Rev.* A **51**, 1420.

Khein, A., Singh, D. J. and Umrigar, C. J. (1995) *Phys. Rev.* B **51**, 4105.

Khriplovich, I. B. and Lamoreaux, S. K. (1997) *CP violation without strangeness: electric dipole moments of particles, atoms, and molecules*. Texts and Monographs in Physics. Springer.

Kibble, T. W. (1966) *Phys. Rev.* **150**, 1060.

Kim, Y.-H., Städele, M. and Martin, R. M. (1999) *Phys. Rev.* A **60**, 3633.

Kim, Y.-K. (1967) *Phys. Rev.* **154**, 17–39.

Kim, Y. K. (1993a) *Commun. At. Mol. Phys.* **28**, 201–210.

Kim, Y. K. (1993b) *Physica Scr.* T **47**, 54–58.

Kim, Y.-K., Baik, D. H., Indelicato, P. and Deslaux, J. P. (1991) *Phys. Rev.* A **44**, 148.

Kim, Y.-K., Parente, F., Marques, J. P., Indelicato, P. and Desclaux, J. P. (1998) *Phys. Rev.* A **58**, 1885–1888.

Kinoshita, T. (1990) *Quantum Electrodynamics*. World Scientific, Singapore.

Klaft, I., Borneis, S., Engel, T., Fricke, B., Grieser, R., Huber, G., Kühl, T., Marx, D., Neumann, R., Schröder, S., Seelig, P. and Völker, L. (1994) *Phys. Rev. Lett.* **73**, 2425.

Klarsfeld, S. (1977) *Phys. Lett.* B **66**, 86.

Kleiner, W. H. (1966) *Phys. Rev.* **142**, 318.

Kleinschmidt, M. and Marian, C. M. (2000) Efficient generation of matrix elements of one-electron spin–orbit operators. (Unpublished.)

Klopper, W. (2000) R12 methods, Gaussian geminals. In Grotendorst (2000), pp. 181–229.

Knöpfle, K. and Sandratskii, L. M. (2000) *Phys. Rev.* B **61**, 014411.

Knöpfle, K., Sandratskii, L. M. and Kübler, J. (2000) *J. Alloys Comp.* **309**, 31.

Knülle, M., Ahlers, D. and Schütz, G. (1995) *Solid State Commun.* **94**, 267.

Kobayashi, S., Asayama, K. and Itoh, J. (1963) *J. Phys. Soc. Jpn* **18**, 1735.

Koelling, D. D. and MacDonald, A. H. (1983) In *Relativistic Effects in Atoms Molecules and Solids* (ed. G. L. Malli), p. 227. Plenum Press, New York.

Kohstall, C., Fritzsche, S., Fricke, B. and Sepp, W.-D. (1998) *At. Data Nucl. Data Tables* **70**, 63–92.

Kołakowska, A. (1997) *J. Phys.* B **30**, 2773–2779.

Kołos, W. and Wolniewicz, L. (1964) *J. Chem. Phys.* **41**, 3663.

Kopalj, A. and Causà, M. (1999) *J. Phys.: Condens. Matter* **11**, 7463.

Korkisch, J. (1989) *Handbook of Ion Exchange Resins; Their Application to Inorganic Analytical Chemistry*, vol. IV. CRC Press.

Körmendi, F. F. and Farkas, G. (1996) *Phys. Rev.* A **53**, R637.

Kortright, J. B. and Kim, S. K. (2000) *Phys. Rev.* B **62**, 12 216.

Kotani, T. (1994) *Phys. Rev.* B **50**, 14 816.

Kotani, T. (1995) *Phys. Rev. Lett.* **74**, 2989.

Kotani, T. (1998) *J. Phys.: Condens. Matter* **10**, 9241.

Kotani, T. and Akai, H. (1996) *Phys. Rev.* B **54**, 16 502.

Krajci, M., Hafner, J. and Mihalkovic, M. (1997) *Phys. Rev.* B **56**, 3072.

Kratz, J. V. (1999) In Greiner and Gupta (1999), pp. 129–193.

Kratz, J. V., Zimmermann, H. P., Scherer, U. W., Schädel, M., Brüchle, W., Gregorich, K. E., Gannett, C. M., Hall, H. L., Henderson, R. A., Lee, D. M., Leyba, J. D., Nurmia, M., Hoffman, D. C., Gäggeler, H. W., Jost, D., Baltensperger, U., Ya, N.-Q., Türler, A. and Lienert, C. (1989) *Radiochim. Acta* **48**, 121.

Krauss, M. and Stevens, W. (1984) *Ann. Rev. Phys. Chem.* **35**, 357.

Kreibich, T., Gross, E. K. U. and Engel, E. (1998) *Phys. Rev.* A **57**, 138.

Kretzschmar, I., Fiedler, A., Harvey, J. N., Schröder, D. and Schwarz, H. (1997) *J. Phys. Chem.* A **101**, 6252.

Kretzschmar, I., Schröder, D., Schwarz, H., Rue, C. and Armentrout, P. B. (1998) *J. Phys. Chem.* A **102**, 10 060.

Krieger, J. B., Li, Y. and Iafrate, G. J. (1990) *Phys. Lett.* A **146**, 256.

Krieger, J. B., Li, Y. and Iafrate, G. J. (1992) *Phys. Rev.* A **45**, 101.

Krieger, R. and Voitländer, J. (1980) *Z. Phys.* B **40**, 39.

Krüger, S., Stener, M., Mayer, M., Nörtemann, F. and Rösch, N. (2000) *J. Mol. Struct (THEOCHEM)* **527**, 63–74.

Krüger, S., Vent, S. and Rösch, N. (1997) *Ber. BunsenGes. Phys. Chem.* **101**, 1640–1643.

Krutzen, B. C. H. and Springelkamp, F. (1989) *J. Phys. Cond. Matt.* **1**, 8369.

Kübler, J., Höck, K.-H., Sticht, J. and Williams, A. R. (1988) *J. Appl. Phys.* **63**, 3482.

Küchle, W., Dolg, M. and Stoll, H. (1997) *J. Phys. Chem.* A **101**, 7128–7133.

Küchle, W., Dolg, M., Stoll, H. and Preuss, H. (1991) *J. Chem. Phys.* **74**, 1245.

Kuck, D. (1990) *Mass Spectrom. Rev.* **9**, 583.

Kulander, K., Schäfer, K. and Krauses, J. (1991) *Phys. Rev. Lett.* **64**, 862.

Kullie, O., Duesterhoeft, C. and Kolb, D. (1999) *Chem. Phys. Lett.* **314**, 307–310.

Kurth, S., Marques, M., Lüders, M. and Gross, E. K. U. (1999) *Phys. Rev. Lett.* **83**, 2628.

Kutzelnigg, W. (1982) *J. Chem. Phys.* **77**, 3081.

Kutzelnigg, W. (1989) *Z. Phys.* D **11**, 15–28.

Kutzelnigg, W. (1990) *Z. Phys.* D **15**, 27–50.

Kutzelnigg, W. (1997) *Chem. Phys.* **225**, 203–222.

Kutzelnigg, W. (1999) *J. Chem. Phys.* **110**, 8283–8294.

Laaksonen, L., Grant, I. P. and Wilson, S. (1988) *J. Phys.* B **21**, 1969–1985.

Labzowsky, L. N. and Mitrushenkov, A. O. (1996) *Phys. Rev.* A **53**, 3029.

Labzowsky, L. N., Johnson, W. R., Soff, G. and Schneider, S. M. (1995) *Phys. Rev.* A **51**, 4597.

Labzowsky, L. N., Klimchitskaya, G. and Dmitriev, Y. (1993) *Relativistic Effects in the Spectra of Atomic Systems*. IOP Publishing, Bristol.

Labzowsky, L., Nefiodov, A., Plunien, G., Soff, G. and Pyykkö, P. (1997) *Phys. Rev.* A **56**, 4508.

Ladik, J. (1959) *Acta. Phys. Acad. Sci. Hung.* **10**, 271.

Ladik, J. J. (1997) *J. Mol. Struct. (THEOCHEM)* **391**, 1–14.

Laerdahl, J. K., Saue, T., Faegri Jr, K. and Quiney, H. M. (1997) *Phys. Rev. Lett* **79**, 1642–1645.

Laerdahl, J. K. and Schwerdtfeger, P. (1999) *Phys. Rev.* A **60**, 4439–4453.

Laerdahl, J. K., Fægri Jr, K., Visscher, L. and Saue, T. (1998) *J. Chem. Phys.* **109**, 10 806–10 817.

Laerdahl, J. K., Saue, T. and Faegri Jr, K. (1997) *Theor. Chem. Acc.* **97**, 177–184.

Laerdahl, J. K., Schwerdtfeger, P. and Quiney, H. M. (2000a) *Phys. Rev. Lett.* **84**, 3811–3814.

Laerdahl, J. K., Wesendrup, R. and Schwerdtfeger, P. (2000b) *Chem. Phys. Chem.* **1**, 60–62.

Lahamer, A. S., Mahurin, S. M., Compton, R. N., House, D., Laerdahl, J. K., Lein, M. and Schwerdtfeger, P. (2000) *Phys. Rev. Lett.* **85**, 4470–4473.

LaJohn, L. A., Christiansen, P. A., Ross, R. B., Atashroo, T. and Ermler, W. C. (1987) *J. Chem. Phys.* **87**, 2812.

Lammertsma, K., v. R. Schleyer, P. and Schwarz, H. (1989a) *Angew. Chem.* **101**, 1313.

Lammertsma, K., v. R. Schleyer, P. and Schwarz, H. (1989b) *Angew. Chem. Int. Ed. Engl.* **28**, 1321.

Landau, A., Eliav, E. and Kaldor, U. (1999) *Chem. Phys. Lett.* **313**, 399–403.

Landau, A., Eliav, E. and Ishikawa, Y. (2001) *J. Chem. Phys.* **114**, 2977–2980.

Landau, L. D. and Lifshitz, E. M. (1934) *Sov. Phys.* **6**, 244.

Lang, P., Stepanyuk, S., Wildberger, K., Zeller, R. and Dederichs, P. (1994) *Solid State Commun.* **92**, 755.

Langhoff, S. R. and Kern, C. W. (1977) Molecular fine structure. In *Modern Theoretical Chemistry* (ed. H. F. Schaefer III), pp. 381–437. Plenum Press, New York.

Langreth, D. C. and Mehl, M. J. (1983) *Phys. Rev.* B **28**, 1809.

Laughlin, C. and Victor, G. (1988) *Adv. At. Mol. Phys.* **25**, 163.

Lawrence, T. W., Szöke, A. and Laughlin, R. B. (1992) *Phys. Rev. Lett.* **69**, 1439.

Lee, C., Yang, W. and Parr, R. G. (1988) *Phys. Rev.* B **37**, 785.

Lee, T. D. and Yang, C. N. (1956) *Phys. Rev.* **104**, 254–258.

Lee, Y. S. and McLean, A. D. (1982) *J. Chem. Phys.* **76**, 735–736.

Leininger, T., Nicklass, A., Küchle, W., Stoll, H., Dolg, M. and Bergner, A. (1996a) *Chem. Phys. Lett.* **255**, 274–280.

Leininger, T., Nicklass, A., Stoll, H., Dolg, M. and Schwerdtfeger, P. (1996b) *J. Phys. Chem.* **105**, 1052–1059.

Ley-Koo, E., Bunge, C. F. and Jáuregui, R. (2000) *J. Mol. Struct. (THEOCHEM)* **527**, 11–25.

Lide, D. R. (ed.) (1996) *Handbook of Chemistry and Physics 1996–97*. CRC Press, Boca Raton.

Lieb, E. H. and Siedentop, H. (2000) *Commun. Math. Phys.* **213**, 673.

Lieb, E. H. and Yau, H.-T. (1988) *Commun. Math. Phys.* **118**, 177.

Lieb, E. H., Loss, M. and Siedentop, H. (1996) *Helv. Phys. Acta* **69**, 974.

Lieb, E. H., Siedentop, H. and Solovej, J. P. (1997a) *J. Statist. Phys.* **89**, 37.

Lieb, E. H., Siedentop, H. and Solovej, J. P. (1997b) *Phys. Rev. Lett.* **79**, 1785.

Liebsch, A. (1979) *Phys. Rev. Lett.* **43**, 1431.

Lihn, H.-T. S., Wu, S., Drew, H. D., Kaplan, S., Li, Q. and Fenner, D. B. (1996) *Phys. Rev. Lett.* **78**, 3810.

Lim, S. P., Price, D. L. and Cooper, B. R. (1991) *IEEE Trans. Magn.* **27**, 3648.

Lindgren, I. (1987) *Physica Scr.* **36**, 591–601.

Lindgren, I. (1989) A relativistic coupled-cluster approach with radiative corrections. In Kaldor (1989), pp. 293–306.

Lindgren, I. (1998) *Mol. Phys.* **94**, 19–28.

Lindgren, I. (2000) *Mol. Phys.* **98**, 1159.

Lindgren, I. and Morrison, J. (1986) *Atomic Many-Body Theory*, vol. 3, 2nd edn. Springer.

Lindgren, I., H. Persson, S. S. and Sunnergren, P. (1998) *Phys. Rev.* A **58**, 1001.

Lindgren, I., Persson, H., Salomonson, S. and Labzowsky, L. (1995) *Phys. Rev.* A **51**, 1167.

Lindgren, I., Persson, H., Salomonson, S., Karasiev, V., Labzowsky, L., Mitrushenkov, A. and Tokman, M. (1993) *J. Phys.* B **26**, L503.

Lindroth, E. (1988) *Phys. Rev.* A **37**, 316–328.

Lindroth, E. and Hvarfner, J. (1992) *Phys. Rev.* A **45**, 2771–2776.

Lindroth, E. and Salomonson, S. (1990) *Phys. Rev.* A **41**, 4659–4669.

Lingott, R. M., Liebermann, H.-P., Alekseyev, A. B. and Buenker, R. J. (1999) *J. Chem. Phys.* **110**, 11 294–11 302.

Lipkowitz, K. B. and Boyd, D. B. (eds) (1993) *Reviews in Computational Chemistry*, vol. 3. VCH, New York.

Liu, K. L., MacDonald, A. H., Daams, J. M., Vosko, S. H. and Koelling, D. D. (1979) *J. Magn. Magn. Mater.* **12**, 43.

Liu, W. and van Wüllen, C. (1999) *J. Chem. Phys.* **110**, 3730–3735.

Liu, W. and van Wüllen, C. (2000) *J. Chem. Phys.* **113**, 2506.

Liu, W., Hong, G., Dai, D., Li, L. and Dolg, M. (1997) *Theor. Chem. Acc.* **96**, 75–83.

Liu, W., Dolg, M. and Li, L. (1998) *J. Chem. Phys.* **108**, 2886–2895.

Liu, W., Franke, R. and Dolg, M. (1999) *Chem. Phys. Lett.* **302**, 231–239.

Llusar, R., Casarrubios, M., Barandiarán, Z. and Seijo, L. (1996) *J. Chem. Phys.* **105**, 5321–5330.

López-Urrutia, J. R. C., Beiersdorfer, P., Savin, D. W. and Widmann, K. (1996) *Phys. Rev. Lett.* **77**, 826.

López-Urrutia, J. R. C., Beiersdorfer, P., Widmann, K., Birkett, B. B., Mårtensson-Pendrill, A.-M. and Gustavsson, M. G. H. (1998) *Phys. Rev.* A **57**, 879.

Louie, S. G., Froyen, S. and Cohen, M. L. (1982) *Phys. Rev.* B **26**, 1738.

Lynch, M. and Hu, P. (2000) *Surf. Sci.* **458**, 1.

Lyons, K. B., Kwo, J., Dillon, J. F., Espinosa, G. P., McGlashan-Powell, M., Ramirez, A. P. and Schneemeyer, L. F. (1990) *Phys. Rev. Lett.* **64**, 2949.

Lyons, K. B., Dillon, J. F., Hellman, E. S., Hartford, E. H. and McGlashan-Powell, M. (1991) *Phys. Rev.* B **43**, 11 408.

MacDonald, A. H. (1983) *J. Phys.* C **16**, 3869.

MacDonald, A. H. and Vosko, S. H. (1979) *J. Phys.* C **12**, 2977.

MacDonald, A. H., Daams, J. M., Vosko, S. H. and Koelling, D. D. (1981) *Phys. Rev.* B **23**, 6377.

MacDonald, A. H., Daams, J. M., Vosko, S. H. and Koelling, D. D. (1982) *Phys. Rev.* B **25**, 713.

McGuire, T. R. and Potter, R. I. (1975) *IEEE Trans. Magn.* **11**, 1018.

McGuire, T. R., Aboaf, J. A. and Klokholm, E. (1984) *J. Appl. Phys.* **55**, 1951.

Mackintosh, A. R. and Andersen, O. K. (1980) The electronic structure of transition metals. In *Electrons at the Fermi Surface* (ed. M. Springford). Cambridge University Press.

McPerson, A., Gibson, G., Jara, H., Johanson, U., Luk, T. S., McIntyre, I., Boyer, K. and Rhoades, C. K. (1897) *J. Opt. Am.* B **4**, 595.

McWeeny, R. (1992) *Methods of Molecular Quantum Mechanics*, 2nd edn. Academic.

Maine, P., Strickland, D., Bado, P., Pessot, M. and Mourou, G. (1988) *IEEE J. Quant. Electron.* **24**, 398.

Malkin, V. G., Malkina, O. L. and Salahub, D. R. (1996) *Chem. Phys. Lett* **261**, 335.

Malkina, O. L., Schimmelpfennig, B., Kaupp, M., Hess, B. A., Chandra, P., Wahlgren, U. and Malkin, V. G. (1998) *Chem. Phys. Lett.* **93**, 296.

Malkina, O. L., Vaara, J., Schimmelpfennig, B., Munzarová, M., Malkin, V. G. and Kaupp, M. (2000) *J. Am. Chem. Soc.* **122**, 9206–9218.

Mallampalli, S. and Sapirstein, J. (1996) *Phys. Rev.* A **54**, 2714.

Mallampalli, S. and Sapirstein, J. (1998) *Phys. Rev.* A **57**, 1548.

Malli, G. L. (ed.) (1994) *Polyatomic Molecular Dirac–Hartree–Fock Calculations with Gaussian Basis Sets: Theory, Implementation and Application*. NATO ASI Series, Series B: Physics, vol. 318. Plenum Press, New York.

Malli, G. L. and Ishikawa, Y. (1998) *J. Chem. Phys.* **109**, 8759–8763.

Malli, G. L. and Styszyński, J. (1994) *J. Chem. Phys.* **101**, 10 736–10 745.

Malli, G. L. and Styszynski, J. (1996) *J. Chem. Phys.* **104**, 1012–1017.

Malli, G. L. and Styszynski, J. (1998) *J. Chem. Phys.* **109**, 4448–4455.

Malli, G. L., Da Silva, A. B. F. and Ishikawa, Y. (1992) *Can. J. Chem.* **70**, 1822–1826.

Malli, G. L., Da Silva, A. B. F. and Ishikawa, Y. (1993) *Phys. Rev.* A **47**, 143–146.

Malli, G. L., Da Silva, A. B. F. and Ishikawa, Y. (1994) *J. Chem. Phys.* **101**, 6829–6833.

Manakov, N. L., Nekipelov, A. A. and Fainstein, A. G. (1989) *Sov. Phys. JETP* **68**, 673.

Marian, C. M. (2000) Spin–orbit coupling in molecules. In *Reviews in Computational Chemistry* (ed. K. Lipkowitz and D. Boyd), pp. 152–278. Wiley–VCH.

Marian, C. M. and Wahlgren, U. (1996) *Chem. Phys. Lett.* **251**, 357–364.

Marques, M., Capelle, K. and Gross, E. K. U. (1999) *Physica* C **317–318**, 508.

Marrs, R. E., Elliott, S. R. and Stöhlker, T. (1995) *Phys. Rev.* A **52**, 3577.

Mårtensson-Pendrill, A.-M., Lindgren, I., Lindroth, E., Salomonson, S. and Staudte, D. S. (1995) *Phys. Rev.* A **51**, 3630–3635.

Maruyama, H., Koizumi, A., Yamazaki, H., Iwazumi, T. and Kawata, H. (1992) *J. Magn. Magn. Mater.* **104**, 2055.

Mason, R. S., Williams, C. M. and Anderson, P. D. J. (1995) *J. Chem. Soc. Chem. Commun.*, p. 1027.

Mateev, A. V. and Rösch, N. (2002) Self-consistent spin–orbit interaction in the Douglas–Kroll approach to relativistic density functional theory. (In preparation.)

Matila, T., Ellingsen, K., Saue, T., Aksela, H. and Gropen, O. (2000) *Phys. Rev.* A **61**, 032712.

Matsumoto, M., Staunton, J. B. and Strange, P. (1990) *J. Phys. Cond. Matt.* **2**, 8365.

Matsuoka, O. (1992) *J. Chem. Phys.* **97**, 2271–2275.

Matsuoka, O. (1993) LINMOL: Dirac–Fock–Roothaan SCF program for linear molecules. In Clementi (1993), pp. 337–343.

Matsuoka, O. and Koga, T. (2001) *Theor. Chem. Acc.* **105**, 473–476.

Mayer, M. (1999) A parallel density functional method: implementation of the two-component Douglas–Kroll–Hess method and applications to relativistic effects in heavy-element chemistry. PhD thesis, Technical University of Munich.

Mayer, M., Häberlen, O. D. and Rösch, N. (1996) *Phys. Rev.* A **54**, 4775–4782.

Mayer, M., Krüger, S. and Rösch, N. (2002) *J. Chem. Phys.* **115**, 4411.

Meier, F. (1985) Polarized electrons in surface physics. In *Spin Polarized Photoemission by Optical Spin Orientation in Semiconductors* (ed. R. Feder). World Scientific, Singapore.

Mendelsohn, L. B., Biggs, F. and Mann, J. B. (1970) *Phys. Rev.* A **2**, 1130.

Mertig, I., Zeller, R. and Dederichs, P. H. (1993) *Phys. Rev.* B **47**, 16 178.

Meskers, S. C. J., Polonski, T. and Dekkers, H. P. J. M. (1995) *J. Phys. Chem.* **99**, 1134–1142.

Metz, B., Schweizer, M., Stoll, H., Dolg, M. and Liu, W. (2000a) *Theor. Chim. Acta* **104**, 22–28.

Metz, B., Stoll, H. and Dolg, M. (2000b) *J. Chem. Phys.* **113**, 2563–2569.

Meyer, J., Sepp, W.-D., Fricke, B. and Rosén, A. (1996) *Comp. Phys. Commun.* **96**, 263.

Meyerhofer, D. D., Knauer, J. P., McNaught, S. J. and Moore, C. I. (1996) *J. Opt. Soc. Am.* B **13**, 113.

Min, B. I. and Jang, Y.-R. (1991) *J. Phys. Cond. Matt.* **3**, 5131.

Minaev, B. F. and Ågren, H. (1996) *Int. J. Quant. Chem.* **57**, 519–532.

Mitas, L. (1994) *Phys. Rev.* A **49**, 4411.

Mitrushenkov, A., Labzowsky, L., Lindgren, I., Persson, H. and Salomonson, S. (1995) *Phys. Lett.* A **51**, 200.

Mittleman, M. H. (1981) *Phys. Rev.* A **24**, 1167.

Mohanty, A. K. (1992) *Int. J. Quant. Chem.* **42**, 627–662.

Mohanty, A. and Clementi, E. (1990) Dirac–Fock self-consistent field calculations for closed-shell molecules with kinetic balance and finite nuclear size. In *Modern Techniques in Computational Chemistry: MOTECC-90* (ed. E. Clementi), pp. 693–730. ESCOM, Leiden.

Mohanty, A. K. and Clementi, E. (1991) *Int. J. Quant. Chem.* **39**, 487–517.

Mohr, P. J. (1974a) *Ann. Phys. (NY)* **88**, 26.

Mohr, P. J. (1974b) *Ann. Phys. (NY)* **88**, 52.

Mohr, P. J. (1992) *Phys. Rev.* A **46**, 4421.

Mohr, P. J. (1996) In *Atomic, Molecular, and Optical Physics Handbook* (ed. G. W. F. Drake). AIP, Woodbury, NY.

Mohr, P. J. and Sapristein, J. (2000) *Phys. Rev.* A **62**, 052501.

Mohr, P. J. and Soff, G. (1993) *Phys. Rev. Lett.* **70**, 158.

Mohr, P. J. and Taylor, B. N. (2000) *Rev. Mod. Phys.* **72**, 351–495.

Mohr, P. J., Plunien, G. and Soff, G. (1998) *Phys. Rep.* **293**, 227–369.

Mokler, P. H., Stöhlker, T., Kozhuharov, C., Moshammer, R., Rymuza, P., Stachura, Z. and Warczak, A. (1995) *J. Phys.* B **28**, 617.

Molzberger, K. and Schwarz, W. H. E. (1996) *Theor. Chim. Acta* **94**, 213–222.

Momberger, K., Belkacem, A. and Sorensen, A. H. (1996) *Phys. Rev.* A **53**, 1605.

Monge Oroz, M., Schier, A. and Schmidbaur, H. (1999) *Z. Naturf.* B **54**, 26.

Moore, C. I., Knauer, J. P. and Meyerhofer, D. D. (1995) *Phys. Rev. Lett.* **74**, 2439.

Morgan, C. G., Kratzer, P. and Scheffler, M. (1999) *Phys. Rev. Lett.* **82**, 4886.

Moroni, E. G., Kresse, G., Hafner, J. and Furthmüller, J. (1997) *Phys. Rev.* B **56**, 15 629.

Morrison, J. D. and Moss, R. E. (1980) *Mol. Phys.* **41**, 491–507.

Moskovits, M. and DiLella, D. P. (1980) *J. Chem. Phys.* **73**, 4917.

Moskovits, M., DiLella, D. P. and Limm, W. (1984) *J. Chem. Phys.* **80**, 626.

Mott, N. F. (1949) *Proc. Phys. Soc.* A **62**, 416.

Moukara, M., Städele, M., Majewski, J. A., Vogl, P. and Görling, A. (2000) *J. Phys.* C **12**, 6783.

Muller, H. G. (1999) *Phys. Rev. Lett.* **83**, 3158.

Muller, H. G. and Gavrila, M. (1993) *Phys. Rev. Lett.* **71**, 1693.

Müller, W., Flesch, J. and Meyer, W. (1984) *J. Chem. Phys.* **80**, 3297.

Mulliken, R. S. (1955) *J. Chem. Phys.* **23**, 1833.

Munro, L. J., Johnson, J. K. and Jordan, K. D. (2001) *J. Chem. Phys.* **114**, 5545–5551.

Münzenberg, G. and Hofmann, S. (1999) In Greiner and Gupta (1999), p. 9.

Murad, E. (1980) *J. Chem. Phys.* **73**, 1381.

Mustre de Leon, J., Rehr, J. J., Zabinsky, S. I. and Albers, R. C. (1991) *Phys. Rev.* B **44**, 4146.

Nakatsuji, H., Takashima, H. and Hada, M. (1995) *Chem. Phys. Lett.* **233**, 95–101.

Narath, A. (1968) *J. Appl. Phys.* **39**, 553.

Nash, C. S., Bursten, B. E. and Ermler, W. C. (1997) *J. Chem. Phys.* **106**, 5133.

Nasluzov, V. A. and Rösch, N. (1996) *Chem. Phys.* **210**, 413–425.

Nefiodov, A. V., Labzowsky, L., Plunien, G. and Soff, G. (1996) *Phys. Lett.* A **222**, 227.

Neyman, K. M., Nasluzov, V. A., Hahn, J., Landis, C. R. and Rösch, N. (1997) *Organometallics* **16**, 995–1000.

Nicklass, A. and Stoll, H. (1995) *Mol. Phys.* **86**, 317.

Nicklass, A., Dolg, M., Stoll, H. and Preuß, H. (1995) *J. Chem. Phys.* **102**, 8942–8952.

Niering, M., Holzwarth, R., Reichert, J., Pokasov, P., Udem, T., Weitz, M., Hänsch, T. W., Lemonde, P., Santarelli, G., Abgrall, M., Laurent, P., Salomon, C. and Clairon, A. (2000) *Phys. Rev. Lett.* **84**, 5496.

Nieuwpoort, W. C., Aerts, P. J. C. and Visscher, L. (1994) Molecular electronic structure calculations based on the Dirac–Coulomb–(Breit) Hamiltonian. In Malli (1994), pp. 59–70.

Nikitin, A. G. (1998) *J. Phys.* A **31**, 3297–3300.

Nomura, Y., Takeuchi, Y. and Nakagawa, N. (1969) *Tetrahedron Lett.*, p. 639.

Nonas, B., Cabria, I., Zeller, R., Dederichs, P. H., Huhne, T. and Ebert, H. (2001) *Phys. Rev. Lett.* **86**, 2146.

Nordström, L. (1991) PhD thesis, Uppsala University.

Nordström, L. and Singh, D. J. (1996) *Phys. Rev. Lett.* **76**, 4420.

Norrington, P. H. and Grant, I. P. (1987) *J. Phys.* B **20**, 4869–4881.

Noya, H., Arima, A. and Horie, H. (1958) *Prog. Theor. Phys.* **8**, 33.

Oda, T., Pasquarello, A. and Car, R. (1998) *Phys. Rev. Lett.* **80**, 3622.

Ogletree, D. F., van Hove, M. A. and Somorjai, G. A. (1986) *Surf. Sci.* **173**, 351.

Ohnishi, S., Freeman, A. J. and Weinert, M. (1983) *Phys. Rev.* B **28**, 6741.

Okada, S., Shinada, M. and Matsuoka, O. (1990) *J. Chem. Phys.* **93**, 5013–5019.

Olah, G. A. (1993a) *Angew. Chem.* **105**, 805.

Olah, G. A. (1993b) *Angew. Chem. Int. Ed. Engl.* **32**, 767.

Olah, G. A. (1995a) *Angew. Chem.* **107**, 1519.

Olah, G. A. (1995b) *Angew. Chem. Int. Ed. Engl.* **34**, 1393.

Oliveira, L. N., Gross, E. K. U. and Kohn, W. (1988) *Phys. Rev. Lett.* **60**, 2430.

Pachucki, K. (1999) *Phys. Rev.* A **60**, 3593.

Pacios, L. F. and Christiansen, P. A. (1985) *J. Chem. Phys.* **82**, 2664.

Paduch, K. and Bieroń, J. (2000) *J. Phys.* B **33**, 303–311.

Park, C.-Y. and Almlöf, J. E. (1994) *Chem. Phys. Lett.* **231**, 269–276.

Parpia, F. A. and Mohanty, A. K. (1995) *Phys. Rev.* A **52**, 962–968.

Parpia, F. A., Mohanty, A. K. and Clementi, E. (1992a) *J. Phys.* B **25**, 1–16.

Parpia, F. A., Tong, M. and Fischer, C. F. (1992b) *Phys. Rev.* A **46**, 3717–3724.

Parpia, F. A., Fischer, C. F. and Grant, I. P. (1996) *Comp. Phys. Commun.* **94**, 249.

Patterson, F. G. and Perry, M. D. (1991) *J. Opt. Soc. Am.* B **8**, 2384.

Patterson, F. G., Gonzales, R. and Perry, M. D. (1991) *J. Opt. Lett.* **16**, 1107.

Pauli, W. and Villars, F. (1949) *Rev. Mod. Phys.* **21**, 434.

Paulus, W., Kratz, J. V., Strub, E., Zauner, S., Brüchle, W., Pershina, V., Schädel, M., Schausten, B., Adams, J. L., Gregorich, K. E., Hoffman, D. C., Lane, M. R., Laue, C., Lee, D. M., McGrath, C. A., Shaughnessy, D. K., Strellis, D. A. and Sylwester, E. R. (1999) *Radiochim. Acta* **84**, 69.

Pavlov, M., Blomberg, M. R. A., Siegbahn, P. E. M., Wesendrup, R., Heinemann, C. and Schwarz, H. (1997) *J. Phys. Chem.* **101**, 1567.

Pendry, J. B. (1974) *Low Energy Electron Diffraction*. Academic.

Perdew, J. P. (1986a) *Phys. Rev.* B **22**, 8822.

Perdew, J. P. (1986b) *Phys. Rev.* B **34**, 7406.

Perdew, J. P. (1991) *Electronic Structure of Solids 1991*. Akademie Verlag, Berlin, p. 11.

Perdew, J. P. and Zunger, A. (1981) *Phys. Rev.* B **23**, 5048.

Perdew, J. P., Chevary, J. A., Vosko, S. H., Jackson, K. A., Pederson, M. R., Singh, D. J. and Fiolhais, C. (1992) *Phys. Rev.* B **46**, 6671.

Perez-Jorda, J. M., Becke, A. D. and San-Fabian, E. (1994) *J. Chem. Phys.* **100**, 6520.

Pernpointner, M. and Visscher, L. (2001a) *J. Chem. Phys.* **114**, 10 389–10 395.

Pernpointner, M. and Visscher, L. (2001b) Parallelization of the relativistic coupled cluster program RELCCSD. NCF Technical Report NRG-1999-08, Vrije Universiteit Amsterdam, The Netherlands.

Pernpointner, M., Seth, M. and Schwerdtfeger, P. (1998) *J. Chem. Phys.* **108**, 6722–6738.

Pernpointner, M., Visscher, L., de Jong, W. A. and Broer, R. (2000) *J. Comp. Chem.* **21**, 1176–1186.

Pershina, V. (1996) *Chem. Rev.* **96**, 1977.

Pershina, V. (1998a) *Radiochim. Acta* **80**, 65.

Pershina, V. (1998b) *Radiochim. Acta* **80**, 75.

Pershina, V. and Bastug, T. (1999) *Radiochim. Acta* **84**, 79–84.

Pershina, V. and Fricke, B. (1993) *J. Chem. Phys.* **99**, 9720–9729.

Pershina, V. and Fricke, B. (1999) In Greiner and Gupta (1999), pp. 194–262.

Pershina, V. and Kratz, J. V. (2001) *Inorg. Chem.* **40**, 776–780.

Pershina, V., Fricke, B., Kratz, J. V. and Ionova, G. V. (1994) *Radiochim. Acta* **64**, 31.

Pershina, V., Johnson, E. and Fricke, B. (1999) *J. Phys. Chem.* A **103**, 8463–8470.

Pershina, V., Bastug, T., Varga, S. and Fricke, B. (2001) *J. Chem. Phys.* **115**, 792–799.

Persson, H., Lindgren, I. and Salomonson, S. (1993a) *Physica Scr.* T **46**, 125.

Persson, H., Lindgren, I., Salomonson, S. and Sunnergren, P. (1993b) *Phys. Rev.* A **48**, 2772.

Persson, H., Lindgren, I., Labzowsky, L. N., Plunien, G., Beier, T. and Soff, G. (1996a) *Phys. Rev.* A **54**, 2805.

Persson, H., Salomonson, S., Sunnergren, P. and Lindgren, I. (1996b) *Phys. Rev. Lett.* **76**, 204.

Persson, H., Schneider, S. M., Greiner, W., Soff, G. and Lindgren, I. (1996c) *Phys. Rev. Lett.* **76**, 1433.

Pfrepper, G., Pfrepper, R., Kronenberg, A., Kratz, J. V., Nähler, A., Brüchle, W. and Schädel, M. (2000) *Radiochim. Acta* **88**, 273.

Pfrepper, G., Pfrepper, R., Yakushev, A. B., Timokhin, S. N. and Zvara, I. (1997) *Radiochim. Acta* **77**, 201.

Philipsen, P. H. T., van Lenthe, E., Snijders, J. G. and Baerends, E. J. (1997) *Phys. Rev.* B **56**, 13 556.

Pickett, W. (1989) *Comput. Phys. Rep.* **9**, 115.

Pisani, L. and Clementi, E. (1993) Dirac–Fock self-consistent field calculations for closed-shell molecules with kinetic balance. In Clementi (1993), pp. 345–360.

Pisani, L. and Clementi, E. (1994a) *J. Comp. Chem.* **15**, 466–474.

Pisani, L. and Clementi, E. (1994b) *J. Chem. Phys.* **101**, 3079–3084.

Pisani, L. and Clementi, E. (1995a) *J. Chem. Phys.* **103**, 9321–9323.

Pisani, L. and Clementi, E. (1995b) Methods and techniques in computational chemistry METECC-95. In Clementi and Corongiu (1995), pp. 219–218.

Pittini, R., Schoenes, J., Vogt, O. and Wachter, P. (1996) *Phys. Rev. Lett.* **77**, 944.

Pitzer, K. (1984) *Int. J. Quant. Chem.* **25**, 131.

Pitzer, K. S. (1979) *Acc. Chem. Res.* **12**, 271–276.

Plante, D. R., Johnson, W. R. and Sapirstein, J. (1994) *Phys. Rev.* A **49**, 3519–3530.

Plehn, H., Wacker, O.-J. and Kümmel, R. (1994) *Phys. Rev.* B **49**, 12 140.
Plunien, G. and Soff, G. (1995) *Phys. Rev.* A **51**, 1119.
Plunien, G., Beier, T., Soff, G. and Persson, H. (1998) *Eur. Phys. J.* D **1**, 177.
Plunien, G., Müller, B., Greiner, W. and Soff, G. (1991) *Phys. Rev.* A **43**, 5853.
Popescu, V., Ebert, H. and Jenkins, A. C. (1999) *J. Synchr. Rad.* **6**, 711.
Preisenberger, M., Schier, A. and Schmidbaur, H. (1998) *Z. Naturf.* B **53**, 781.
Preisenberger, M., Pyykkö, P., Schier, A. and Schmidbaur, H. (1999a) *Inorg. Chem.* **38**, 5870.
Preisenberger, M., Schier, A. and Schmidbaur, H. (1999b) *J. Chem. Soc., Dalton Trans.*, p. 1645.
Pulay, P. (1983) *J. Chem. Phys.* **78**, 5044.
Pulm, F., Schramm, J., Lagies, H., Hormes, J., Grimme, S. and Peyerimhoff, S. D. (1997) *Chem. Phys.* **224**, 143.
Purdum, H., Montana, P. A., Shenoy, G. K. and Morrison, T. (1982) *Phys. Rev.* B **25**, 4412.
Pyper, N. C. (1983) *Chem. Phys. Lett.* **96**, 204.
Pyper, N. C. (1988) *Mol. Phys.* **64**, 933.
Pyykkö, P. (1978) *Adv. Quant. Chem.* **11**, 353–409.
Pyykkö, P. (1983) *Chem. Phys.* **74**, 1.
Pyykkö, P. (1986) *Relativistic Theory of Atoms and Molecules*, vol. I. Springer.
Pyykkö, P. (1988) *Chem. Rev.* **88**, 563.
Pyykkö, P. (1993) *Relativistic Theory of Atoms and Molecules*, vol. II. Springer.
Pyykkö, P. (1995) *J. Am. Chem. Soc.* **117**, 2067.
Pyykkö, P. (1997) *Chem. Rev.* **97**, 597.
Pyykkö, P. (2000) *Relativistic Theory of Atoms and Molecules*, vol. III. Springer.
Pyykkö, P. (2001) Database 'RTAM' (relativistic quantum chemistry database) 1915–1998; http://www.csc.fi/lul/rtam/rtamquery.html.
Pyykkö, P. and Desclaux, J. P. (1979) *Acc. Chem. Res.* **12**, 276–281.
Pyykkö, P. and Seth, M. (1997) *Theor. Chem. Acc.* **96**, 92–104.
Pyykkö, P. and Stoll, H. (2000) In *Chemical Modelling, Applications and Theory* (ed. A. Hinchliffe), pp. 239–305. RSC Special Periodical Report, vol. 1. The Royal Society of Chemistry, Cambridge, UK.
Pyykkö, P., Pajanne, E. and Inokuti, M. (1973) *Int. J. Quant. Chem.* **7**, 785.
Pyykkö, P., Görling, A. and Rösch, N. (1987) *Mol. Phys.* **61**, 195.
Pyykkö, P., Runeberg, N., Straka, M. and Dyall, K. G. (2000) *Chem. Phys. Lett.* **328**, 415–419.
Quiney, H. M. and Grant, I. P. (1993) *Physica Scr.* T **46**, 132.
Quiney, H. M. and Grant, I. P. (1994) *J. Phys.* B **27**, L299.
Quiney, H. M., Grant, I. P. and Wilson, S. (1989a) *J. Phys.* B **22**, L15–L19.
Quiney, H. M., Grant, I. P. and Wilson, S. (1989b) On the relativistic many-body perturbation theory of atomic and molecular electronic structure. In Kaldor (1989), pp. 307–344.
Quiney, H. M., Skaane, H. and Grant, I. P. (1997) *J. Phys.* B **30**, L829–L834.
Quiney, H. M., Laerdahl, J. K., Fægri Jr, K. and Saue, T. (1998a) *Phys. Rev.* A **57**, 920–944.
Quiney, H. M., Skaane, H. and Grant, I. P. (1998b) *Adv. Quant. Chem.* **32**, 1–49.
Quiney, H. M., Skaane, H. and Grant, I. P. (1998c) *Chem. Phys. Lett.* **290**, 473–480.
Quiney, H. M., Skaane, H. and Grant, I. P. (1998d) *J. Phys.* B **38**, L85–L95.
Racah, G. (1937) *Nuovo Cim.* **14**, 93.
Racah, G. (1942) *Phys. Rev.* **42**, 438.
Rafelski, J., Fulcher, L. P. and Klein, A. (1978) *Phys. Rep.* C **38**, 227.

Rajagopal, A. K. (1978) *J. Phys.* C **11**, L943.

Rajagopal, A. K. and Callaway, J. (1973) *Phys. Rev.* B **7**, 1912.

Rakowitz, F. (1999) Entwicklung, Implementierung und Anwendung effizienter Methoden in der relativistischen Elektronenstrukturtheorie. PhD thesis, University of Bonn. Available at http://www.thch.uni-bonn.de/tc/.

Rakowitz, F. and Marian, C. M. (1996) *Chem. Phys. Lett.* **257**, 105–110.

Rakowitz, F. and Marian, C. M. (1997) *Chem. Phys.* **225**, 223–238.

Rakowitz, F., Casarrubios, M., Seijo, L. and Marian, C. M. (1998) *J. Chem. Phys.* **108**, 7980–7987.

Rakowitz, F., Marian, C. and Seijo, L. (1999a) *J. Chem. Phys.* **111**, 10 436.

Rakowitz, F., Marian, C., Seijo, L. and Wahlgren, U. (1999b) *J. Chem. Phys.* **110**, 3678.

Rakowitz, F., Marian, C. M. and Schimmelpfennig, B. (2000) *Phys. Chem. Chem. Phys.* **2**, 2481.

Ramana, M. V. and Rajagopal, A. K. (1979) *J. Phys.* C **12**, L845.

Ramana, M. V. and Rajagopal, A. K. (1981a) *J. Phys.* C **14**, 4291.

Ramana, M. V. and Rajagopal, A. K. (1981b) *Phys. Rev.* A **24**, 1689.

Ramana, M. V., Rajagopal, A. K. and Johnson, W. R. (1982) *Phys. Rev.* A **25**, 96.

Rampe, A., Hartmann, D. and Güntherodt, G. (1996) In Ebert and Schütz (1996), p. 49.

Reiher, M. (1998) Development and implementation of numerical algorithms for the solution of multi-configuration self-consistent field equations for relativistic atomic structure calculations. PhD thesis, Fakultät für Chemie, Universität Bielefeld, Germany.

Reiher, M. and Hess, B. A. (2000) Relativistic electronic-structure calculations for atoms and molecules. In Grotendorst (2000), pp. 451–477.

Reiher, M. and Hinze, J. (1999) *J. Phys.* B **32**, 5489–5505.

Reiher, M. and Kind, C. (2001) *J. Phys.* B **34**, 3133–3156.

Reinert, T. (2000) Anwendungsgrenzfälle der Dichtefuntionaltheorie. PhD thesis, Technical University of Clausthal.

Reiss, H. R. (1990) *J. Opt. Soc. Am.* B **7**, 574.

Reiss, H. R. (1996) *Phys. Rev.* A **54**, R1765.

Rhoades-Brown, M. J. and Weneser, J. (1991) *Phys. Rev.* A **44**, 330.

Richards, W. G., Trivedi, H. P. and Cooper, D. L. (1981) *Spin–Orbit Coupling in Molecules.* Clarendon Press, Oxford.

Richter, M. and Eschrig, H. (1989) *Solid State Commun.* **72**, 263.

Riegel, D., Barth, H. J., Büermann, L., Haas, H. and Stenzel, C. (1986) *Phys. Rev. Lett.* **57**, 388.

Rieger, M. M. and Vogl, P. (1995) *Phys. Rev.* A **52**, 282.

Rodrigues, G. C., Ourdane, M. A., Bieroń, J., Indelicato, P. and Lindroth, E. (2000) *Phys. Rev.* A **63**, 012510–1–10.

Röhrl, N. (2000) PhD thesis, Universität München.

Rösch, N., Krüger, S., Mayer, M. and Nasluzov, V. A. (1996) The Douglas–Kroll–Hess approach to relativistic density functional theory: Methodological aspects and applications to metal complexes and clusters. In *Recent Developments and Applications of Modern Density Functional Theory* (ed. J. M. Seminario), pp. 497–566. Elsevier.

Rose, M. E. (1961) *Relativistic Electron Theory.* Wiley.

Rosén, A. and Ellis, D. E. (1975) *J. Chem. Phys.* **62**, 3039.

Ross, R. B., Gayen, S. and Ermler, W. C. (1994) *J. Chem. Phys.* **100**, 8145.
Ross, R. B., Powers, J. M., Atashroo, T., Ermler, W. C., LaJohn, L. A. and Christiansen, P. A. (1990) *J. Chem. Phys.* **93**, 6654.
Roszak, S. and Balasubramanian, K. (1995) *J. Chem. Phys.* **103**, 1043.
Rue, C., Armentrout, P. B., Kretzschmar, I., Schröder, D., Harvey, J. N. and Schwarz, H. (1999) *J. Chem. Phys.* **110**, 7858.
Rutkowski, A. (1986a) *J. Phys.* B **19**, 149–158.
Rutkowski, A. (1986b) *J. Phys.* B **19**, 3431–3441.
Rutkowski, A. (1986c) *J. Phys.* B **19**, 3443–3455.
Rutkowski, A. and Schwarz, W. H. E. (1990) *Theor. Chim. Acta* **76**, 391–410.
Rutkowski, A., Rutkowska, D. and Schwarz, W. H. E. (1992) *Theor. Chim. Acta* **84**, 105–114.
Rutkowski, A., Schwarz, W. H. E. and Koslowski, R. (1993) *Theor. Chim. Acta* **87**, 75–87.
Safronova, M. S., Derevianko, A. and Johnson, W. R. (1998) *Phys. Rev.* A **58**, 1016–1028.
Safronova, M. S., Johnson, W. R. and Derevianko, A. (1999a) *Phys. Rev.* A **60**, 4476–4487.
Safronova, M. S., Johnson, W. R. and Safronova, U. I. (1996) *Phys. Rev.* A **53**, 4036–4053.
Safronova, M. S., Johnson, W. R. and Safronova, U. I. (1997) *J. Phys.* B **30**, 2375–2393.
Safronova, U. I., Derevianko, A., Safronova, M. S. and Johnson, W. R. (1999b) *J. Phys.* B **32**, 3527–3545.
Safronova, U. I., Johnson, W. R. and Albritton, J. R. (2000a) *Phys. Rev.* A **62**, 052505–1–14.
Safronova, U. I., Johnson, W. R. and Berry, H. G. (2000b) *Phys. Rev.* A **61**, 052503–1–11.
Safronova, U. I., Johnson, W. R., Kato, D. and Ohtani, S. (2001) *Phys. Rev.* A **63**, 032518–1–11.
Sahni, V., Gruenebaum, J. and Perdew, J. P. (1982) *Phys. Rev.* B **26**, 4371.
Sakurai, J. J. (1967) *Advanced Quantum Mechanics*. Addison-Wesley, Reading.
Salamin, Y. I. and Faisal, F. H. M. (1996) *Phys. Rev.* A **54**, 4383.
Salamin, Y. I. and Faisal, F. H. M. (1997) *Phys. Rev.* A **55**, 3678.
Salamin, Y. I. and Faisal, F. H. M. (1999) *Phys. Rev.* A **60**, 2505.
Salamin, Y. I. and Faisal, F. H. M. (2000) *Phys. Rev.* A **61**, 043801–1.
Samzow, R. and Hess, B. A. (1991) *Chem. Phys. Lett.* **184**, 491–495.
Samzow, R., Hess, B. A. and Jansen, G. (1992) *J. Chem. Phys.* **96**, 1227–1231.
Sändig, N. and Koch, W. (1998) *Organometallics* **17**, 2344.
Sandratskii, L. M. (1998) *Adv. Phys.* **47**, 91.
Sandratskii, L. M. and Kübler, J. (1995) *Phys. Rev. Lett.* **75**, 946.
Santos, J. P., Marques, J. P., Parente, F., Lindroth, E., Boucard, S. and Indelicato, P. (1998) *Eur. Phys. J.* D **1**, 149–163.
Santos, J. P., Marques, J. P., Parente, F., Lindroth, E., Indelicato, P. and Desclaux, J. P. (1999) *J. Phys.* B **32**, 2089–2097.
Sapirstein, J. (1993) *Physica Scr.* T **46**, 52–60.
Sapirstein, J. (1998) *Rev. Mod. Phys.* **70**, 55–76.
Sarachik, E. S. and Schappert, G. T. (1970) *Phys. Rev.* D **1**, 2738.
Satoko, C. (1981) *Chem. Phys. Lett.* **83**, 111.
Satoko, C. (1984) *Phys. Rev.* B **30**, 1754.
Saue, T. and Jensen, H. J. A. (1999) *J. Chem. Phys.* **111**, 6211–6222.
Saue, T., Faegri, K. and Gropen, O. (1996) *Chem. Phys. Lett.* **263**, 360–366.
Saue, T., Fægri, K., Helgaker, T. and Gropen, O. (1997) *Mol. Phys.* **91**, 937–950.
Schädel, M., Brüchle, W. and Haefner, B. (1988) *Nucl. Instrum. Meth. Phys. Res.* A **264**, 308.

Schädel, M., Brüchle, W., Jäger, E., Schimpf, E., Kratz, J. V., Scherer, U. W. and Zimmermann, H. P. (1989) *Radiochim. Acta* **88**, 273.

Schädel, M., Brüchle, W., Dressler, R., Eichler, B., Gäggeler, H. W., Günther, R., Gregorich, K. E., Hoffman, D. G., Hübener, S., Jost, D. T., Kratz, J. V., Paulus, W., Schumann, D., Timokhin, S., Trautmann, N., Türler, A., Wirth, G. and Yakuschev, A. (1997a) *Nature* **388**, 55.

Schädel, M., Brüchle, W., Schausten, B., Schimpf, E., Jäger, E., Wirth, G., Günther, R., Kratz, J. V., Paulus, W., Seibert, A., Thörle, P., Trautmann, N., Zauner, S., Schumann, D., Andrassy, M., Misiak, R., Gregorich, K. E., Hoffman, D. G., Lee, D. M., Sylwester, E. R., Nagame, Y. and Oura, Y. (1997b) *Radiochim. Acta* **77**, 149.

Schädel, M., Brüchle, W., Jäger, E., Schausten, B., Wirth, G., Paulus, W., Günther, R., Eberhardt, K., Kratz, J. V., Seibert, A., Strub, E., Thörle, P., Trautmann, N., Waldek, A., Zauner, S., Schumann, D., Kirbach, U., Kubica, B., Misiak, R., Nagame, Y. and Gregorich, K. E. (1998) *Radiochim. Acta* **83**, 163.

Schadler, G., Albers, R. C., Boring, A. M. and Weinberger, P. (1987) *Phys. Rev.* B **35**, 4324.

Scherbaum, F., Grohmann, A., Huber, B., Krüger, C. and Schmidbaur, H. (1988a) *Angew. Chem.* **100**, 1602.

Scherbaum, F., Grohmann, A., Huber, B., Krüger, C. and Schmidbaur, H. (1988b) *Angew. Chem. Int. Ed. Engl.* **27**, 1544.

Scherff, H. L. and Herrmann, G. (1966) *Radiochim. Acta* **6**, 53.

Scheunemann, T., Halilov, S. V., Henk, J. and Feder, R. (1994) *Solid State Commun.* **91**, 487.

Schier, A., Grohmann, A., de Luzuriaga, J. M. L. and Schmidbaur, H. (2000) *Inorg. Chem.* **39**, 547.

Schimmelpfennig, B. (1996) Atomic spin–orbit Mean-Field Integral program AMFI. Stockholms Universitet.

Schlegel, H. B. (1987) In *Ab Initio Methods in Quantum Chemistry* (ed. K. P. Lawley), p. 249. Wiley.

Schleyer, v. R. P. (ed.) (1998) *Encyclopedia of Computational Chemistry*. Wiley.

Schmid, R. N. (2000) PhD thesis, University of Frankfurt.

Schmid, R. N., Engel, E., Dreizler, R. M., Blaha, P. and Schwarz, K. (1998) *Adv. Quant. Chem.* **33**, 209.

Schmid, R. N., Engel, E., Dreizler, R. M., Blaha, P. and Schwarz, K. (1999) *Adv. Quantum. Chem.* in press.

Schmidbaur, H. (1995) *Chem. Soc. Rev.*, p. 391.

Schmidbaur, H. (2000) *Gold Bull.* **33**, 3.

Schmidbaur, H., Hofreiter, S. and Paul, M. (1995) *Nature* **377**, 503.

Schmidt, J. M., Soff, G. and Mohr, P. J. (1989) *Phys. Rev.* A **40**, 2176.

Schmidt, K. and Springborg, M. (2000) *Chem. Phys. Lett.* **320**, 1–5.

Schmiedeskamp, B., Vogt, B. and Heinzmann, U. (1988) *Phys. Rev. Lett.* **60**, 651.

Schneider, C. M., Schuster, P., Hammond, M. S. and Kirschner, J. (1991) *Europhys. Lett.* **16**, 689.

Schneider, S. M., Greiner, W. and Soff, G. (1993a) *J. Phys.* B **26**, L529.

Schneider, S. M., Schaffner, J., Greiner, W. and Soff, G. (1993b) *J. Phys.* B **26**, L581.

Schneider, S. M., Greiner, W. and Soff, G. (1994) *Phys. Rev.* A **50**, 118.

Schröder, D., Hrušák, J., Tornieporth-Oetting, I. C., Klapötke, T. M. and Schwarz, H. (1994a) *Angew. Chem.* **106**, 223.

Schröder, D., Hrušák, J., Tornieporth-Oetting, I. C., Klapötke, T. M. and Schwarz, H. (1994b) *Angew. Chem. Int. Ed. Engl.* **33**, 212.

Schröder, D., Hrušák, J., Hertwig, R. H., Koch, W., Schwerdtfeger, P. and Schwarz, H. (1995) *Organometallics* **14**, 312.

Schröder, D., Harvey, J. N. and Schwarz, H. (1998a) *J. Phys. Chem.* A **102**, 3639.

Schröder, D., Heinemann, C., Schwarz, H., Harvey, J. N., Dua, S., Blanskby, S. J. and Bowie, J. H. (1998b) *Chem. Eur. J.* **4**, 2550.

Schröder, D., Schwarz, H., Hrušák, J. and Pyykkö, P. (1998c) *Inorg. Chem.* **37**, 624.

Schröder, D., Diefenbach, M., Klapötke, T. M. and Schwarz, H. (1999a) *Angew. Chem.* **111**, 206.

Schröder, D., Diefenbach, M., Klapötke, T. M. and Schwarz, H. (1999b) *Angew. Chem. Int. Ed.* **38**, 137.

Schröder, D., Brown, R., Schwerdtfeger, P. and Schwarz, H. (2000a) *Int. J. Mass Spectrom.* **203**, 155.

Schröder, D., Shaik, S. and Schwarz, H. (2000b) *Acc. Chem. Res.* **33**, 139.

Schröder, D., Wesendrup, R., Hertwig, R. H., Dargel, T., Grauel, H., Koch, W., Bender, B. R. and Schwarz, H. (2000c) *Organometallics* **19**, 2608.

Schütz, G. and Ahlers, D. (1997) *J. Phys. (Paris)* **7**, C2 59.

Schütz, G., Wagner, W., Wilhelm, W., Kienle, P., Zeller, R., Frahm, R. and Materlik, C. (1987) *Phys. Rev. Lett.* **58**, 737.

Schütz, G., Frahm, R., Mautner, P., Wienke, R., Wagner, W., Wilhelm, W. and Kienle, P. (1989) *Phys. Rev. Lett.* **62**, 2620.

Schütz, G., Knülle, M. and Ebert, H. (1993) *Physica Scr.* T **49**, 302.

Schwacke, M. (2000) Relativistische *Ab-initio*-Berechnungen der Stoßanregungsquerschnitte mittelschwerer bis schwerer Ionen mit Hilfe von R-Matrix-Verfahren. PhD thesis, Fakultät für Physik und Astronomie, Ruhr-Universität Bochum, Germany.

Schwarz, H. and Schröder, D. (2001) *Pure Appl. Chem.* **72**, 2319.

Schwarz, J., Heinemann, C., Schröder, D., Schwarz, H. and Hrušák, J. (1996a) *Helv. Chim. Acta* **79**, 1.

Schwarz, J., Heinemann, C., Schröder, D., Schwarz, H. and Hrušák, J. (1996b) *Helv. Chim. Acta* **79**, 1110.

Schwarz, W. H. E. (1987) *Physica Scr.* **36**, 403–411.

Schwarz, W. H. E. (1991) Fundamentals of relativistic effects in chemistry. In *The Concept of the Chemical Bond* (ed. Z. B. Maksić), pp. 593–643. Vol. 2 of *Theoretical Models of Chemical Bonding*. Springer.

Schwarz, W. H. E., van Wezenbeek, E. M., Baerends, E. J. and Snijders, J. G. (1989) *J. Phys.* B **22**, 1515–1530.

Schwarz, W. H. E., Rutkowski, A. and Collignon, G. (1991) Nonsingular relativistic perturbation theory and relativistic changes of molecular structure. In Wilson *et al.* (1991), pp. 135–147.

Schwarz, W. H. E., Rutkowski, A. and Wang, S. G. (1996c) *Int. J. Quant. Chem.* **57**, 641–653.

Schweppe, J., Belkacem, A., Blumenfeld, L., Claytor, N., B. Feinberg, B., Gould, H., Kostroun, V. E., Levy, L., Misawa, S., Mowat, J. R. and Prior, M. H. (1991) *Phys. Rev. Lett.* **66**, 1434.

Schwerdtfeger, P., McFeaters, J. S., Stephens, R. L., Liddell, M. J., Dolg, M. and Heß, B. A. (1994) *Chem. Phys. Lett.* **218**, 362.

Schwerdtfeger, P., Fischer, T., Dolg, M., Igel-Mann, G., Nicklass, A., Stoll, H. and Haaland, A. (1995a) *J. Chem. Phys.* **102**, 2050–2062.

Schwerdtfeger, P., McFeaters, J. S., Liddell, M. J., Hrušák, J. and Schwarz, H. (1995b) *J. Chem. Phys.* **103**, 245.

Schwerdtfeger, P., Brown, J. R., Laerdahl, J. K. and Stoll, H. (2000) *J. Chem. Phys.* **113**, 7110–7118.

Schwerdtfeger, P., Wesendrup, R., Moyano, G. E., Sadlej, A. J., Greif, J. and Hensel, F. (2001) *J. Chem. Phys.* **115**, 7401–7412.

Seebauer, E. G., Kong, A. C. F. and Schmidt, L. D. (1982) *Surf. Sci.* **123**, 164.

Seelig, P., Borneis, S., Dax, A., Engel, T., Faber, S., Gerlach, M., Holbrow, C., Kühl, G. H. T., Marx, D., Meier, K., Merz, P., Quint, W., Schmitt, F., Tomaselli, M., Völker, L., Winter, H., Würtz, M., Beckert, K., Franzke, B., Nolden, F., Reich, H., Steck, M. and Winkler, T. (1998) *Phys. Rev. Lett.* **81**, 4824.

Seewald, G., Hagn, E., Zech, E., Forkel-Wirth, D. and Buchard, A. (1999) *Phys. Rev. Lett.* **82**, 1024.

Seewald, G., Hagn, E., Zech, E., Forkel-Wirth, D., Buchard, A. and ISOLDE Collaboration (1997) *Phys. Rev. Lett.* **78**, 1795.

Segal, G. A., Wetmore, R. W. and Wolf, K. (1978) *Chem. Phys.* **30**, 269.

Segev, B. and Wells, J. C. (1998) *Phys. Rev.* A **57**, 1849.

Seidel, S. and Seppelt, K. (2000) *Science* **290**, 117.

Seidl, M., Perdew, J. P. and Kurth, S. (2000) *Phys. Rev. Lett.* **84**, 5070.

Seifert, G., Guttierrez, R. and Schmidt, R. (1996) *Phys. Lett.* A **211**, 357.

Seijo, L. (1995) *J. Chem. Phys.* **102**, 8078.

Seijo, L. and Barandiarán, Z. (1999) In *Computational Chemistry: Reviews of Current Trends* (ed. J. Leszczynski), pp. 55–152. World Scientific, Singapore.

Seijo, L., Barandiarán, Z. and Harguindey, E. (2001) *J. Chem. Phys.* **114**, 118–129.

Senoussi, S., Campbell, I. A. and Fert, A. (1977) *Solid State Commun.* **21**, 269.

Sepp, W.-D., Kolb, D., Sengler, W., Hartung, H. and Fricke, B. (1986) *Phys. Rev.* A **33**, 3679–3687.

Seth, M. and Schwerdtfeger, P. (2000) *Chem. Phys. Lett.* **318**, 314–318.

Seth, M., Dolg, M., Fulde, P. and Schwerdtfeger, P. (1995) *J. Am. Chem. Soc.* **117**, 6597–6598.

Seth, M., Schwerdtfeger, P. and Dolg, M. (1997) *J. Chem. Phys.* **106**, 3623–3632.

Seth, M., Schwerdtfeger, P., Dolg, M., Fægri, K., Hess, B. A. and Kaldor, U. (1996) *Chem. Phys. Lett* **250**, 461–465.

Severin, L., Brooks, M. S. S. and Johansson, B. (1993) *Phys. Rev. Lett.* **71**, 3214.

Severin, L., Richter, M. and Steinbeck, L. (1997) *Phys. Rev.* B **55**, 9211.

Shabaev, V. M. (1985) *Theor. Math. Phys.* **63**, 588.

Shabaev, V. M. (1990a) *Theor. Math. Phys.* **82**, 57.

Shabaev, V. M. (1990b) *Sov. Phys. J.* **33**, 660.

Shabaev, V. M. (1991) *J. Phys.* A **24**, 5665.

Shabaev, V. M. (1993) *J. Phys.* B **26**, 1103.

Shabaev, V. M. (1994) *J. Phys.* B **27**, 5825.

Shabaev, V. M. (1998) Hyperfine structure of highly charged ions. In *Atomic Physics with Heavy Ions* (ed. H. F. Beyer and V. P. Shevelko), p. 138. Springer.

Shabaev, V. M., Tomaselli, M., Kühl, T., Artemyev, A. N. and Yerokhin, V. A. (1997) *Phys. Rev.* A **56**, 252.

Shabaev, V. M., Artemyev, A. N., Beier, T., Plunien, G., Yerokhin, V. A. and Soff, G. (1998a) *Phys. Rev.* A **57**, 4235.

Shabaev, V. M., Shabaeva, M. B., Tupitsyn, I. I. and Yerokhin, V. A. (1998b) *Hyperfine Interactions* **114**, 129.

Shabaev, V. M., Shabaeva, M. B., Tupitsyn, I. I., Yerokhin, V. A., Artemyev, A. N., Kühl, T., Tomaselli, M. and Zherebtsov, O. M. (1998c) *Phys. Rev.* A **58**, 149.

Shabaev, V. M., Artemyev, A. N. and Yerokhin, V. A. (2000a) *Physica Scr.* T **86**, 7.

Shabaev, V. M., Artemyev, A. N., Zherebtsov, O. M., Yerokhin, V. A., Plunien, G. and Soff, G. (2000b) *Hyperfine Interactions* **127**, 279.

Shabaev, V. M., Artemyev, A. N., Zherebtsov, O. M., Yerokhin, V. A. and Soff, G. (2001) *Phys. Rev. Lett* **86**, 3959.

Shadwick, B. A., Talman, J. D. and Norman, M. R. (1989) *Comput. Phys. Commun.* **54**, 95.

Shaik, S., Danovich, D., Fiedler, A., Schröder, D. and Schwarz, H. (1995) *Helv. Chim. Acta* **78**, 1393.

Shaik, S., Filatov, M., Schröder, D. and Schwarz, H. (1998) *Chem. Eur. J.* **4**, 193.

Sham, L. J. (1985) *Phys. Rev.* B **32**, 3876.

Shi, Q., Zhang, S., Cho, H., Xu, K., Li, J.-M. and Kais, S. (1998) *J. Phys.* B **31**, 4123–4135.

Shick, A. B. and Gubanov, V. A. (1994) *Phys. Rev.* B **49**, 12860.

Shick, A. B., Novikov, D. L. and Freeman, A. J. (1997) *Phys. Rev.* B **56**, R14 259–R14 262.

Shick, A. B., Gornostyrev, Y. N. and Freeman, A. J. (1999) *Phys. Rev.* B **60**, 3029–3032.

Shiota, Y. and Yoshizawa, K. (2000) *J. Am. Chem. Soc.* **122**, 12 317.

Sienkiewicz, J. E. (1997) *J. Phys.* B **30**, 1261–1267.

Silver, D. M. (1980) *Phys. Rev.* A **21**, 1106.

Sinha, H. K., Abou-Zied, O. K. and Steer, R. P. (1993) *Chem. Phys. Lett.* **201**, 433–436.

Sinzig, J., deJongh, L. J., Ceriotti, A., della Pergola, R., Longoni, G., Stener, M., Albert, K. and Rösch, N. (1998) *Phys. Rev. Lett.* **81**, 3211–3214.

Sjøvoll, M., Fagerli, H., Gropen, O., Almlöf, J., Saue, T., Olsen, J. and Helgaker, T. (1997) *J. Chem. Phys.* **107**, 5496–5501.

Skudlarski, P. and Vignale, G. (1993) *Phys. Rev.* B **48**, 8547.

Smit, J. (1951) *Physica* **16**, 612.

Smith, F. C. and Johnson, W. R. (1967) *Phys. Rev.* **160**, 136–142.

Soff, G. (1980) Electron–positron pair creation and K-shell ionisation in relativistic heavy ion collisions. In *Proc. XVIII Winter School, Selected Topics in Nuclear Structure, Bielsko-Biala, Polen*, p. 201.

Soff, G. (1989) *Z. Phys.* D **11**, 29.

Soff, G. (1993) *Physica Scr.* T **46**, 266.

Soff, G. and Mohr, P. (1988) *Phys. Rev.* A **38**, 5066.

Solovyev, I. V., Shik, A. B., Antropov, V. P., Liechtenstein, A. I., Gubanov, V. A. and Andersen, O. K. (1989) *Sov. Phys. Solid State* **31**, 1285.

Sousa, C., de Jong, W. A., Broer, R. and Nieuwpoort, W. C. (1997) *J. Chem. Phys.* **106**, 7162–7169.

Soven, P. (1967) *Phys. Rev.* **156**, 809.

Springborg, M. (2000) *Methods of Electronic-Structure Calculations*. Wiley.

Städele, M., Majewski, J. A., Vogl, P. and Görling, A. (1997) *Phys. Rev. Lett.* **79**, 2089.
Stähler, S., Knülle, M., Schütz, G., Fischer, P., Welzel-Gerth, S. and Buchholz, B. (1993) *J. Appl. Phys.* **73**, 6063.
Stanton, R. E. and Havriliak, S. (1984) *J. Chem. Phys.* **81**, 1910–1918.
Staude, U., Bosselmann, P., Büttner, R., Horn, D., Schartner, K.-H., Folkmann, F., Livingston, A. E., Ludziejewski, T. and Mokler, P. H. (1998) *Phys. Rev.* A **58**, 3516.
Staunton, J. B. (1982) PhD thesis, University of Bristol.
Stearns, M. B. (1987) *Magnetic properties of 3d, 4d and 5d elements, alloys and compounds*. Vol. III/19a of *Landolt–Börnstein, New Series*. Springer.
Steih, T. (1999). (Unpublished.)
Steinbrenner, U., Bergner, A., Dolg, M. and Stoll, H. (1994) *Mol. Phys.* **82**, 3.
Stener, M., Albert, K. and Rösch, N. (1999) *Inorg. Chim. Acta* **286**, 30–36.
Stepanyuk, V., Hergert, W., Wildberger, K., Zeller, R. and Dederichs, P. (1996) *Phys. Rev.* B **53**, 2121.
Stevens, W. J., Basch, H. and Krauss, M. (1984) *J. Chem. Phys.* **81**, 6026.
Stevens, W. J., Kraus, M., Basch, H. and Jasien, P. G. (1992) *Can. J. Chem.* **70**, 612.
Sticht, J. and Kübler, J. (1985) *Solid State Commun.* **53**, 529.
Stöhlker, T., Elliott, S. R. and Marrs, R. E. (1996) *Hyperfine Interactions* **99**, 217.
Stöhlker, T., Mokler, P. H., Geissel, H., Moshammer, R., Rymuza, P., Bernstein, E. M., Cocke, C. L., Kozhuharov, C., Münzenberg, G., Nickel, F., Scheidenberger, C., Stachura, Z., Ullrich, J. and Warczak, A. (1992) *Phys. Lett.* A **168**, 285.
Stöhlker, T., Mokler, P. H., Bosch, F., Dunford, R. W., Franzke, F., Klepper, O., Kozhuharov, C., T, L., Nolden, F., Reich, H., Rymuza, P., Stachura, Z., Steck, M., Swiat, P. and Warczak, A. (2000) *Phys. Rev. Lett.* **85**, 3109.
Stöhr, J., Scholl, A., Regan, T. J., Anders, S., Luning, J., Scheinfein, M. R., Padmore, H. A. and White, R. L. (1999) *Phys. Rev. Lett.* **83**, 1862.
Stoll, H., Metz, B. and Dolg, M. (2001) *J. Comput. Chem.* submitted.
Strange, P., Ebert, H., Staunton, J. B. and Gyorffy, B. L. (1989) *J. Phys. Cond. Matt.* **1**, 2959.
Strange, P., Staunton, J. B. and Gyorffy, B. L. (1984) *J. Phys.* C **17**, 3355.
Stranz, D. D. and Khanna, R. K. (1981) *J. Chem. Phys.* **74**, 2116.
Strayer, M. R., Bottcher, C., Oberacker, V. E. and Umar, A. S. (1990) *Phys. Rev.* A **254**, 1399.
Strub, E., Kratz, J. V., Kronenberg, A., Brüchle, W., Li, Z., Jäger, E., Schädel, M., Schausten, B., Schimpf, E., Gäggeler, H. W., Gärtner, M. and Jost, D. (1999) GSI Scientific Report 1998, GSI.
Strub, E., Kratz, J. V., Kronenberg, A., Nähler, A., Thörle, P., Zauner, S., Brüchle, W., Jäger, E., Schädel, M., Schausten, B., Schimpf, E., Li, Z., Kirbach, U., Schumann, D., Jost, D., Türler, A., Asai, M., Nagame, Y., Sakama, M., Tsukada, K., Gäggeler, H. W. and Glatz, J. P. (2000) *Radiochim. Acta* **88**, 265.
Střda, P. and Smrčka, L. (1975) *Physica Status Solidi* (b) **70**, 537.
Styszyński, J., Cao, X., Malli, G. L. and Visscher, L. (1997) *Int. J. Quant. Chem.* **18**, 601–608.
Sucher, J. (1957) *Phys. Rev.* **107**, 1448.
Sucher, J. (1980) *Phys. Rev.* A **22**, 348–362.
Sundholm, D. (1987) *Chem. Phys. Lett.* **149**, 251–256.
Sundholm, D. (1994) *Chem. Phys. Lett.* **223**, 469–473.
Sundholm, D., Pyykkö, P. and Laaksonen, L. (1987) *Physica Scr.* **36**, 400–402.

Sunnergren, P., Persson, H., Salomonson, S., Schneider, S. M., Lindgren, I. and Soff, G. (1998) *Phys. Rev.* A **58**, 1055.

Szasz, L. (1992) *The Electronic Structure of Atoms*. Wiley.

Szmytkowski, R. (1997) *J. Phys.* B **30**, 825–861.

Szmytkowski, R. (2001) *Phys. Rev.* A **63**, 062704–1–14.

Szmytkowski, R. and Hinze, J. (1996a) *J. Phys.* B **29**, 761–777.

Szmytkowski, R. and Hinze, J. (1996b) *J. Phys.* A **29**, 6125–6141.

Szunyogh, L. and Weinberger, P. (1999) *J. Phys. Cond. Matt.* **11**, 10 451.

Szunyogh, L., Újfalussy, B., Weinberger, P. and Kollar, J. (1994) *Phys. Rev.* B **49**, 2721.

Szunyogh, L., Újfalussy, B. and Weinberger, P. (1995) *Phys. Rev.* B **51**, 9552.

Szymanski, M., Steer, R. P. and Maciejewski, A. (1987) *Chem. Phys. Lett.* **135**, 243–248.

Taherian, M. R. and Maki, A. H. (1983) *Chem. Phys. Lett.* **96**, 541—546.

Talman, J. D. (1986) *Phys. Rev. Lett.* **57**, 1091–1094.

Tamura, E. (1992) *Phys. Rev.* B **45**, 3271.

Tamura, E., Piepke, W. and Feder, R. (1987) New spin-polarization effect in photoemission from nonmagnetic surfaces. *Phys. Rev. Lett.* **59**, 9347.

Tatchen, J. (1999) Anwendung von mean-field Operatoren zur Beschreibung von Spin-Bahn-Effekten in organischen Molekülen. Diploma thesis, University of Bonn. Available at http://www.thch.uni-bonn.de/tc/.

Tatchen, J. and Marian, C. M. (1999) *Chem. Phys. Lett.* **313**, 351–357.

Tatchen, J., Waletzke, M., Marian, C. M. and Grimme, S. (2001) *Chem. Phys.* **264**, 245–254.

Tatewaki, H. and Matsuoka, O. (1997) *J. Chem. Phys.* **106**, 4558–4565.

Tatewaki, H. and Matsuoka, O. (1998) *Chem. Phys. Lett.* **283**, 161–166.

Taylor, W. S., Campbell, A. S., Barnas, D. F., Babcock, L. M. and Linder, C. B. (1997) *J. Phys. Chem.* A **101**, 2654.

Teichteil, C. and Spiegelmann, F. (1983) *Chem. Phys.* **180**, 1.

Teichteil, C. H., Pélissier, M. and Spiegelmann, F. (1983) *Chem. Phys.* **81**, 273–282.

Temmerman, W. M. and Sterne, P. A. (1990) *J. Phys. Cond. Matt.* **2**, 5529.

Temmerman, W. M., Szotek, Z. and Gehring, G. A. (1997) *Phys. Rev. Lett.* **79**, 3970.

Tenzer, R., Busic, O., Gail, M., Grün, N. and Scheid, W. (2000a) In *Nuclear Matter, Hot and Cold, Proc. Symp. in Memory of J. M. Eisenberg, Tel Aviv University*, p. 249.

Tenzer, R., Grün, N. and Scheid, W. (2000b) *Eur. Phys. J.* D **11**, 347.

Terakura, K., Hamada, N., Oguchi, T. and Asada, T. (1982) *J. Phys.* F **12**, 1661.

te Velde, G. and Baerends, E. J. (1992) *J. Comp. Phys.* **99**, 84.

Thaller, B. (1992) *The Dirac Equation*. Springer.

Theileis, V. and Bross, H. (2000) *Phys. Rev.* B **62**, 13 338–13 346.

Thiel, J., Grün, N. and Scheid, W. (1994) *Hot and Dense Nuclear Matter*, p. 595. NATO ASI Series B: Physics, vol. 335. Plenum, New York.

Thole, B. T., Carra, P., Sette, F. and van der Laan, G. (1992) *Phys. Rev. Lett.* **68**, 1943.

Thyssen, J. (2001) Development and application of methods for correlated relativitic calculations of molecular properties. PhD thesis, Department of Chemistry, University of Southern Denmark, Odense, Denmark.

Thyssen, J., Laerdahl, J. K. and Schwerdtfeger, P. (2000) *Phys. Rev. Lett* **85**, 3105–3108.

Tilson, J. L., Ermler, W. C. and Pitzer, R. M. (2000) *Comp. Phys. Commun.* **128**, 128.

Tischer, M., Hjortstam, O., Arvanitis, D., Dunn, J. H., May, F., Baberschke, K., Trygg, J., Wills, J. M., Johansson, B. and Eriksson, O. (1995) *Phys. Rev. Lett.* **75**, 1602.

Tomaselli, M., Schneider, S. M., Kankeleit, E. and Kühl, T. (1995) *Phys. Rev.* C **51**, 2989.

Tomaselli, M., Kühl, T., Seelig, P., Holbrow, C. and Kankeleit, E. (1997) *Phys. Rev.* C **58**, 1524.

Tordoir, X., Biémont, E., Garnir, H. P., Dumont, P.-D. and Träbert, E. (1999) *Eur. Phys. J.* D **6**, 1–7.

Trautmann, N. (1995) *Radiochim. Acta* **70/71**, 237.

Triguero, L., Pettersson, L. G. M., Minaev, B. and Ågren, H. (1998) *J. Chem. Phys.* **108**, 1193.

Tripathi, U. M., Schier, A. and Schmidbaur, H. (1998a) *Z. Naturf.* B **53**, 171.

Tripathi, U. M., Wegner, G. L., Schier, A., Jokisch, A. and Schmidbaur, H. (1998b) *Z. Naturf.* B **53**, 939.

Troullier, N. and Martins, J. L. (1991) *Phys. Rev.* B **43**, 1993.

Trygg, J., Johansson, B., Eriksson, O. and Wills, J. M. (1995) *Phys. Rev. Lett.* **75**, 2871.

Tterlikkis, L., Mahanti, S. D. and Das, T. P. (1968) *Phys. Rev.* **176**, 10.

Tytko, K. H. and Glemser, O. (1976) *Adv. Inorg. Chem. Radiochem.* **19**, 239–315.

Tzeng, B.-C., Schier, A. and Schmidbaur, H. (1999a) *Inorg. Chem.* **38**, 3987.

Tzeng, B.-C., Zank, J., Schier, A. and Schmidbaur, H. (1999b) *Z. Naturf.* B **54**, 825.

Ueda, K. and Rice, T. M. (1985) *Phys. Rev.* B **31**, 7149.

Újfalussy, B., Szunyogh, L. and Weinberger, P. (1997) In *Properties of Complex Inorganic Solids* (ed. A. Gonis, A. Meike and P. E. A. Turchi), p. 181. Plenum Press, New York.

Umrigar, C. J. and Gonze, X. (1994) *Phys. Rev.* A **50**, 3827.

v. Kopylow, A. and Kolb, D. (1998) *Chem. Phys. Lett.* **295**, 439–446.

v. Kopylow, A., Heinemann, D. and Kolb, D. (1998) *J. Phys.* B **31**, 4743–4754.

Vaara, J., Malkina, O. L., Stoll, H., Malkin, V. G. and Kaupp, M. (2001) *J. Chem. Phys.* **114**, 61–71.

Vallet, V., Maron, L., Teichteil, C. and Flament, J.-P. (2000) *J. Chem. Phys.* **113**, 1391–1402.

van der Laan, G. (1998) *J. Phys. Cond. Matt.* **10**, 3239.

van der Laan, G., Thole, B. T., Sawatzky, G. A., Goedkoop, J. B., Fuggle, J. C., Esteva, J.-M., Karnatak, R., Remeika, J. P. and Dabkowska, H. A. (1986) *Phys. Rev.* B **34**, 6529.

van Leeuwen, R., van Lenthe, E., Baerends, E. J. and Snijders, J. G. (1994) *J. Chem. Phys.* **101**, 1271–1281.

van Lenthe, E., Baerends, E. J. and Snijders, J. G. (1993) *J. Chem. Phys.* **99**, 4597–4610.

van Lenthe, E., Baerends, E. J. and Snijders, J. G. (1994) *J. Chem. Phys.* **101**, 9783–9792.

van Lenthe, E., van Leeuwen, R., Baerends, E. J. and Snijders, J. G. (1995) In *New Challenges in Computational Quantum Chemistry* (ed. R. Broer, P. J. C. Aerts and P. S. Bagus), pp. 93–111.

van Lenthe, E., van Leeuwen, R., Baerends, E. J. and Snijders, J. G. (1996) *Int. J. Quant. Chem.* **57**, 281–293.

van Schilfgaarde, M., Abrikosov, I. A. and Johansson, B. (1999) *Nature* **400**, 46.

van Wüllen, C. (1998) *J. Chem. Phys* **109**, 392–400.

Varga, S., Engel, E., Sepp, W.-D. and Fricke, B. (1999) *Phys. Rev.* A **59**, 4288.

Varga, S., Fricke, B., Hirata, M., Baştuğ, T., Pershina, V. and Fritzsche, S. (2000a) *J. Phys. Chem.* A **104**, 6495–6498.

Varga, S., Fricke, B., Nakamatsu, H., Mukoyama, T., Anton, J., Geschke, D., Heitmann, A., Engel, E. and Baştuğ, T. (2000b) *J. Chem. Phys.* **112**, 3499–3506.

Varga, S., Rosén, A., Sepp, W.-D. and Fricke, B. (2001) *Phys. Rev.* A **63**, 022510.

Versluis, L. and Ziegler, T. (1988) *J. Chem. Phys* **88**, 322.

Vignale, G. (1995) *Current Density Functional Theory and Orbital Magnetism*. Nato ASI Series, Series B, vol. 337, p. 485. Plenum Press, New York.

Vignale, G. and Rasolt, M. (1987) *Phys. Rev. Lett.* **59**, 2360.

Vignale, G. and Rasolt, M. (1988) *Phys. Rev.* B **37**, 10 685.

Vignale, G. and Rasolt, M. (1989) *Phys. Rev. Lett.* **62**, 115.

Vilkas, M. J., Koc, K. and Ishikawa, Y. (1997) *Chem. Phys. Lett.* **280**, 167–176.

Vilkas, M. J., Ishikawa, Y. and Koc, K. (1998a) *Phys. Rev.* E **58**, 5096–5110.

Vilkas, M. J., Ishikawa, Y. and Koc, K. (1998b) *Int. J. Quant. Chem.* **70**, 813–823.

Vilkas, M. J., Koc, K. and Ishikawa, Y. (1998c) *Chem. Phys. Lett.* **296**, 68–76.

Vilkas, M. J., Ishikawa, Y. and Koc, K. (1999) *Phys. Rev.* A **60**, 2808–2821.

Vilkas, M. J., Ishikawa, Y. and Hirao, K. (2000) *Chem. Phys. Lett.* **321**, 243–252.

Visscher, L. (1997) *Theor. Chem. Acc.* **98**(2/3), 68–70.

Visscher, L. and Dyall, K. G. (1996) *J. Chem. Phys.* **104**, 9040–9046.

Visscher, L. and Dyall, K. G. (1997) *At. Data Nucl. Data Tables* **67**, 207–224.

Visscher, L. and Saue, T. (2000) *J. Chem. Phys.* **113**, 3996–4002.

Visscher, L. and van Lenthe, E. (1999) *Chem. Phys. Lett.* **306**, 357–365.

Visscher, L., Aerts, P. J. C. and Visser, O. (1991a) General contraction in four-component relativistic Hartree–Fock calculations. In Wilson *et al.* (1991), pp. 197–205.

Visscher, L., Aerts, P. J. C., Visser, O. and Nieuwpoort, W. C. (1991b) *Int. J. Quant. Chem.: Quant. Chem. Symp.* **25**, 131–139.

Visser, O., Aerts, P. J. C. and Visscher, L. (1991c) Open Shell Relativistic Molecular Dirac–Hartree–Fock SCF-Program. In Wilson *et al.* (1991), pp. 185–195.

Visser, O., Visscher, L., Aerts, P. J. and Nieuwpoort, W. C. (1992a) *Theor. Chim. Acta* **81**, 405–416.

Visser, O., Visscher, L., Aerts, P. J. C. and Nieuwpoort, W. C. (1992b) *J. Chem. Phys.* **96**, 2910–2919.

Visscher, L., Saue, T., Nieuwpoort, W. C., Fægri, K. and Gropen, O. (1993) *J. Chem. Phys.* **99**, 6704–6715.

Visscher, L., Visser, O., Aerts, P. J. C., Merenga, H. and Nieuwpoort, W. C. (1994) *Comp. Phys. Commun.* **81**, 120–144.

Visscher, L., de Jong, W. A., Visser, O., Aerts, P. J. C., Merenga, H. and Nieuwpoort, W. C. (1995a) Relativistic Quantum Chemistry: the MOLFDIR program package. In Clementi and Corongiu (1995), pp. 169–218.

Visscher, L., Dyall, K. G. and Lee, T. J. (1995b) *Int. J. Quant. Chem. Quant. Chem. Symp.* **29**, 411–419.

Visscher, L., Lee, T. J. and Dyall, K. G. (1996a) *J. Chem. Phys.* **105**, 8769–8776.

Visscher, L., Styszynski, J. and Nieuwpoort, W. C. (1996b) *J. Chem. Phys.* **105**, 1987–1994.

Visscher, L., Saue, T. and Oddershede, J. (1997) *Chem. Phys. Lett.* **274**, 181–188.

Visscher, L., Enevoldsen, T., Saue, T. and Oddershede, J. (1998) *J. Chem. Phys.* **109**, 9677–9684.

Visscher, L., Enevoldsen, T., Saue, T., Jensen, H. J. A. and Oddershede, J. (1999) *J. Comp. Chem.* **20**, 1262–1273.

von Barth, U. and Hedin, L. (1972) *J. Phys.* C **5**, 1629.

Vosko, S. H., Wilk, L. and Nusair, M. (1980) *Can. J. Phys.* **58**, 1200.

Wadt, W. R. and Hay, P. J. (1985) *J. Chem. Phys.* **82**, 284.

Wagmann, A. D. (1982) *J. Phys. Chem. Rev. Data* **11**, 2.

Wagner, R., Chen, S. Y., Maksimuhuk, A. and Umstadter, D. (1997) *Phys. Rev. Lett.* **78**, 3125.

Wallace, N. M., Blaudeau, J.-P. and Pitzer, R. (1991) *Int. J. Quant. Chem.* **40**, 789.

Wang, D. S., Wu, R. Q., Zhong, L. P. and Freeman, A. J. (1995) *J. Magn. Magn. Mater.* **144**, 643.

Wang, X., Zhang, X.-G., Butler, W. H., Stocks, G. M. and Harmon, B. N. (1992) *Phys. Rev.* B **46**, 9352–9358.

Wang, Y., Flad, H.-J. and Dolg, M. (2000) *Phys. Rev.* B **61**, 2362.

Watanabe, Y. and Matsuoka, O. (1997) *J. Chem. Phys.* **107**, 3738–3739.

Watanabe, Y. and Matsuoka, O. (1998) *J. Chem. Phys.* **109**, 8182–8187.

Watson, K. M. (1953) *Phys. Rev.* **89**, 575.

Weber, H. J., Weitbrecht, D., Brach, D., Shelankov, A. L., Keiter, H., Weber, W., Wolf, T., Geerk, J., Linker, G., Roth, G., Splittgerber-Hünnekes, P. C. and Güntherodt, G. (1990) *Solid State Commun.* **76**, 511.

Weeks, J., Hazi, A. and Rice, S. (1969) *Adv. Quant. Chem.* **16**, 283.

Weigend, F. and Häser, M. (1997) *Theor. Chem. Acc.* **97**, 331.

Weigend, F., Häser, M., Patzelt, H. and Ahlrichs, R. (1998) *Chem. Phys. Lett.* **294**, 143–152.

Weil, D. A. and Wilkins, C. L. (1985) *J. Am. Chem. Soc.* **107**, 7316.

Weinberger, P. (1990) *Electron Scattering Theory for Ordered and Disordered Matter*. Oxford University Press.

Weinberger, P., Levy, P. M., Banhart, J., Szunyogh, L. and Újfalussy, B. (1996) *J. Phys. Cond. Matt.* **8**, 7677.

Weinstock, B. and Goodman, G. L. (1965) *Adv. Chem. Phys.* **9**, 169.

Wesendrup, R. and Schwarz, H. (1995a) *Angew. Chem.* **107**, 2176.

Wesendrup, R. and Schwarz, H. (1995b) *Angew. Chem. Int. Ed. Engl.* **34**, 2033.

Wesendrup, R., Schröder, D. and Schwarz, H. (1994a) *Angew. Chem.* **106**, 1232.

Wesendrup, R., Schröder, D. and Schwarz, H. (1994b) *Angew. Chem. Int. Ed. Engl.* **33**, 1174.

Wesendrup, R., Schalley, C. A., Schröder, D. and Schwarz, H. (1995) *Chem. Eur. J.* **1**, 608.

Wetmore, R. W. and Segal, G. A. (1975) *Chem. Phys. Lett.* **36**, 478.

Wichmann, E. H. and Kroll, N. M. (1956) *Phys. Rev.* **101**, 843.

Wick, G. C. (1950) *Phys. Rev.* **80**, 268.

Wienke, R., Schütz, G. and Ebert, H. (1991) *J. Appl. Phys.* **69**, 6147.

Wildman, S. A., DiLabio, G. A. and Christiansen, P. A. (1997) *J. Chem. Phys.* **107**, 9975.

Wilson, S. (2001) *J. Mol. Struct. (THEOCHEM)* **547**, 279–291.

Wilson, S., Grant, I. P. and Gyorffy, B. (eds) (1991) Plenum Press, New York.

Winter, H., Borneis, S., Dax, A., Faber, S., Kühl, T., Marx, D., Schmitt, F., Seelig, P., Seelig, W., Shabaev, V. M., Tomaselli, M. and Würtz, M. (1999) Bound electron g-factor in hydrogen-like bismuth. GSI Scientific Report 1998, GSI, DE-64291 Darmstadt, Germany.

Wiśniewski, P., Gukasov, A. and Henkie, Z. (1999) *Phys. Rev.* B **60**, 6242.

Wood, C. S., Bennett, S. C., Cho, D., Masterson, B. P., Roberts, J. L., Tanner, C. E. and Wieman, C. E. (1997) *Science* **275**, 1759–1763.

Wood, J. H. and Boring, A. M. (1978) *Phys. Rev.* B **18**, 2701–2711.

Woon, D. E. (1994) *J. Chem. Phys.* **100**, 2838.

Wu, J., Derrickson, J. H., Parnell, T. A. and Strayer, M. R. (1999) *Phys. Rev.* A **60**, 3722.

Wu, S., Kaplan, S., Lihn, H.-T. S., Drew, H. D., Hou, S. Y., Philips, J. M., Barbour, J. C., Venturini, E. L., Li, Q. and Fenner, D. B. (1996) *Phys. Rev.* B **54**, 13 343.

Xiao, C., Krüger, S. and Rösch, N. (1999) *Int. J. Quant. Chem.* **74**, 405–416.

Xu, B. X., Rajagopal, A. K. and Ramana, M. V. (1984) *J. Phys.* C **17**, 1339.

Yabushita, S., Zhang, Z. Y. and Pitzer, R. M. (1999) *J. Phys. Chem.* A **103**, 5791.

Yamagami, H. (2000) *Phys. Rev.* B **61**, 6246.

Yamagami, H., Mavromaras, A. and Kübler, J. (1997) *J. Phys.: Condens. Matter* **9**, 10 881.

Yanai, T., Nakajima, T., Ishikawa, Y. and Hirao, K. (2001) *J. Chem. Phys.* **114**, 6526–6538.

Yang, L., Heinemann, D. and Kolb, D. (1992) *Chem. Phys. Lett.* **192**, 499–502.

Yarkony, D. R. (2001) *J. Phys. Chem.* A **105**, 6277.

Yasui, M. and Shimizu, M. (1985) *J. Phys.* F **15**, 2365.

Yeo, Y. Y., Vattuone, L. and King, D. A. (1997) *J. Chem. Phys.* **106**, 392.

Yerokhin, V. A. and Shabaev, V. M. (2001) *Phys. Rev.* A **64**, 062507.

Yerokhin, V. A., Artemyev, A. N. and Shabaev, V. M. (1997a) *Phys. Lett.* A **234**, 361.

Yerokhin, V. A., Shabaev, V. M. and Artemyev, A. N. (1997b) *JETP Lett.* **66**, 18.

Yerokhin, V. A., Artemyev, A. N., Beier, T., Plunien, G., Shabaev, V. M. and Soff, G. (1999) *Phys. Rev.* A **60**, 3522.

Yerokhin, V. A., Artemyev, A. N., Shabaev, V. M., Sysak, M. N., Zherebtsov, O. M. and Soff, G. (2000) *Phys. Rev. Lett.* **85**, 4699.

Yerokhin, V. A., Artemyev, A. N., Shabaev, V. M., Sysak, M. M., Zherebtsov, O. M. and Soff, G. (2001) *Phys. Rev.* A **64**, 032109.

Ynnerman, A. and Froese Fischer, C. (1995) *Phys. Rev.* A **51**, 2020–2030.

Ynnerman, A., James, J., Lindgren, I., Persson, H. and Salomonson, S. (1994) *Phys. Rev.* A **50**, 4671.

Yu, M. and Dolg, M. (1997) *Chem. Phys. Lett.* **273**, 329–336.

Yuan, X., Dougherty, R. W., Das, T. P. and Andriessen, J. (1995) *Phys. Rev.* A **52**, 3563–3571.

Zank, J., Schier, A. and Schmidbaur, H. (1998) *J. Chem. Soc., Dalton Trans.*, p. 323.

Zank, J., Schier, A. and Schmidbaur, H. (1999) *J. Chem. Soc., Dalton Trans.*, p. 415.

Zecha, C. (1997) Diploma thesis, University of Munich.

Zeller, R., Dederichs, P. H., Újfalussy, B., Szunyogh, L. and Weinberger, P. (1995) *Phys. Rev.* B **52**, 8807.

Zemach, A. C. (1956) *Phys. Rev.* **104**, 1771.

Zhang, X.-G., Liyanage, R. and Armentrout, P. B. (2001) *J. Am. Chem. Soc.* **123**, 5563.

Zimmermann, H. P., Gober, M. K., Kratz, J. V., Schädel, M., Brüchle, W., Schimpf, E., Gregorich, K. E., Türler, A., Czerwinski, K. R., Hannink, N. J., Kadkhodayan, B., Lee, D. M., Nurmia, M., Hoffman, D. C., Gäggeler, H. W., Jost, D., Kovacs, J., Scherer, U. W. and Weber, A. (1993) *Radiochim. Acta* **60**, 11.

Zou, Y. and Froese Fischer, C. (2000) *Phys. Rev.* A **62**, 062505–1–6.

Zumbro, J. D., Shera, E. B., Tanaka, Y., Bemis, C. E., Naumann, R. A., Hoehn, M. V., Reuter, W. and Steffen, R. M. (1984) *Phys. Rev. Lett.* **53**, 1888.

Index